Advances in
INORGANIC CHEMISTRY

Volume 60

ADVISORY BOARD

Advances in
INORGANIC CHEMISTRY

EDITED BY

Rudi van Eldik

Department of Chemistry and Pharmacy
University of Erlangen-Nürnberg
91058 Erlangen
Germany

VOLUME 60

AMSTERDAM • BOSTON • HEIDELBERG • LONDON • NEW YORK • OXFORD
PARIS • SAN DIEGO • SAN FRANCISCO • SINGAPORE • SYDNEY • TOKYO
ACADEMIC PRESS IS AN IMPRINT OF ELSEVIER

Academic Press is an imprint of Elsevier
32 Jamestown Road, London NW1 7BY, UK
Radarweg 29, PO Box 211, 1000 AE Amsterdam, The Netherlands
30 Corporate Drive, Suite 400, Burlington, MA 01803, USA
525 B Street, Suite 1900, San Diego, CA 92101-4495, USA

First edition 2008

Library of Congress Cataloging-in-Publication Data
A catalog record for this book is available from the Library of Congress

British Library Cataloguing in Publication Data
A catalogue record for this book is available from the British Library

ISBN: 978-0-12-373977-3
ISSN: 0898-8838

For information on all Academic Press publications
visit our website at elsevierdirect.com

Printed and bound in USA

08 09 10 11 12 10 9 8 7 6 5 4 3 2 1

CONTENTS

Tripodal *N,N,O*-Ligands for Metalloenzyme Models and Organometallics

NICOLAI BURZLAFF

Hydroxypyranones, Hydroxypyridinones, and their Complexes

JOHN BURGESS and MARIA RANGEL

Late Transition Metal-Oxo Compounds and Open-Framework Materials that Catalyze Aerobic Oxidations

RUI CAO, JONG WOO HAN, TRAVIS M. ANDERSON, DANIEL A. HILLESHEIM, KENNETH I. HARDCASTLE, ELENA SLONKINA, BRITT HEDMAN, KEITH O. HODGSON, MARTIN L. KIRK, DJAMALADDIN G. MUSAEV, KEIJI MOROKUMA, YURII V. GELETII and CRAIG L. HILL

PREFACE

Most of the contributions presented in this volume originate from the 2nd Erlangen Symposium on Redox-Active Metal Complexes held in 2006. The authors were requested to prepare an 'Accounts of Chemical Research' type of presentation in which they focused on the work presented at the symposium. Within this context the first five contributions of this issue focus on small molecule activation at electron-rich uranium and transition metal centers (Karsten Meyer and co-worker), the use of tripodal N,N,O ligands for metalloenzymes and organometallics (Nicolai Burzlaff), the application of β-cyclodextrin -linked ruthenium complexes in oxidation and reduction processes (Wolf Woggon and co-workers), the use of late transition metal-oxo compounds and open framework materials to catalyze aerobic oxidation reactions (Craig Hill and co-workers), and the catalytic dismutation and reversible binding of superoxide (Ivana Ivanovic-Burmazovic). These contributions give a representative flavor of the type of topics covered at the symposium that was attended by *ca.* 150 participants.

As final contribution to this volume, John Burgess and Maria Rangel present a detailed account of the chemistry of hydroxypyranones and hydroxypyridinones, and their transition metal complexes.

I trust that you will all agree that the topics covered in this issue are very stimulating and that the contributions report the most recent advances in inorganic and bioinorganic chemistry that especially deals with the activation of small molecules and molecular recognition. The next volume of this series will be a thematic issue on 'Metal Ion Controlled Reactivity'.

Rudi van Eldik
University of Erlangen-Nürnberg
Germany
May 2008

TRIPODAL CARBENE AND ARYLOXIDE LIGANDS FOR SMALL-MOLECULE ACTIVATION AT ELECTRON-RICH URANIUM AND TRANSITION METAL CENTERS

KARSTEN MEYER and SUZANNE C. BART

University of Erlangen-Nürnberg, Department of Chemistry and Pharmacy, Inorganic Chemistry, Egerlandstr. 1, 91058 Erlangen, Germany

I. Introduction

Chelating ligands that induce tripodal configurations at coordinated metal centers are known to provide powerful platforms for small-molecule activation. These ligands consist of an atom or small molecule which acts as an anchor, and holds three pendant arms capable of coordinating to a metal in a trigonal conformation (Fig. 1). Typically, these ligands hold several advantages over monodentate and bidentate ligands. Due to the enhanced chelating effects, tripodal ligands often bind to metal ions very strongly and can stabilize reactive intermediate species with unusual electronic and geometric structures. In addition, the steric bulk of both the anchor and pendant arms is highly modular, providing the electronic and structural flexibility to effectively block undesired side- and decomposition reactions, more effectively controlling reactivity at the metal center. Due to these distinctive benefits, the design and development of new tripodal ligand systems has been an active area in inorganic and organometallic coordination chemistry (*1–4*).

ADVANCES IN INORGANIC CHEMISTRY
VOLUME 60 ISSN 0898-8838 / DOI: 10.1016/S0898-8838(08)00001-9

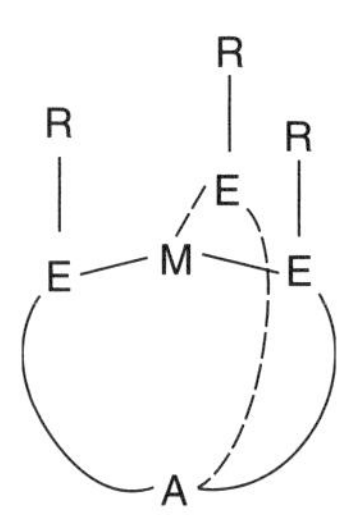

Fig. 1. Schematic diagram of tripodal ligand with sterically encumbering R groups oriented perpendicular to the ME_3 coordination plane.

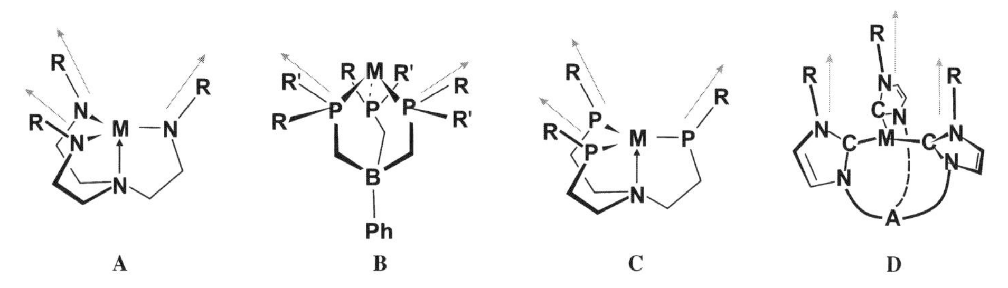

Fig. 2. Tripodal ligand scaffolds with arrows indicating the direction of steric bulk.

The chemistry presented herein has been presented on the occasion of the SFB symposium "Redox active metal complexes – Control of Reactivity via Molecular Architecture."

Commonly used tripodal ligands such as tripodal tris(amido)amine (Fig. 2, **A**) and tris-phosphine ligands (Fig. 2, **B**) have drawn much attention recently for supporting metal centers capable of small-molecule activation (*1*). Metal complexes supported by tris(pyrazolyl)-borate (Tp) ligands promote catalytic transformations, including C–H activation (*5*), C–C (*6*), C–O (*7*), and C–N (*8*) bond formation, assist dioxygen activation (*9*) and serve as structural mimics of metal-containing enzymes (*10*). The tetradentate tris(amido)amine framework, composed of three negatively charged "hard" (*11*) amido donors, coordinates to transition metal (*2,12*) and main group elements (*13*) in 3+ or higher oxidation states. The resulting metal complexes have a pair of degenerate π-type frontier orbitals that aid in metal–ligand multi-bond formation (*2*) and have proven essential for supporting well-defined catalytic reactions such as dinitrogen reduction at a single molybdenum center (*14,15*). In contrast, tripodal ligands bearing "soft" donor atoms such as sulfur (*16*) and phosphorus (*4,17,18*) are more suitable for stabilizing electron-rich, low-valent metal centers. A prime example is the work by Sacconi (*19*) which shows the development of the tris(phosphino)amine ligand (Fig. 2, **C**).

Using this rich chemistry as inspiration, the development of tripodal N-heterocyclic carbene (NHC) analogues was explored (Fig. 2, **D**). These chelators should mimic the properties of monodentate NHC ligands, producing distinct beneficial synthetic, electronic, and steric properties over these previously known tripodal ligand systems, including reduction of oxidative ligand degradation which can occur with air-sensitive phosphine ligands. Recent studies show that the seemingly "soft" NHCs can coordinate to both "soft", electron-rich metal fragments and "hard", electron-deficient metal centers (*20*), resulting in their coordination to virtually every metal in the periodic table with a range of oxidation states (*20a*). The newly developed tripodal NHC ligands thus are complementary to both the tris(amido)amine and the tris(phosphino)borate ligand systems.

The geometries of tripodal ligand systems are determined by the type of pendant arm and the number of atoms in the linker to the anchor (Fig. 2, Types **A** and **B**).

Depending on these variables, the sterically encumbering substituents may be directed away from the metal center, leaving the reactive core wide open or be directed towards the metal center in order to protect it (see arrows in Fig. 2). Protecting the metal center prevents binuclear decomposition of reactive species. For instance, Peters *et al.* report that the bulky isopropyl derivatized tris(phosphino)-borate ligand does not prevent dimerization of the unique yet highly reactive terminal nitrido complex $[(PhBP^{iPr})_3Fe{\equiv}N]$ to form a dinitrogen-bridged dinuclear species $[(PhBP_3^{iPr})Fe]_2(\mu\text{-}N_2)$ (Fig. 3, **E**) (*21*). To prevent $[(^{Ar}N_3N)Mo{\equiv}N]$ from forming a similar dinuclear dinitrogen complex $[\{(^{Ar}N_3N)Mo\}_2\ (\mu{:}\eta^1,\eta^1\text{–}N_2)]$, Schrock *et al.* had to introduce three extremely bulky hexaisopropyl terphenyl substituents at the tris(amido)amine ligand (Fig. 3, **D**) (*14*). The synthesis of these sterically bulky ligand derivatives is both challenging and time-consuming. In contrast, the sterics of tripodal NHC ligands are controlled by the alkyl or aryl substituents at the sp^2-hybridized ring

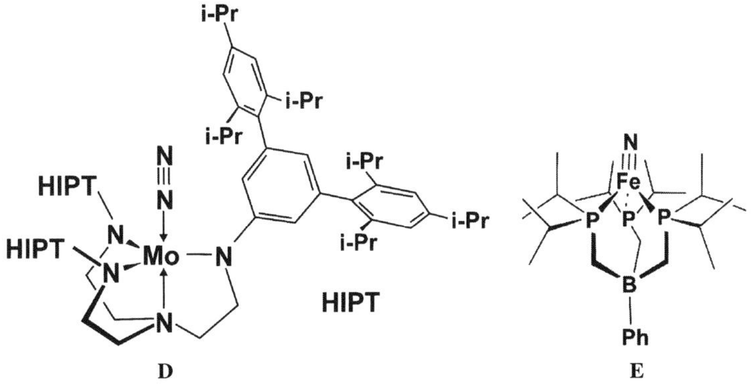

Fig. 3. Complexes containing sterically bulky tripodal ligands.

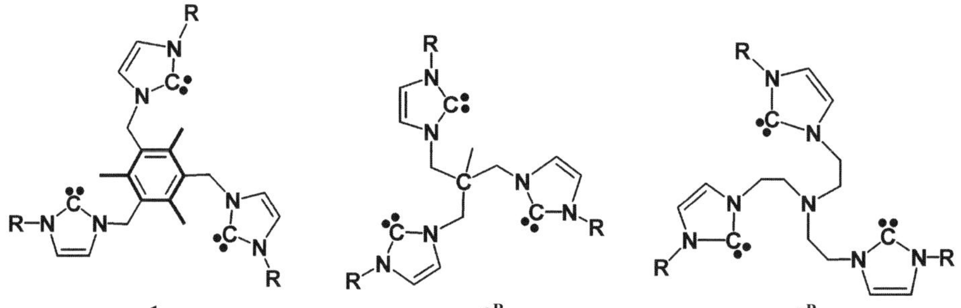

1 $\mathbf{2^R}$ $\mathbf{3^R}$

FIG. 4. Tripodal N-heterocyclic carbene chelators with mesitylene (mes-carbene, left), carbon (**TIME$^{\mathbf{R}}$**, center), and nitrogen (**TIMEN$^{\mathbf{R}}$**, right) anchoring units.

nitrogen (N3), allowing perpendicular alignment of the steric bulk to the plane of the pendant arms forming a deep (5–6 Å) well-protected cavity for ligand binding to the metal center (Fig. 2, Type **C**).

Prior to our work, only two tripodal NHC ligand systems were known. The mesitylene-anchored tris(carbene) ligand (Fig. 4, **1**) was unique (*22*), yet metal complexation had not been achieved with this ligand. Thorough investigation by Nakai *et al.* of the coordination chemistry of **1** and its derivatives showed that the cavity of this ligand system can only host exceptionally large metal ions, such as the thallium(I) cation (*23*). Attempts to synthesize transition metal complexes of derivatives of **1** have been unsuccessful up to this point. The development of new tripodal NHC ligand systems for stabilization of a single transition metal center in a coordinatively unsaturated ligand environment to allow the binding and activation of small molecules in a controlled manner is discussed here. In particular, we describe the synthesis of two new classes of tripodal NHC ligands TIME^{R} (1,1,1-*t*ris(3-alkyl*i*midazol-2-ylidene)*m*ethyl]*e*thane) ($\mathbf{2^R}$) and TIMEN^{R} (tris[2-(3-alkylimidazol-2-ylidene)ethyl]amine) ($\mathbf{3^R}$) and their coordination chemistry.

The coordination chemistry of uranium centers with a classic Werner-type polyamine tripodal chelator is also being investigated (Fig. 5) with a goal of identifying and isolating uranium complexes with enhanced reactivity towards binding, activation, and functionalization of small molecules. Because of the large size of the uranium center, instead of using a single atom anchor, the small molecule 1,4,7-triazacyclononane is used. Enacting a small, weakly binding molecule in this position protects one side of the uranium center from unwanted side and decomposition reactions. Each nitrogen contains an alkyl-substituted aryloxide pendant arm which coordinates strongly to the uranium center in a distorted trigonal planar fashion. Because the polyamine chelator is a *weak* ligand for uranium ions, the metal orbitals do not participate in strong metal–ligand interactions, thus creating a more electron-rich uranium center for small-molecule

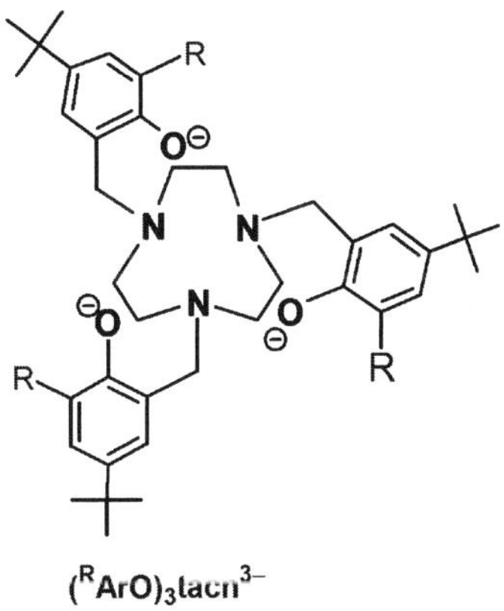

(^{R}ArO)$_3$tacn^{3-}

4^R

FIG. 5. Tris-aryloxide triazacyclononane ligand for uranium coordination chemistry.

activation trans to the tacn anchor. This coordination geometry places the aliphatic ortho substitutents (R) perpendicular to the plane formed by the aryloxides, making a protective cavity around the uranium ion similar to that for the NHC ligand system. Specifically, we show herein that the introduction of hexadentate tris-anionic 1,4,7-tris(3,5-alkyl-2-hydroxybenzylate)-1,4,7-triazacyclononane derivatives ((^{R}ArO)$_3$tacn^{3-} with R = *tert*-butyl (*t*-Bu) (*24*) and 1-adamantyl (Ad) (*25*)) to redox-active uranium centers results in formation of stable, coordinatively unsaturated core complexes with a single axial coordination site (L) available for ligand binding, substitution reactions, and redox events associated with small-molecule activation and functionalization.

The molecular architecture of the axial binding site is dependant on the type of substituent R used for both ligand systems and thus must be chosen carefully. In the case of the NHC ligands, flat xylyl and mesityl groups are used to form the cavity, while for the polyamine ligand, 3D *tert*-butyl and adamantyl substituents were chosen. Although using the alkyl groups may form a narrower cavity, groups such as these also provide the possibility for van der Waals interactions with each other and incoming small molecules, which might be beneficial in view of alkane binding and C–H activation events.

II. Synthesis and Characterization of Ligand Precursors and Low-Valent Metal Complexes for Small-Molecule Activation

A. SYNTHESIS AND METALATION OF (TIME^{ME})

A single carbon atom anchors the three pendant arms of the neopentane based tripodal NHC chelator 1,1,1-tris

[(3-alkylimidazol-2-ylidene)methyl]ethane (TIME^{R}) producing a binding cavity with the appropriate size to host a transition metal ion.

Access to the imidazolium precursors of the free monodentate carbenes was carried out according to the procedures reported by Dias and Jin in their synthesis of the mesitylene-anchored tris(carbene) ligand **1** (Fig. 6) (*22*). Attempts to isolate the free tris(carbene) ligand, TIME^{Me}, were unsuccessful and consequently only formed *in situ* by deprotonation of the [H_3TIME^{Me}]$(PF_6)_3$ salt with a strong base. Complexation of transition metals with the *in situ* generated carbene had limited success (*26*). Metal complexes of TIME^{Me} were made successfully through the transmetalation method developed by Lin *et al.* (*26*). Treatment of [H_3TIME^{Me}]Br_3 with Ag_2O in DMSO affords the trinuclear silver complex [(TIME^{Me})$_2Ag_3$]$(PF_6)_3$ (**2^{Me}-Ag**), which serves as a carbene transfer agent and reacts with copper(I) bromide and (dimethylsulfido)gold(I) chloride to form the corresponding copper(I) and gold(I) complexes [(TIME^{Me})$_2$Cu]$(PF_6)_3$ (**2^{Me}-Cu**) and [(TIME^{Me})$_2Au_3$]$(PF_6)_3$ (**2^{Me}-Au**) (Scheme 1) (*26,27*).

X-ray diffraction analyses revealed isostructural geometries for all trinuclear complexes, where three metal ions are coordinated to two ligand molecules (Fig. 7). Each metal ion is coordinated to two carbenoid carbon atoms each from a different ligand. The structure exhibits a D_3 symmetry with the 3-fold axis passing through the anchoring C atoms of the two ligands.

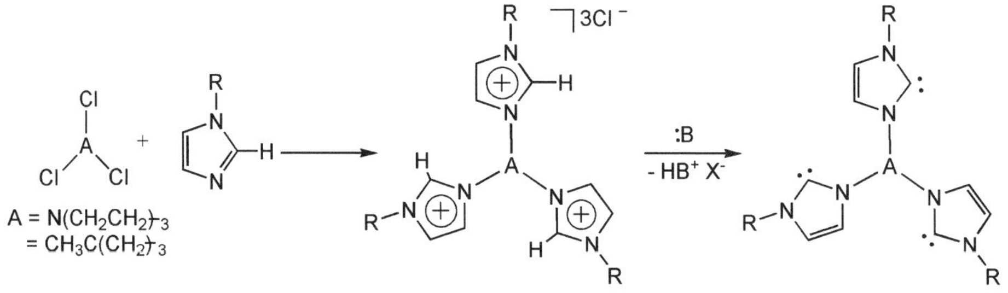

FIG. 6. General synthesis of tripodal NHC ligands.

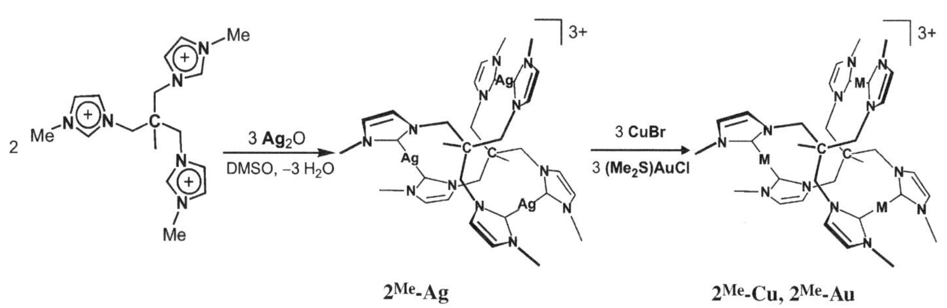

SCHEME 1. Synthesis of group 11 metal complexes of **TIME^{Me}**.

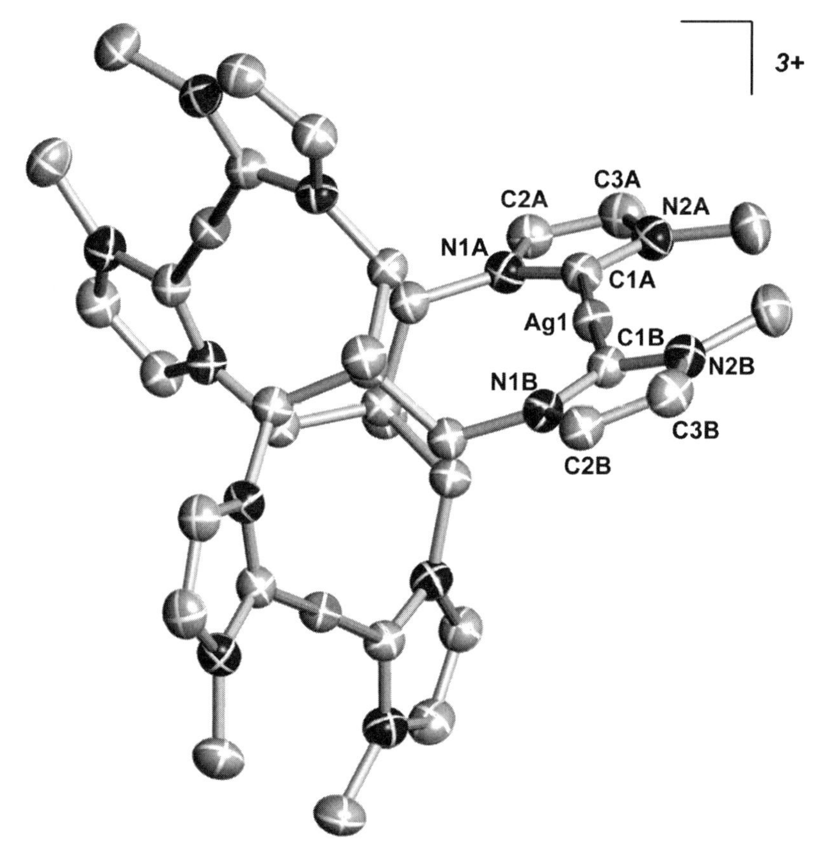

FIG. 7. Solid-state molecular structure of $\mathbf{[(TIME^{Me})_2Ag_3]^{3+}}$ ($\mathbf{2^{Me}}$**-Ag**).

The metal–carbene bond distances in this family of complexes (2.082 (2) Å for Ag, 1.9124 (16) Å for Cu, and 2.035 (12) Å for Au) are within the range of reported values for typical group 11 metal NHC complexes (*23*). The metal carbene units are almost linear, with a C–M–C bond angle of 178.56 (13)°, 177.70 (9)°, and 177.7 (6)° for Ag, Cu, and Au, respectively. The imidazole units for $\mathbf{2^{Me}}$**-Ag**, $\mathbf{2^{Me}}$**-Cu**, $\mathbf{2^{Me}}$**-Au** exhibit structural parameters typically observed for coordinated NHC ligands. There are no inter- or intramolecular metal–metal interactions in these complexes.

B. SYNTHESIS AND METALATION OF TIME$^{t\text{-Bu}}$

The formation of trinuclear complexes $\mathbf{2^{Me}}$**-Ag**, $\mathbf{2^{Me}}$**-Cu**, $\mathbf{2^{Me}}$**-Au** shows the tendency of group 11 metal ions to form linear, two-coordinate complexes. As a result, the metal centers of these complexes are not situated in a well-protected cavity as planned.

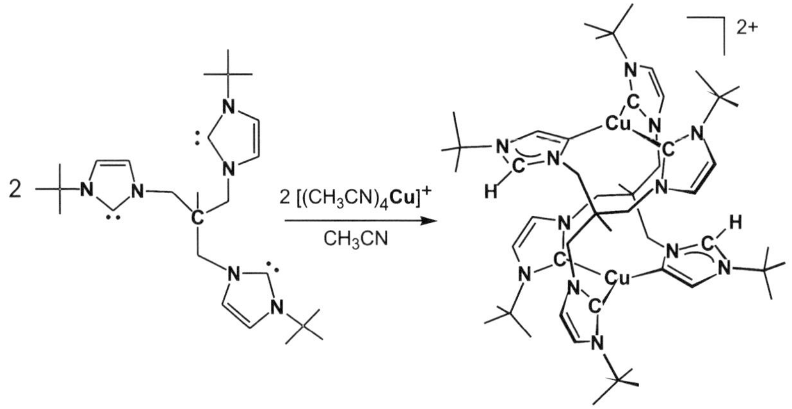

SCHEME 2. Synthesis of the bis(carbene)alkenyl complex **[(TIME$^{t\text{-Bu}}$)$_2$Cu$_2$](PF$_6$)$_2$ (2$^{t\text{-Bu}}$-Cu)**.

To prevent the formation of polynuclear species, the steric bulk at the TIME ligand was increased by synthesizing the *tert*-butyl derivative TIME$^{t\text{-Bu}}$.

Deprotonation of the imidazolium salt [H$_3$TIME$^{t\text{-Bu}}$](PF$_6$)$_3$ with three equivalents of potassium *tert*-butoxide produced the free carbene TIME$^{t\text{-Bu}}$ (**2$^{t\text{-Bu}}$**), which was isolated and fully characterized (*28*). Addition of one equivalent of **2$^{t\text{-Bu}}$** to [(CH$_3$CN)$_4$Cu](PF$_6$) affords the copper carbene complex [(TIME$^{t\text{-Bu}}$)$_2$Cu$_2$](PF$_6$)$_2$ (**2$^{t\text{-Bu}}$-Cu**) (Scheme 2).

The X-ray crystallographic analysis of **2$^{t\text{-Bu}}$-Cu** reveals a dimeric structure, where two copper ions are coordinated to two ligand molecules (Fig. 8). Each copper ion is situated in a trigonal planar carbene/alkenyl carbon ligand environment, coordinated to two carbene arms from one chelator and a third carbon from the pendant arm of a second chelator. The average Cu–C bond distance is 1.996 (1) Å, consistent with that of other reported Cu(I) carbene complexes (*29*).

Interestingly, the X-ray structure of **2$^{t\text{-Bu}}$-Cu** reveals the existence of an "alkenyl" binding mode within the Cu–C entity. Two of the three carbon chelators coordinated to the copper ion are "normal" diamino carbene centers, i.e., the C2 carbon of the imidazole rings (Scheme 2). The third carbon chelator formally is the C5 backbone atom of the imidazole ring structure, and can be considered an alkenyl functionality, also known as "abnormal" carbene binding. This formulation is further supported by NMR spectroscopy (*28*). Although this type of metal binding to N-heterocyclic carbene ligands was previously reported for complexes synthesized directly from imidazolium salts (*30*), this was the first example of this binding mode generated by metal coordination to the free carbene starting material.

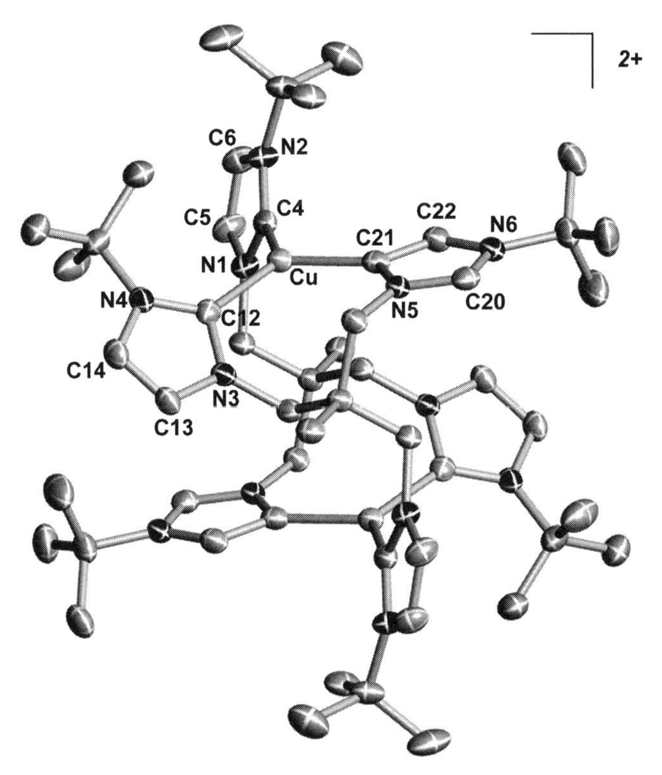

FIG. 8. Solid-state molecular structure of $[(\mathbf{TIME}^{t\text{-}Bu})_2\mathbf{Cu}_2]^{2+}$ ($\mathbf{2}^{t\text{-}Bu}$-**Cu**).

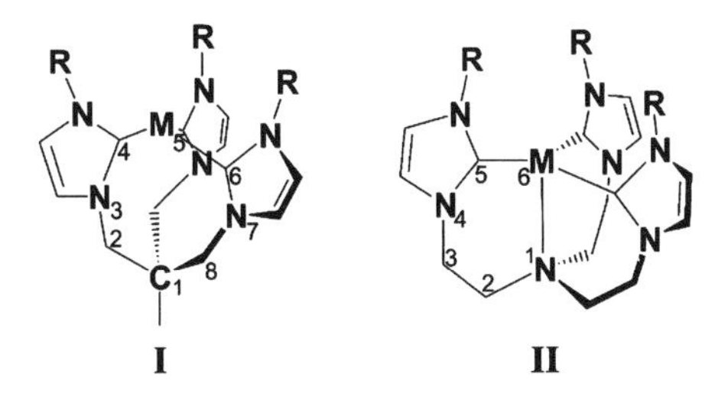

FIG. 9. Structural models of carbon- and nitrogen-anchored tris(carbene) metal complexes.

C. SYNTHESIS AND METALATION OF TIMEN^R LIGAND

Examination of the backbones of the neopentane-based, carbon-anchored TIME^{Me} and bulky TIME$^{t\text{-}Bu}$ tripodal ligands reveals possible reasons for the lack of 1:1 complex formation with metal centers. A hypothetical κ^3 complex of TIME would contain three eight-membered rings (Fig. 9, **I**) with much lower stability than the more commonly observed five, six, or seven-membered rings formed, for instance, by poly(pyrazolyl)borate or poly(phosphine) ligands. In order to overcome this, a coordinating atom at the anchoring position of the carbene tripod was incorporated, such that 1:1 metal complexation was favored by forming three stable six-membered rings (Fig. 9, **II**). Accordingly, the nitrogen-anchored tris(carbene) ligand system tris[2-(3-alkylimidazol-2-ylidene)ethyl]amine (TIMEN^{alkyl}) was synthesized. The imidazolium precursors $[H_3TIMEN^R]^{3+}$ were prepared by quaternization of functionalized N-alkylimidazoles with tris-(2-chloroethyl)amine. Deprotonation of the imidazolium salts with potassium *tert*-butoxide yields the free carbenes TIMEN^{alkyl} (**3^R**).

Monomeric copper(I) tris(carbene) complexes were prepared by reacting TIMEN^{alkyl} with copper(I) salts. The X-ray structures of $[(TIMEN^R)Cu]^+$ complexes (R = *tert*-butyl (**3$^{t\text{-}Bu}$-Cu**), benzyl (Bz, **3^{Bz}-Cu**)) show that the TIMEN^R ligand coordinates to the copper(I) ion via the three carbenoid carbons in the predicted κ^3-fashion (Fig. 10, left) (*31*). The Cu–C distances of 1.95–2.00 Å compare well with bond distances found for other Cu(I) NHC complexes (*29*). The cuprous center is located in the plane of the three carbenes, with the sum of the three C–Cu–C angles close to 360°. The Cu–N distances range from 2.4–2.6 Å, suggesting the interaction between the anchoring nitrogen atom and the copper center is non-significant. The space filling model shows that the copper center in complex $[(TIMEN^{t\text{-}Bu})Cu]^+$ is well shielded by the three sterically encumbering *tert*-butyl groups. This steric congestion is largely reduced in the analogous complex $[(TIMEN^{Bz})Cu]^+$.

Accordingly, the complex $[(TIMEN^{Bz})Cu]^{2+}$ could be synthesized, isolated, and spectroscopically characterized. Electrochemical measurements show that the complexes exhibit a reversible Cu(I)/Cu(II) couple at 0.11 V and −0.1 V vs. Fc/Fc$^+$ for **3$^{t\text{-}Bu}$-Cu** and **3^{Bz}-Cu**, respectively. However, all Cu(I) complexes $[(TIMEN^{alkyl})Cu]^+$ have

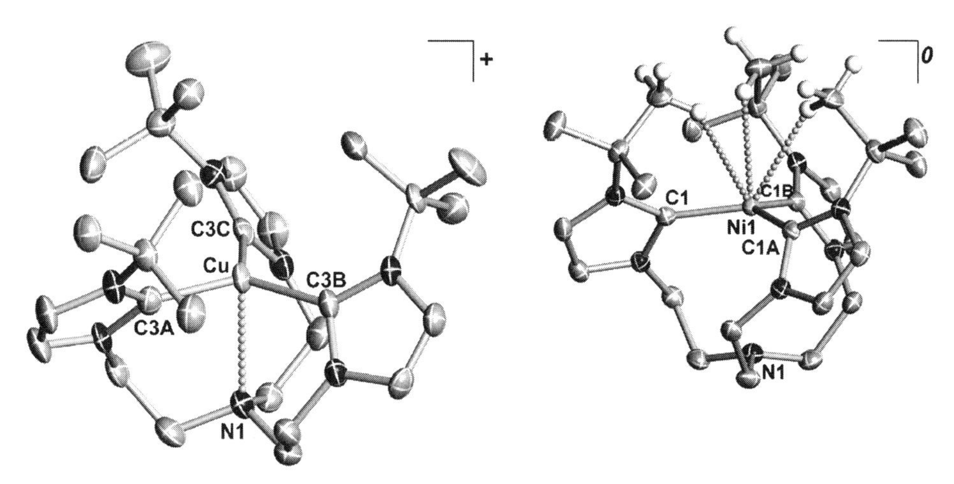

FIG. 10. Solid-state molecular structure of **$[(TIMEN^{t\text{-}Bu})Cu]^+$ (3$^{t\text{-}Bu}$-Cu)** (left) and **$[(TIMEN^{t\text{-}Bu})Ni]$ (3$^{t\text{-}Bu}$-Ni)** (right).

proven to be inert towards additional ligand binding and activation of small molecules. The inactivity is probably due to the full 18 electron count of the copper(I) ion in these complexes.

Treating TIMEN$^{t\text{-Bu}}$ with Ni(COD)$_2$ (*32*) produced [(TIMEN$^{t\text{-Bu}}$)Ni] (**3$^{t\text{-Bu}}$-Ni**) (Fig. 10, right), which forms the Ni(I) complex [(TIMEN$^{t\text{-Bu}}$)Ni]Cl (**3$^{t\text{-Bu}}$-NiCl**) upon exposure to CH_2Cl_2. Interestingly, the molecular structures of these two nickel complexes **3$^{t\text{-Bu}}$-Ni** and **3$^{t\text{-Bu}}$-NiCl** differ significantly. The TIMEN$^{t\text{-Bu}}$ ligand of **3$^{t\text{-Bu}}$-Ni** is tridentate and coordinates to the nickel center only through the three carbene chelators. In complex **3$^{t\text{-Bu}}$-NiCl**, however, TIMEN$^{t\text{-Bu}}$ coordinates in a tetradentate fashion and additionally binds to the nickel center via the anchoring nitrogen atom (Ni–N = 2.22 Å). This observation strongly supports our theory that the central nitrogen can function as an electron sink supplying two additional electrons for metal centers in higher oxidation states. The difference between the Ni–C bond distance of 1.892 (1) Å for Ni(0) (**3$^{t\text{-Bu}}$–Ni**) and 1.996 (4) Å for Ni(I) (**3$^{t\text{-Bu}}$-NiCl**) is also noteworthy. The short Ni–C distance of 1.892 (1) Å for the bigger Ni(0) center indicates a substantially higher degree of π-backbonding in this complex.

The cyclic voltammogram of complex **3$^{t\text{-Bu}}$-Ni** shows two quasi-reversible Ni(0)/(I) and Ni(I)/(II) redox waves at −2.5 V and −1.1 V vs. Fc/Fc$^+$, respectively. Neither of the two oxidized complexes was isolable, most likely due to the masking of the low-valent nickel(0) center by the three *tert*-butyl substituents on the carbene ligand. The search for alternative routes to these complexes using less sterically bulky NHC ligands, such as the TIMEN^{ipr} isopropyl derivative, is under investigation.

The chemistry of aryl-substituted TIMEN ligands was investigated by studying the coordination chemistry of TIMEN^{Ar} with low-valent cobalt ions. Treatment of TIMEN^{Ar} with a suitable cobalt(I) precursor, i.e., Co(PPh$_3$)$_3$Cl, yields the [(TIMEN^{Ar})Co]Cl (Ar = xylyl (**3^{xyl}-CoCl**), mesityl (**3mes-CoCl**)) target complexes (Fig. 11) (*33*). In both compounds, the cobalt(I) centers are coordinated in a distorted trigonal–pyramidal fashion, with the anchoring nitrogen coordinated

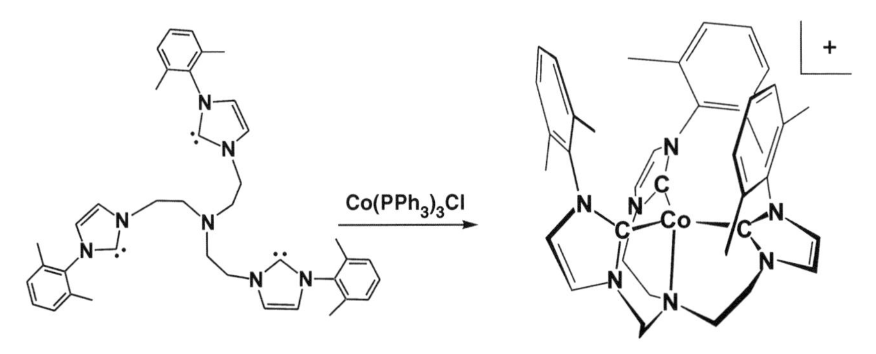

FIG. 11. Synthesis of **[(TIMEN^{Ar})Co]Cl**. Anion was omitted for clarity.

as well. The cobalt centers are coordinatively unsaturated and their reactivity with a variety of small molecules will be discussed later. Both cobalt(I) complexes have high-spin electronic configurations, possesses magnetic moments of about 3.6 μ_B at ambient temperature, and display well-defined, 3-fold symmetrical, paramagnetic ^{1}H NMR spectra.

D. Synthesis and Metalation of Triazacyclononane Derivatives

Initial synthesis of the ligand derivatives of hexadentate tris-anionic 1,4,7-tris(3,5-alkyl-2-hydroxybenzylate)-1,4,7-triazacyclononane derivatives $((^{R}ArO)_3tacn$ *(34)* with R = *tert*-butyl (*t*-Bu) and 1-adamantyl (Ad)) was performed according to literature procedures. Synthesis of the *tert*-butyl derivative was achieved using a Mannich reaction with commercially available 2,4-di-*tert*-butyl phenol, paraformaldehyde, and triazacyclononane with a catalytic amount of hydrochloric acid in methanol (Fig. 12). The ligand is isolated in high yield as a precipitate from the methanol solution. ^{1}H NMR spectra acquired in $CDCl_3$ were identical to those reported in the literature. The adamantyl derivative was made by first synthesizing the 2-adamantyl-4-*tert*-butyl phenol from commercially available 1-adamantyl chloride and 4-*tert*-butylphenol, followed by a similar Mannich reaction without the acid catalyst (Fig. 12). This ligand was isolated on multigram quantities, and characterized by ^{1}H NMR spectroscopy.

The first successful isolation of a mononuclear uranium–tacn complex was achieved by treatment of the free 1,4,7-tris(3,5-di-*tert*-butyl-2-hydroxybenzyl)-1,4,7-triazacyclononane $((^{t\text{-}Bu}ArOH)_3tacn)$ ($\mathbf{4}^{t\text{-}\mathbf{Bu}}$) *(25)* with one equivalent of $[U(N(SiMe_3)_2)_3]$ *(35)* in cold pentane

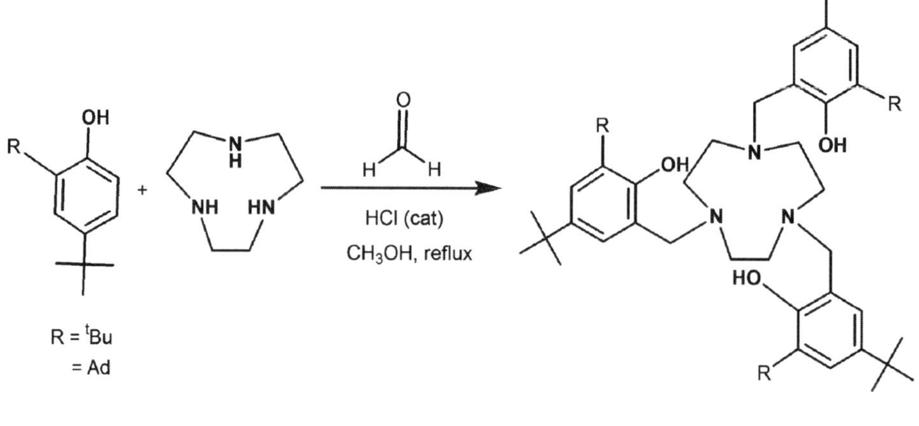

FIG. 12. Synthesis of triazacyclononane ligand derivatives.

solution. This protocol yields the six-coordinate uranium(III) complex [((${}^{t\text{-Bu}}$ArO)$_3$tacn)U] (**4$^{t\text{-Bu}}$-U**, Scheme 3) as a microcrystalline precipitate on a multi-gram scale (*22*). The ^{1}H NMR spectrum of trivalent **4$^{t\text{-Bu}}$-U** (an f^3 species) recorded in benzene-d$_6$ at 20°C displays 10 paramagnetically shifted and broadened resonances between –22 and +13 ppm. The signal pattern is in agreement with an idealized C$_3$ symmetry, with the three pendant arms arranged in a twisted propeller-like arrangement. This splits the signals of the tacn backbone into four, and each methylene linkage into two diastereotopic hydrogens.

Complex **4$^{t\text{-Bu}}$-U** is stable in a dry N$_2$-atmosphere, and very soluble in hydrocarbon solvents. High quality single-crystals suitable for an X-ray diffraction study were obtained from a pentane:phenylsilane (90:10) solution (Fig. 13). Attempts to re-crystallize **4$^{t\text{-Bu}}$-U** from

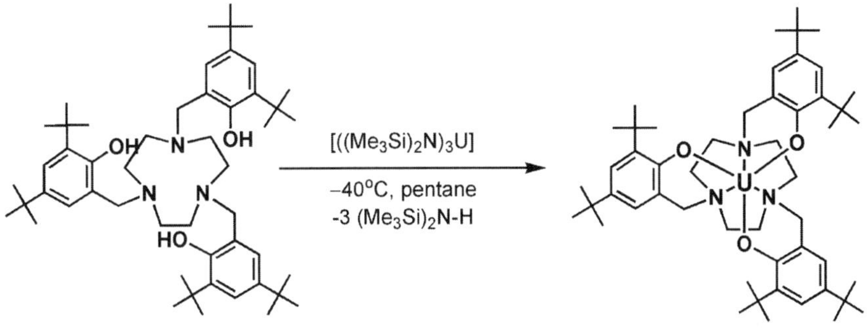

[((Me$_3$Si)$_2$N)$_3$U]

–40°C, pentane
-3 (Me$_3$Si)$_2$N-H

4^{tBu} 4^{tBu}-U

SCHEME 3. Synthesis of complex **[((${}^{t\text{-Bu}}$ArO)$_3$tacn)U] (4$^{t\text{-Bu}}$-U)**.

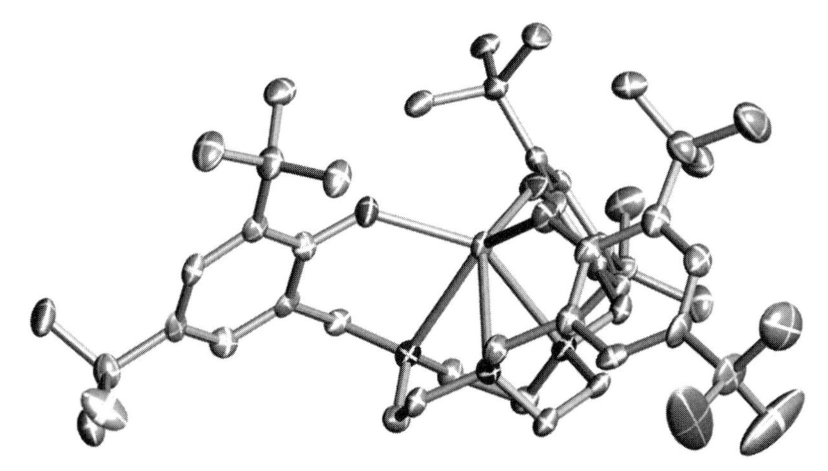

FIG. 13. Molecular Structure of **[((${}^{t\text{-Bu}}$ArO)$_3$tacn)U] (4$^{t\text{-Bu}}$-U)**.

ethereal solvents, such as Et_2O and THF at room or low temperature resulted in activation of the solvent.

Our initial results on the *tert*-butyl derivatized $[((^{t\text{-Bu}}ArO)_3tacn)U]$-system have suggested that undesired side reactions *trans* to the reactive site are effectively eliminated due to shielding by the triazacyclononane fragment. However, it is evident that the *ortho*-functionalized *tert*-butyl groups of the three aryloxide pendant arms do not form a protective cavity at the reactive, electron-rich uranium center. Instead, the tilted aryloxides force the *tert*-butyl groups to form a bowl-shaped cavity (Fig. 14) with little protection to prevent decomposition reactions, such as ligand and solvent degradation and formation of dinuclear complexes. It seemed clear that if the reactive uranium center could be more efficiently protected, a uranium(III) complex of the general $[((ArO)_3tacn)U]$-type would provide a powerful platform for reactivity studies at the apical position. In order to prevent dimerization, a sterically more encumbering derivative was required. Accordingly, protocols for the synthesis of the adamantane-functionalized ligand 1,4,7-tris(3-adamantyl-5-*tert*-butyl-2-hydroxybenzyl)1,4,7-triazacyclononane, $(^{Ad}ArOH)_3tacn$, and its corresponding U(III) precursor complex $[((^{Ad}ArO)_3tacn)U]$ (**4^{Ad}-U**) were developed (*25*).

Similar to the preparation of **$4^{t\text{-Bu}}$-U**, treating $(^{Ad}ArOH)_3tacn$ (**4^{Ad}-U**) with one equivalent of $[U(N(SiMe_3)_2)_3]$ in benzene produced the six-coordinate U(III) complex $[((^{Ad}ArO)_3tacn)U]$ (**4^{Ad}-U**) as a red-brown powder in multi-gram quantities (*20*). In striking contrast to **$4^{t\text{-Bu}}$-U**, complex **4^{Ad}-U** is stable in ethereal solutions and thus, can be re-crystallized from Et_2O. Single crystals of **4^{Ad}-U** were studied by X-ray crystallography (*25*) and its molecular core structure was compared to its parent complex, $[((^{t\text{-Bu}}ArO)_3tacn)U]$ (Fig. 12).

A comparison of the core molecular structures in **$4^{t\text{-Bu}}$-U** and **4^{Ad}-U** (Fig. 14) show comparable U–O(ArO) and U–N(tacn) bond distances (see Table I). The most striking difference in these structures is the

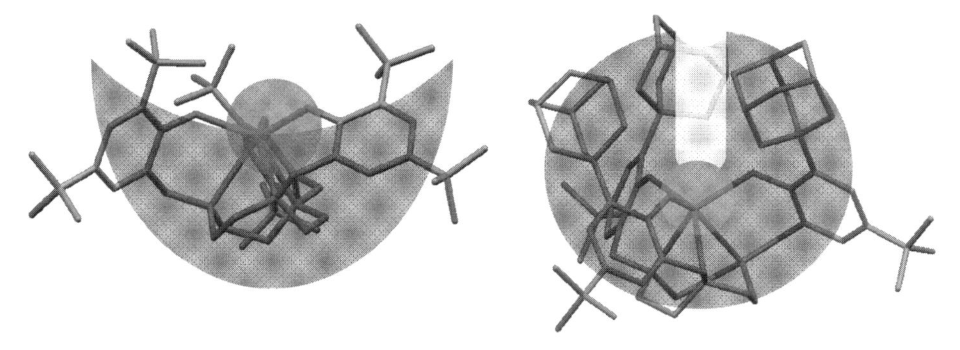

FIG. 14. Molecular structures and schematic representation of steric characteristics in **$[((^{t\text{-Bu}}ArO)_3tacn)U]$** (**$4^{t\text{-Bu}}$-U**, left) and **$[((^{Ad}ArO)_3\text{-}tacn)U]$** (**$4^{Ad}$-U**, right).

TABLE I

SELECTED CORE STRUCTURAL PARAMETERS FOR COMPLEXES [((${}^{t\text{-Bu}}$ArO)$_3$tacn)U] (**4$^{t\text{-Bu}}$-U**) AND [((${}^{\text{Ad}}$ArO)$_3$tacn)U] (**4^{Ad}-U**, TWO INDEPENDENT MOLECULES)

Structural parameters in **Å**	**4^{tBu}-U**	**4^{Ad}-U**
$d(\text{U–N1}_{\text{tacn}})$	2.676(5)	2.633(9)/2.648(8)
$d(\text{U–N2}_{\text{tacn}})$	2.669(5)	2.609(10)/2.669(10)
$d(\text{U–N3}_{\text{tacn}})$	2.659(5)	2.677(11)/2.628(10)
$d(\text{U–N}_{\text{av}})$	2.67	2.64
$d(\text{U–O1}_{\text{ArO}})$	2.230(4)	2.214(10)/2.219(8)
$d(\text{U–O2}_{\text{ArO}})$	2.240(4)	2.218(9)/2.220(9)
$d(\text{U–O3}_{\text{ArO}})$	2.245(4)	2.239(9)/2.248(8)
$d(\text{U–O}_{\text{av}})$	2.24	2.23
$\text{U}_{\text{out-of-plane shift}}$	−0.75	−0.85

displacement of the U ion from the trigonal aryloxide plane. While the out-of-plane shift in 4$^{t\text{-Bu}}$–U is −0.75 Å, the uranium ion in **4^{Ad}-U** was found to be −0.88 Å below the aryloxide plane.

The U ion displacement from the aryloxide plane in addition to the increased steric bulk provided by the adamantane substituents of **4^{Ad}-U** lead to a narrow and approximately ~4.7 Å deep cylindrical cavity. This cavity provides restricted access of an incoming ligand to the uranium ion, thereby protecting the uranium center from bimolecular decomposition reactions (Fig. 14).

III. Small-Molecule Activation

A. REACTIVITY OF [(TIMEN^{R})Co] COMPLEXES

The previously mentioned coordinatively and electronically unsaturated cobalt tris-carbene complexes were studied for small-molecule activation, with a focus on oxygen and nitrogen atom transfer reactions. Initially, complexes of the type [(TIMEN^{xyl})Co(X)]$^+$ (X = CO, Cl$^-$, CH_3CN) containing axial carbonyl, chloro, and acetonitrile ligands were prepared (Fig. 15). X-ray crystallographic analysis shows that the cobalt centers in these complexes again have the distorted trigonal–pyramidal coordination geometry, but in contrast to the previously discussed Co(I) precursors, the nitrogen is no longer coordinated. Instead the axial ligand completes the coordination sphere. Magnetic moments at ambient temperature are approximately 4.2–4.8 μ_B for these cobalt(II) complexes.

We found that a solution of [(TIMEN^{xyl})Co]Cl (**3^{xyl}-Co**) reacts cleanly with dioxygen at room temperature to form the 1:1 cobalt dioxygen species [(TIMEN^{xyl})Co(O_2)]$^+$ (Fig. 16, 3^{xyl}, **-Co(O_2)**). Infrared, NMR, and X-ray crystallography establish the side-on binding

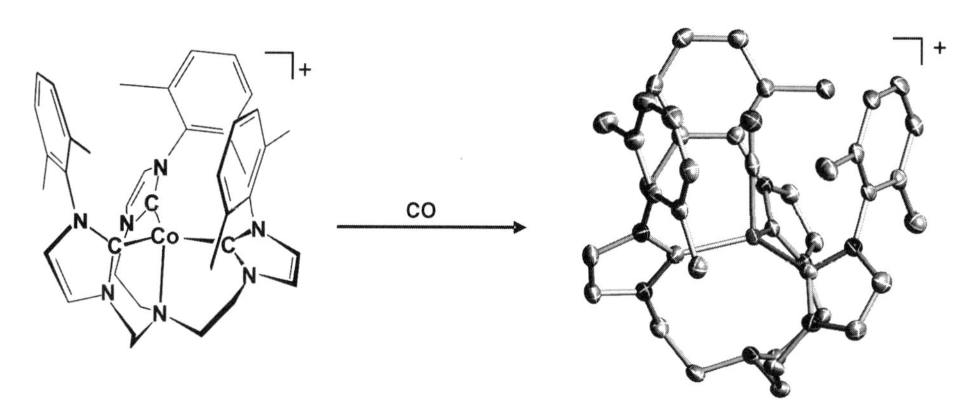

FIG. 15. Synthesis of **[(TIMEN^{xyl})Co(CO)]$^+$** with molecular structure.

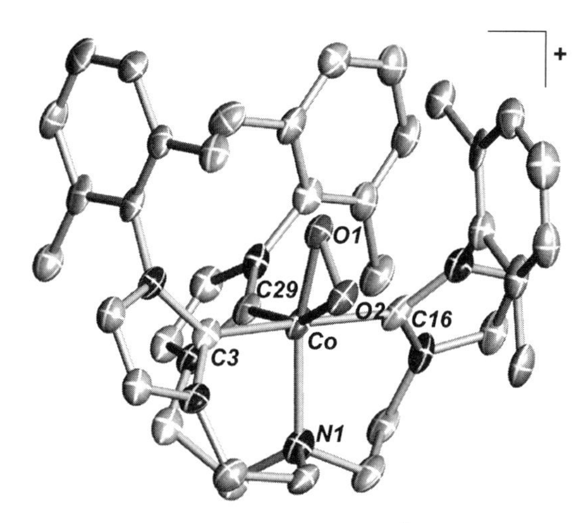

FIG. 16. Molecular structure of **[(TIMEN^{xyl})Co(O$_2$)]$^+$** (3^{xyl}-Co(O$_2$)).

mode of the coordinated peroxo ligand. The ^{16}O–^{16}O stretching frequency of 890 cm^{-1} is shifted to 840 cm^{-1} in an $^{18}O_2$ substituted sample. The molecular structure reveals a six-coordinate cobalt(III) ion in a pseudo-octahedral geometry. The dioxygen ligand is bound in an η^2 side-on fashion, with an O–O distance of 1.429 (3) Å. The coordination sphere is completed by the three carbenoid carbons and the anchoring nitrogen atom. The formation of this octahedral complex with TIMEN is noteworthy as it highlights the great structural flexibility of this type of ligand system.

The coordinated dioxygen in **3^{xyl}-Co(O$_2$)** is nucleophilic and reacts with electron-poor organic substrates such as benzoyl chloride, malonitrile, and tetracyanoethylene to transfer a single oxygen atom. A mechanism involving a homolytic Co–O bond cleavage has been proposed (*33*). The nucleophilic character of **3^{xyl}-Co(O$_2$)** was

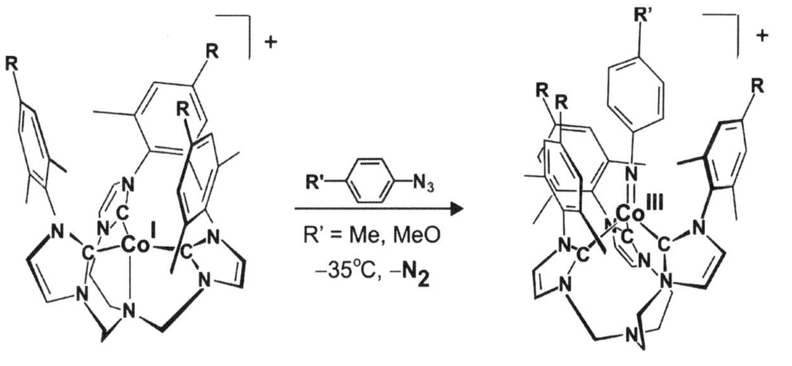

R = H (3^{xyl}-Co)
R = Me (3mes-Co)

R = H: R' = OMe (**3^{xyl}-Co(N(p-PhOMe))**; R' = Me (**3^{xyl}-Co(N(p-PhMe)**)
R = Me: R' = OMe (**3mes-Co(N(p-PhOMe))**; R' = Me (**3mes-Co(N(p-PhMe)**)

FIG. 17. Synthesis of cobalt(III) imide complexes.

supported by DFT calculations, which showed that the HOMO mainly consists of the dioxygen π^* orbital.

The cobalt(I) complexes also react with organic aryl azides to form the terminal cobalt(III) imido complexes [(TIMEN^{Ar})Co(NAr^{R})]Cl (Ar = xyl, mes, R = *p*-PhMe, *p*-PhOMe) at −35°C (Fig. 17). These deep-green complexes are fully characterized, including ^{1}H NMR, IR, UV–Vis spectroscopy, and combustion analysis (*10*). These are diamagnetic (d^6 low-spin, S = 0), and the ^{1}H NMR spectra suggest a C_3-symmetry of these molecules in solution.

The molecular structure of [(TIMEN^{mes})Co(N(*p*-PhOMe))]$^+$ (**3mes-Co (N(*p*-PhOMe)**), presented in Fig. 18, shows a pseudo-tetrahedral geometry featuring a short Co–N_{imido} bond (1.675 (2) Å) (*36*). This short distance, together with an almost linear Co–N–C angle of 168.6(2)°, indicates strong multiple bond character within the Co–NAr entity. According to DFT studies, this bond is most accurately formulated as a double bond (*36*).

B. Uranium: Metal–Alkane Coordination

Aware of the enhanced reactivity of trivalent [($^{t\text{-Bu}}$ArO)$_3$tacn)U] towards "non-innocent" solvents, such as ethers and chlorinated solvents, the reactivity of this molecule was challenged by exposure to more inert solvents like alkanes. Remarkably, recrystallization of **$4^{t\text{-Bu}}$-U** from pentane solutions containing various cycloalkanes, i.e., methylcyclohexane, afforded the coordination of one cycloalkane molecule to the electron-rich U center (Scheme 4) (*37*).

The X-ray diffraction analysis of these complexes (**$4^{t\text{-Bu}}$-Ua–$4^{t\text{-Bu}}$-Ue**) revealed atom positions and connectivities of one molecule of alkane in the coordination sphere of the uranium(III) center and a second molecule of cycloalkane co-crystallized in the lattice. Molecules **$4^{t\text{-Bu}}$-Ua–$4^{t\text{-Bu}}$-Ue** are isostructural and isomorphous. The molecular

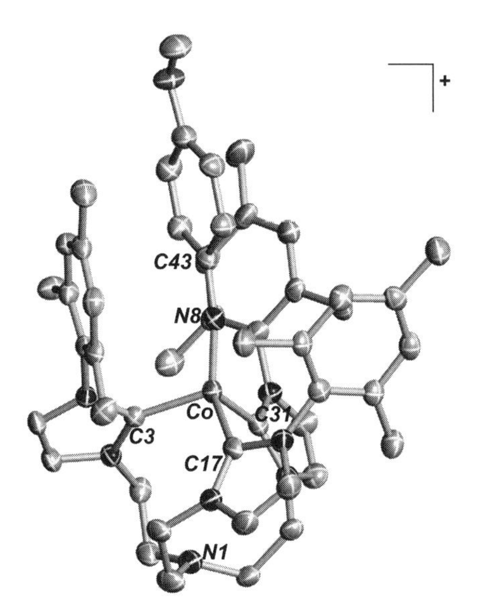

FIG. 18. Molecular structure of **[(TIMEN^{mes})Co(N(*p*-PhOMe))]$^+$** (3mes-Co(N(*p*-PhOMe)).

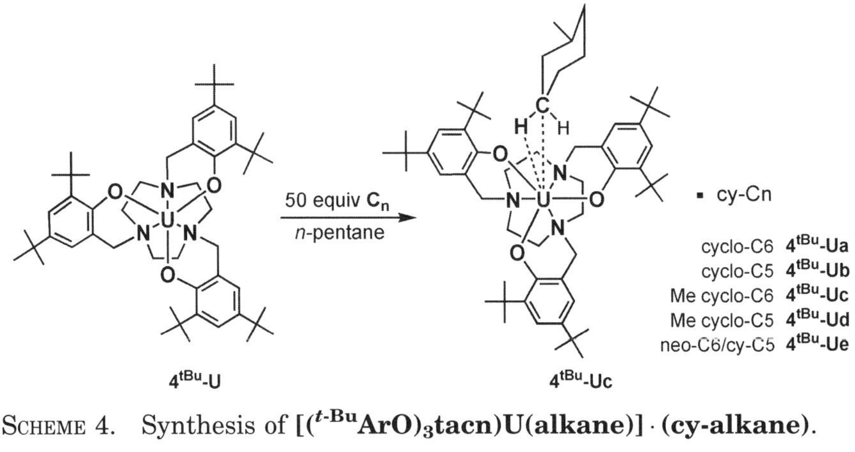

SCHEME 4. Synthesis of **[($^{t\text{-Bu}}$ArO)$_3$tacn)U(alkane)]** · **(cy-alkane)**.

structure of **4$^{t\text{-Bu}}$–Uc** is presented in Fig. 19, and is representative of the series of uranium-alkane complexes. The U–O(ArO) and U–N ligand distances of all alkane-adducts **4$^{t\text{-Bu}}$-Ua–4$^{t\text{-Bu}}$-Ue** vary from 2.24 to 2.26 Å and 2.67 to 2.69 Å, respectively (*37*), and thus, are similar to those found for the uranium(III) starting materials.

Most interestingly, the U–carbon bond distances to the axial cycloalkane *d*(U–C1S) in **4$^{t\text{-Bu}}$-Uc** and **4$^{t\text{-Bu}}$-Ud** were determined to be 3.864 and 3.798 Å, with the shortest U–carbon bond distance of 3.731 Å found in the solid-state structure of complex **4$^{t\text{-Bu}}$-Ue**. Considering that the sum of the van der Waals radii for a U–CH_2 or

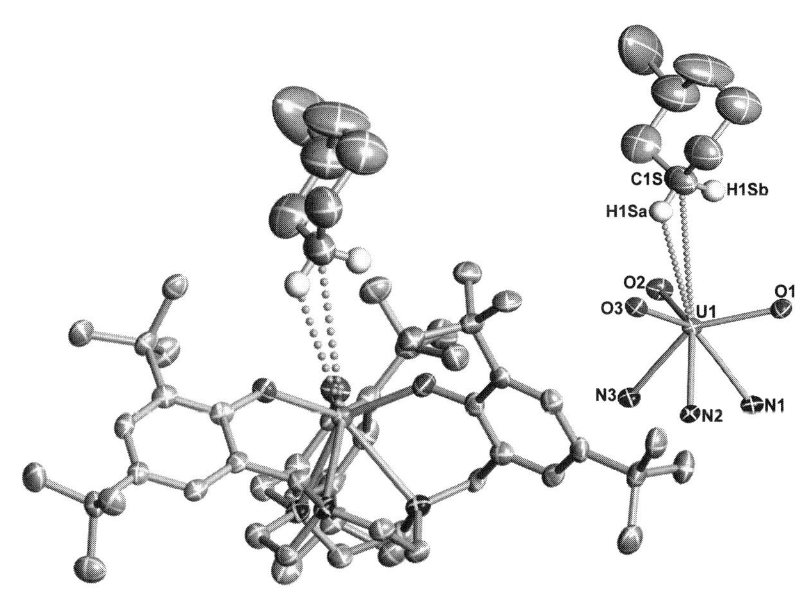

FIG. 19. Molecular representation of **[((${}^{t\text{-Bu}}$ArO)$_3$tacn)U(^{Me}cy-C6)], (4${}^{t\text{-Bu}}$-Uc)**. The dotted lines emphasize the η^2 H,C alkane coordination.

U–CH_3 contact is 3.9 Å (*38,39*),[1] the shorter U–C distances found in complexes of **4${}^{t\text{-Bu}}$-Ua** to **4${}^{t\text{-Bu}}$-Ue** are indicative of a weak but significant orbital interaction. Upon closer inspection, the structures of **4${}^{t\text{-Bu}}$-Ua–4${}^{t\text{-Bu}}$-Ue** exhibit short contacts between the peripheral *tert*-butyl groups and the axial alkane ligand (2.12 and 2.71 Å), thus, the observed alkane coordination may be additionally supported by van der Waals interactions. X-ray diffraction analysis of complexes **4${}^{t\text{-Bu}}$-Uc–4${}^{t\text{-Bu}}$-Ue** allows for calculation of the hydrogen atoms in proximity to the uranium center (calculated positions; d(C–H) = 0.96 Å). For all structures, an η^2-H,C coordination mode is observed that seems to be favored for the metal-alkane binding (*37*).

As with the starting complex, **4${}^{t\text{-Bu}}$-U**, the uranium ion is displaced below the plane formed by the aryloxide ligands by 0.66 Å for complexes **4${}^{t\text{-Bu}}$-Uc–4${}^{t\text{-Bu}}$-Ue**. This out-of-plane shift varies with U oxidation state and axial ligand strength. A more in depth discussion of this trend will be presented later.

C. URANIUM: C1 COORDINATION AND ACTIVATION CHEMISTRY

The greenhouse gas CO_2 is a valuable and renewable carbon source for the production of fine chemicals and fuels because it is readily

[1] The methyl group, CH_3, as a whole is assigned the van der Waals radius 2.0 Å. According to Pauling, the methylene group, CH_2, can be assigned the same value. The van der Waals radius for uranium was taken as 1.9 Å.

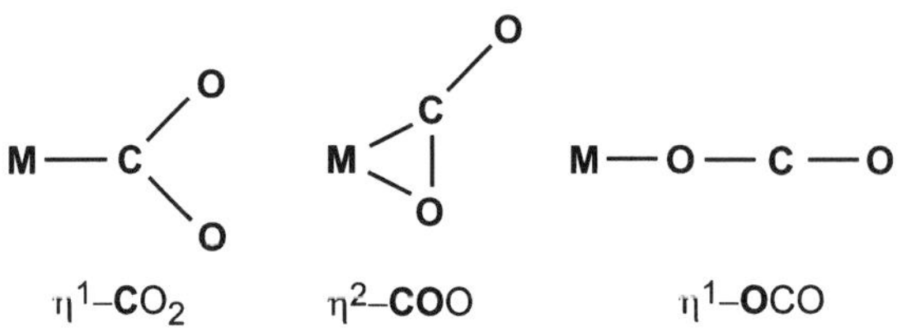

η^1–CO_2 η^2–COO η^1–OCO

FIG. 20. Coordination modes in mononuclear M-CO_2 complexes.

available and inexpensive. There has been much interest in metal-mediated multi-electron reduction of carbon dioxide. Coordinating carbon dioxide and carbon monoxide to a highly reducing metal such as uranium offers the possibility for metal-mediated multi-electron reduction of these molecules. For instance, the molecule [$(Cp)_3U(CO)$], first synthesized in 1986 by Andersen *et al.*, was crystallographically characterized (*40*) as derivatives in 1995 (*41*) and 2003 (*42*). Encouraged by the remarkable reactivity of our complexes and multiple coordination possibilities for CO_2 (Fig. 20) and CO, we set out to explore this possibility for CO_2 and CO activation and reduction with our highly reactive, sterically protected uranium(III) complexes.

C.1. Carbon monoxide and carbon dioxide activation assisted by [((${}^{t\text{-Bu}}$ArO)$_3$tacn)U]

A pentane solution of coordinatively unsaturated, trivalent **$4^{t\text{-Bu}}$-U** cleanly reacts with carbon monoxide by binding and activating the CO, resulting in rapid and quantitative formation of μ-CO bridged diuranium species [{((${}^{t\text{-Bu}}$ArO)$_3$tacn)U}$_2$(μ-CO)] (**$4^{t\text{-Bu}}$-U(CO)**), Scheme 4) (*43*). Infrared vibrational spectra of this material reproducibly show a band at 2092 cm^{-1} (in Nujol) suggestive of a CO ligand. However, this frequency appears to be rather high for a coordinated and activated CO molecule. Despite numerous attempts, CO isotopomers of **$4^{t\text{-Bu}}$-U(CO)** could not be synthesized. This puzzling lack of success in isotopomer synthesis is likely due to impurities in commercially available sources of CO isotopes that, among other impurities, contain up to 20 ppm CO_2 and O_2, which by themselves react rapidly with the U(III) precursors.

The molecular structure of **$4^{t\text{-Bu}}$-U(CO)** was characterized crystallographically and revealed an isocarbonyl-bonding motif, which is unique in actinide chemistry (Fig. 21) (*43*). A representative structure [{((${}^{t\text{-Bu}}$ArO)$_3$tacn)U}$_2$(μ-CO)] in crystals of **$4^{t\text{-Bu}}$-U(CO)** · 3 C_6H_6, (Fig. 20) was modeled by employing an asymmetric U–CO–U entity, with one short U–C bond and a longer U–O isocarbonyl interaction, disordered on two positions at the inversion center (rhombohedral space group *R*-3). The molecular structure of **5** exhibits two staggered [((${}^{t\text{-Bu}}$ArO)$_3$tacn)U]-fragments linked via a linearly bridged CO ligand in a $\mu:\eta^1,\eta^1$ fashion. The resolution of the data is limited and,

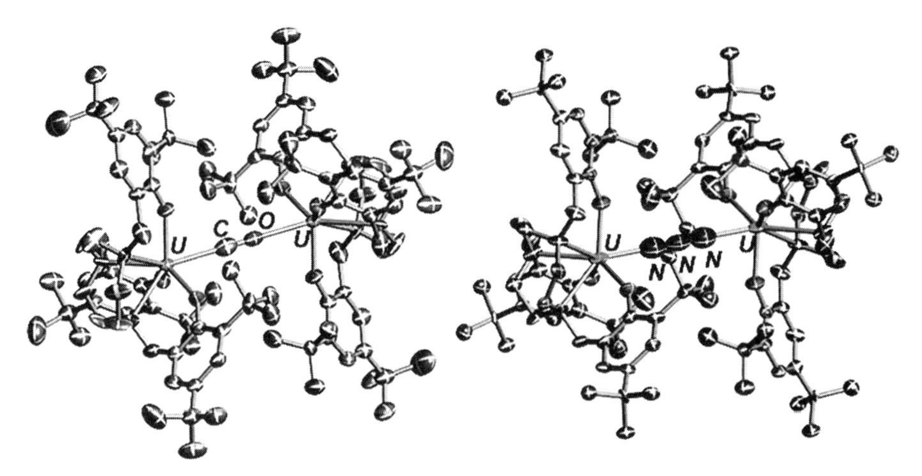

FIG. 21. Molecular representation of **[{((${}^{t\text{-Bu}}$ArO)$_3$tacn)U}$_2$(μ-CO)] (4$^{t\text{-Bu}}$-U(CO))** and **[{((${}^{t\text{-Bu}}$ArO)$_3$tacn)U}$_2$(μ-N$_3$)] (4$^{t\text{-Bu}}$-U(N$_3$)U)**.

therefore, no reliable bond distances between the CO ligand and the U center can be provided. Considering the unusual frequency of the ν(CO) stretch and the crystallographic disorder, it should be emphasized that the corresponding dinitrogen-bridged species could not be synthesized, neither under 1 atm nor an overpressure (80 psi) of N_2 gas, rendering the assignment of the disordered bridging CO-atoms unambiguous.

The U–O(ArO) and U–N(tacn) distances in **4$^{t\text{-Bu}}$-U(CO)** were determined to be 2.185 (5) and 2.676 (4) Å. This U–N(tacn) bond distance is very similar to that found in the [((${}^{t\text{-Bu}}$ArO)$_3$tacn)U] fragments of [((${}^{t\text{-Bu}}$ArO)$_3$tacn)U(alkane)] (d(U–N(tacn)) = 2.676 (6) Å) (*29*) and [((${}^{t\text{-Bu}}$ArO)$_3$tacn)U(NCCH$_3$)] (d(U–N(tacn)) = 2.699 (6) Å) (*37*). In contrast, the average U–O(ArO) bond distance in **4$^{t\text{-Bu}}$-U(CO)** is significantly shorter than those found in structurally related trivalent [((${}^{t\text{-Bu}}$ArO)$_3$tacn)U] complexes.

The displacement of the uranium ion out of the idealized trigonal aryloxide plane towards the coordinated triazacyclononane polyamine chelator in **4$^{t\text{-Bu}}$-U(CO)** was determined to be only –0.377 Å. A diagram depicting this structural parameter for known complexes of the [((${}^{t\text{-Bu}}$ArO)$_3$tacn)U(L$_{ax}$)] type clearly illustrates a linear correlation of higher oxidation states with smaller out-of-plane shifts (Table II) (*37,43,44*). Based on this correlation and in agreement with the crystallographic disorder, **4$^{t\text{-Bu}}$-U(CO)** can be assigned an average oxidation state of +3.5, suggestive of a mixed-valent uranium(III/IV) species. We suggested that **4$^{t\text{-Bu}}$-U(CO)** forms via a charge-separated U(IV)–CO$^{\bullet -}$ intermediate, which reacts with excess **4$^{t\text{-Bu}}$-U** to yield the formally mixed-valent U(IV)–CO–U(III) species **4$^{t\text{-Bu}}$-U(CO)**.

In this context, the μ-azido-bridged U(III/IV) species [{(L)U}$_2$(μ-N$_3$)] (4$^{t\text{-Bu}}$-U(N$_3$)U) was synthesized by reacting [(L)U^{IV}(N$_3$)] with [LU^{III}]

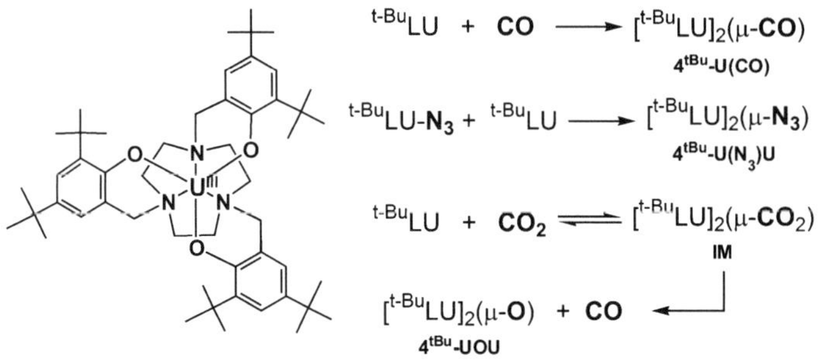

$^{t\text{-Bu}}$LU + CO ⟶ [$^{t\text{-Bu}}$LU]$_2$(μ-CO)
4^{tBu}-U(CO)

$^{t\text{-Bu}}$LU-N_3 + $^{t\text{-Bu}}$LU ⟶ [$^{t\text{-Bu}}$LU]$_2$(μ-N_3)
4^{tBu}-U(N_3)U

$^{t\text{-Bu}}$LU + CO_2 ⇌ [$^{t\text{-Bu}}$LU]$_2$(μ-CO_2)
IM

[$^{t\text{-Bu}}$LU]$_2$(μ-O) + CO
4^{tBu}-UOU

SCHEME 5. Synthesis of **[{((**$^{t\text{-Bu}}$**ArO)**$_3$**tacn)U}**$_2$**(μ-O)] (**$4^{t\text{-Bu}}$**-UOU), [{((**$^{t\text{-Bu}}$**ArO)**$_3$**tacn)U}**$_2$ **(μ-CO)] (**$4^{t\text{-Bu}}$**-U(CO)),** and **[((**$^{t\text{-Bu}}$**ArO)**$_3$**tacn)U (μ-**N_3**)] (**$4^{t\text{-Bu}}$**-U(**N_3**)U).**

to serve as an isostructural (and isomorphous, rhombohedral space group R_3) analogue of triatomic-bridged intermediate IM (see later) as well as an electronic model for mixed-valent **$4^{t\text{-Bu}}$-U(CO)** (Scheme 5). The out-of-plane shift found in mixed-valent **$4^{t\text{-Bu}}$-U(N_3)U** (–0.368 Å) is virtually identical to **$4^{t\text{-Bu}}$-U(CO)** (Table II) and thus supports the charge-separation proposed for mixed-valent **$4^{t\text{-Bu}}$-U(CO)**.

Addition of CO_2-saturated pentane to the deeply colored solution of red-brown **$4^{t\text{-Bu}}$-U** in pentane affords a colorless solution and subsequent formation of a pale-blue solution and CO gas. The pale-blue material was identified as the known μ-oxo bridged diuranium (IV/IV) complex [{(($^{t\text{-Bu}}$ArO)$_3$tacn)U}$_2$(μ-O)] (**$4^{t\text{-Bu}}$-UOU**, see earlier) (*43*). The driving force for this remarkable $2e^-$cleavage reaction of the thermodynamically stable CO_2 molecule likely is the concerted two ion U(III) to U(IV) oxidation, the most stable oxidation state in this system. Additionally, this reaction is sterically facilitated by the ligand environment. Attempts to isolate this colorless intermediate via solvent evaporation in vacuum resulted in recovery of **$4^{t\text{-Bu}}$-U** (Scheme 5). We suggest a dinuclear CO_2-bridged diuranium species **IM** as a possible intermediate. The reaction of **$4^{t\text{-Bu}}$-U** with CO_2 is reminiscent of the reductive cleavage of COS by [(Cp′)$_3$U] (Cp′ = MeC_5H_4), proceeding via a COS-bridged intermediate (*45,46*). Accordingly, complete $2e^-$reduction of CO_2 to yield CO and **$4^{t\text{-Bu}}$-UOU** likely proceeds stepwise via a fleeting CO_2-bridged intermediate, colorless **IM**, that is in equilibrium with **$4^{t\text{-Bu}}$-U** and CO_2.

C.2. Carbon dioxide coordination and activation assisted by [((^{Ad}ArO)$_3$tacn)U]

Due to the dimerization products formed by complex **$4^{t\text{-Bu}}$-U**, it seemed likely that the *tert*-butyl derivatized ligand did not provide sufficient steric bulk to obstruct a complete $2e^-$ reduction of CO_2 to CO. To further study this unique CO_2 activation at the electron-rich uranium center, the adamantyl compound, **$[((^{Ad}ArO)_3tacn)U]$ (4^{Ad}-U)**, was used as it provides greater steric bulk and allows for better protection of the open uranium coordination site.

Exposure of intensely colored $[((^{Ad}ArO)_3tacn)U]$ (**4^{Ad}-U**) in toluene or solid-state to CO_2 gas (1 atm) results in instantaneous discoloration of the samples (*44*). Colorless crystals of $[((^{Ad}ArO)_3tacn)U(CO_2)]$ (**4^{Ad}-U(CO_2)**) was isolated from a saturated CH_2Cl_2/Et_2O solution. The infrared spectrum (nujol) exhibits a distinct vibrational band centered at 2188 cm^{-1}, indicative of a coordinated and activated CO_2 ligand, which shifts to 2128 cm^{-1} upon exposure to 13 CO. The $^{12}C/^{13}C$ isotopic ratio R(2188/2128) of 1.0282 is close to that of free CO_2 gas (R 2349/2284 = 1.0284), suggesting the same linear geometry as free CO_2 and the same carbon motion in the ν_3 (ν_{as}OCO) vibrational mode. While the assignment of this vibrational band is unambiguous, we note that the ν_{as}(OCO) found in **4^{Ad}-U(CO_2)** is significantly lower than frequencies observed for other mononuclear M–CO_2 complexes with carbon-(η^1-CO_2) and carbon–oxygen bound (η^2-**OCO**) bent CO_2 ligands, in which ν_{as}(OCO) were reportedly found between 1550 and 1750 cm^{-1} (*47*).

In mononuclear complexes, such as Aresta's archetypal $[(Cy_3P)_2Ni(CO_2)]$ (Cy = cyclohexyl) (*48,49*) and Herskowitz's $[(diars)_2M(CO_2)(Cl)]$ (diars = o-phenylene bis(dimethylarsine); M = Ir, Rh) (*42*), the CO_2 ligand is coordinated in a bent η^1-**CO_2** or η^2-**OCO** fashion, the only previously known coordination modes for M-CO_2 complexes (Fig. 22). Among the few structurally characterized M-CO_2 complexes (*47*), only $[(Cy_3P)_2Ni(CO_2)]$ (*50*) and $[(bpy)_2Ru(CO_2)(CO)]$ (*51*) are shown here for illustrative purposes.

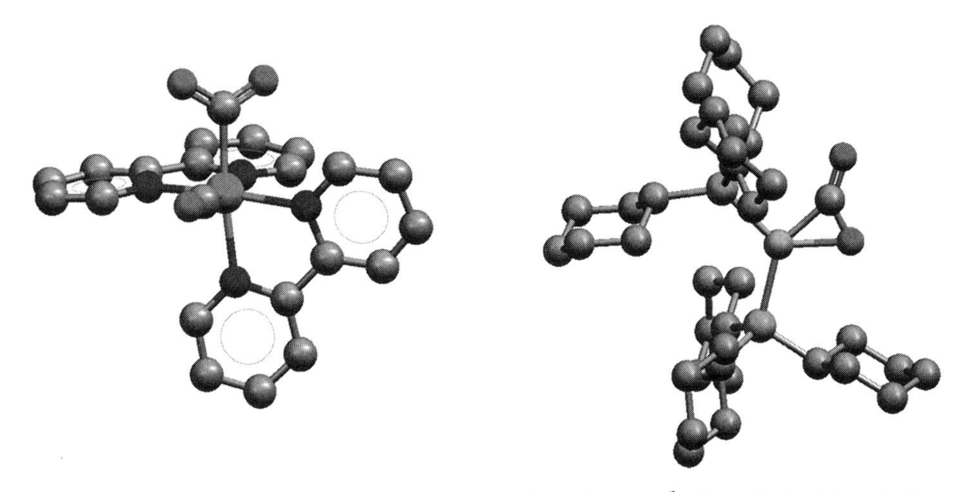

FIG. 22. Structural representations of the bent **η^1-CO_2** (left, **$[(bpy)_2Ru(CO_2)(CO)]$**) and η^2-**OCO** (right, $[(Cy_3P)_2Ni(CO_2)]$) coordination modes (CCDC codes: VUDKIO (left), DAJCUM (right)).

The linear metal–CO_2 coordination mode has been implicated in biological processes such as photosynthesis and previously had been suggested for the crystal structure of the iron-containing enzyme α-ketoglutarate reductase (*52*). Therefore it appears likely that end-on O-coordination is critical for binding, activation, and, most importantly, C-functionalization of the bound CO_2 molecule.

Analysis by X-ray diffraction of the colorless single-crystals obtained from the reaction of **4^{Ad}-U** with CO_2 confirmed the presence of a linearly coordinated η^1-**O**CO fashion and, more importantly, significantly activated CO_2 ligand. The CO_2 ligand in $[((^{Ad}ArO)_3tacn)U(CO_2)]$ 2.5 Et_2O (**4^{Ad}-U(CO_2)** 2.5 Et_2O) is presented in Fig. 23 (*44*). This CO_2 coordination mode is previously unprecedented in synthetic coordination chemistry and is likely facilitated by the adamantyl substituents of the aryloxide pendant arms. The U–OCO group has a U–O bond length of 2.351 (3) Å, a neighboring C–O bond length of 1.122 (4) Å, and a terminal C–O bond length of 1.277 (4) Å. The respective U–O–C and O–C–O angles of 171.1(2)° and 178.0(3)° are close to linear.

These metric parameters along with the redshifted frequency of the vibrational bands (ν_3: $\nu_{O^{12}CO} = 2188\,cm^{-1}$, $\nu_{O^{13}CO} = 2128\,cm^{-1}$) strongly suggest a molecular structure with charge-separated resonance structures $U(IV){=}O{=}C^{\bullet}{-}O^{-} \leftrightarrow U(III){-}^{+}O{\equiv}C{-}O^{-}$. The U(III) ion is either coordinated to a charge-separated CO_2 ligand or oxidized to U(IV) causing the CO_2 ligand to be reduced by one electron and resulting in activation of the inert C=O double bond.

It is interesting to compare structural and spectroscopic features in a systematic study, by comparison of the unique U–CO_2 complex (**4^{Ad}-U(CO_2)**) to analogous complexes. In the following section, we will compare **4^{Ad}-U(CO_2)** to the series of complexes $[((^{Ad}ArO)_3tacn)U^n(L)]^{m+}$ (n = III, IV, V, VI; m = 0, 1; and L = CH_3NC, N_3^-, CH_3NCN^-, OCN^-, and RN^{2-}) and discuss whether or not the bound CO_2 ligand

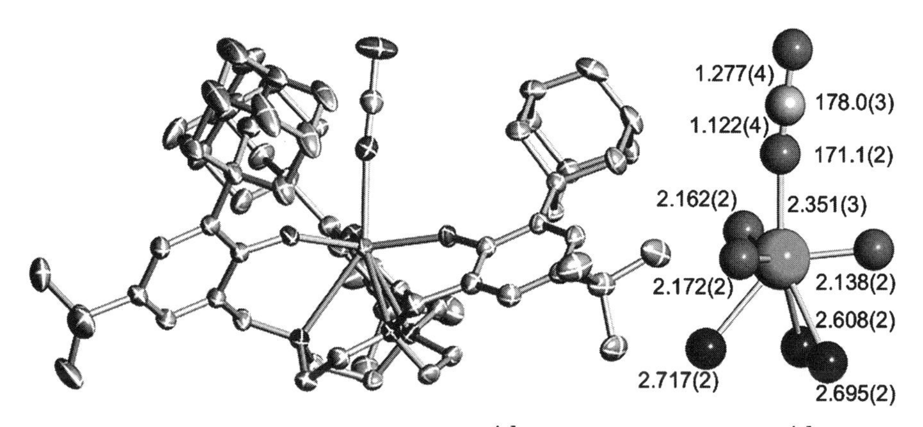

Fig. 23. Molecular representation of **$[((^{Ad}ArO)_3tacn)U(CO_2)]$** (**$4^{Ad}$-U($CO_2$)**, left) with core structure and metric parameters (right) in Å and deg.

in $[((^{Ad}ArO)_3tacn)U(CO_2)]$ is “activated” or “reduced” and if so, to what degree.

It was emphasized earlier that the U(III) starting complexes are deep red-brown in color. Upon addition of anionic ligands such as azide and isocyanate, however, the color bleaches and the result is colorless U(IV) complexes. Exposure of toluene solutions or even solid samples of deeply red-colored **4^{Ad}-U** to CO_2 gas resulted in instantaneous discoloration and, eventually, colorless crystals were obtained. Its solution UV/vis/NIR electronic absorption spectrum is strikingly similar to all other U(IV) complexes synthesized in this study. All other spectroscopic evidence, including advanced techniques, such as single-crystal diffraction, X-ray absorption, and SQUID magnetization studies, as well as the standard laboratory spectroscopy techniques that were accumulated so far suggest that the U ion in **4^{Ad}-U(CO_2)** is oxidized by $1e^-$ and thus, the CO_2 ligand is reduced to a $CO_2^{\bullet-}$ radical anion.

The molecular structure of **4^{Ad}-U(CO_2)** already revealed bond distances of the coordinated CO_2 ligand that were quite different from those of the symmetrical free CO_2 and thus, suggested a significant degree of ligand reduction. The uranium ion's displacement from the idealized trigonal plane of the three aryloxide ligators further implies the U ions oxidation upon CO_2 binding. While the out-of-plane shift in precursor **4^{Ad}-U** was determined to be −0.88 Å, the U ion in **4^{Ad}-U(CO_2)** and *all other* U(IV) heterocumulene complexes of the $[((^{Ad}ArO)_3tacn)U^{IV}(X)]$ system is displaced only 0.29–0.32 Å below the plane (Fig. 24) (*53*). In fact, an extrapolation of all available out-of-plane shifts vs. oxidation state places the two complexes with ambiguous oxidation states – $[((^{Ad}ArO)_3tacn)U(CO_2)]$ and $[\{((^{t\text{-}Bu}ArO)_3tacn)U\}_2(\mu\text{-}CO)]$ – correctly at +4 and +3.5.

TABLE II

SUMMARY OF OUT-OF-PLANE SHIFT CORRELATION WITH OXIDATION STATE

Oxidation state	Out-of-plane shift (in Å)	Complex (L_{ax})	
3	–0.66		$4^{t\text{-}Bu}$-Uc
3	–0.442	**4**	CH_3CN
3.5	–0.377	**5**	μ–CO
3.5	–0.368	**6**	$\mu\text{–}N_3$
4	–0.308	**7**	OCO
4	–0.318	**9**	N_3
4	–0.290	**14**	Cl
4	–0.301	**12**	NCO
4	–0.318	**13**	NCNMe
4	–0.292	**10**	N_3
5	–0.151	**8**	NsiMe
5	–0.147	**8b**	$NCPh_3$
5	–0.188	**11**	NsiMe

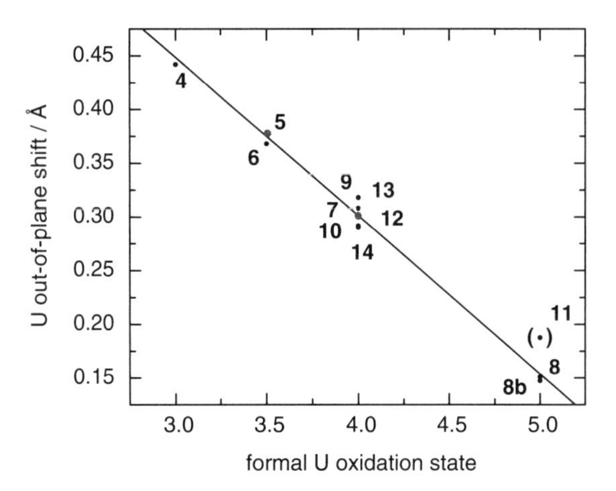

FIG. 24. Plot of out-of-plane shifts vs. oxidation state for complexes **[((^RArO)₃tacn)U(L_ax)]**.

Spectroscopic data further support an intramolecular redox reaction upon CO_2 coordination to **4^{Ad}-U**. The vibrational spectrum of **4^{Ad}-U(CO_2)** exhibits a band at 2188 cm^{-1} that shifts to 2128 cm^{-1} upon ^{13}C isotope labeling. Although this band can be assigned unambiguously to the asymmetric stretching vibration of the coordinated CO_2 ligand, a significantly higher redshift is expected for a $1e^-$ reduced CO_2 ligand. Accordingly, upon initial observation, a comparison of CO_2 stretching frequencies to those of known M–CO_2 complexes, which feature signals $\nu(CO_2)$ between 1600 and 1750 cm^{-1}, suggests that the activation found in **4^{Ad}-U(CO_2)** cannot be a "complete" one-electron reduction. However, considering the linear η^1-OCO coordination mode in **4^{Ad}-U(CO_2)**, which is unprecedented, a comparison of vibrational frequencies with complexes that possess bent C- (η^1-**COO**) or C,O-bound (η^2-**OCO**) CO_2 ligands may not be valid.

SQUID magnetization measurements of **4^{Ad}-U(CO_2)** were recorded and compared to the large number of similar complexes (Fig. 25). The magnetic moment μ_{eff} of **4^{Ad}-U(CO_2)** was determined to be 2.89 μ_B at 300 K and 1.51 μ_B at 5 K. Although the room-temperature moment of **4^{Ad}-U(CO_2)** is close to the magnetic moment found for the azide complex **4^{Ad}-U(N_3)**, the low-temperature value is similar to that of the U(III) (f^3) starting material **4^{Ad}-U**(1.73 μ_B at 5 K), which has a doublet ground state at low temperatures. As mentioned earlier, the magnetic moments of U(III) (f^3) and U(IV) (f^2) complexes at room temperature are generally very similar and often do not allow for an unambiguous assignment of the complexes' oxidation state. The temperature dependence of μ_B in the range 4–300 K, however, shows a curvature reminiscent of data obtained for all closely related U(IV) complexes of this type. Although U(IV) complexes possess a singlet ground state, which typically results in magnetic moments of *ca.* 0.5–0.8 μ_B, the magnetic moment of **4^{Ad}-U(CO_2)** at low temperatures is significantly

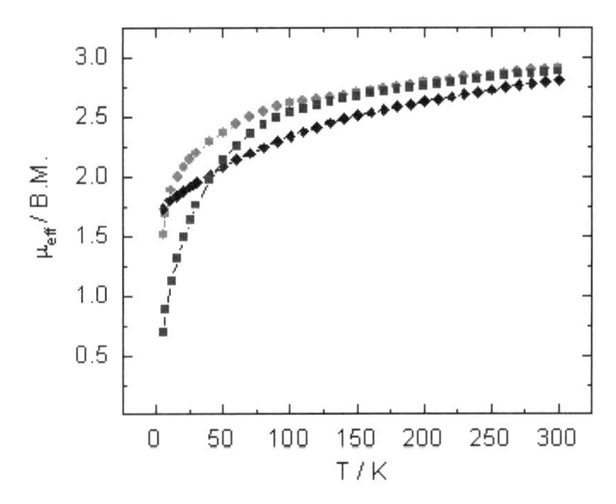

FIG. 25. Temperature dependant SQUID magnetization data for the U(III) and U(IV) complexes **[((AdArO)$_3$tacn)U] (4^{Ad}-U, -♦-)**, **[((AdArO)$_3$tacn)U(N$_3$)]** (**4^{Ad}-U(N$_3$)**, -■-), and **[((AdArO)$_3$tacn)U(CO$_2$)]** (**4^{Ad}-U(CO$_2$)**, -●-).

higher, suggesting that the open-shell $CO_2^{\bullet -}$, unlike the closed-shell N_3^- ligand, likely contributes to the observed increased magnetic moment of **4^{Ad}-U(CO$_2$)** at low temperatures. The temperature dependence and low-temperature value of **4^{Ad}-U(CO$_2$)** are in agreement with the description of the CO_2 ligand as a one-electron reduced $CO_2^{\bullet -}$ radical anion coordinated to a U(IV) ion.

Finally, in order to unambiguously determine the uranium ion's +IV oxidation state in **4^{Ad}-U(CO$_2$)**, UL$_3$ edge energy XANES measurements of the isostructural complexes $[((^{Ad}ArO)_3tacn)U^m(L)]^{n+}$ (with L = CH$_3$CN, N_3^-, and Me$_3$SiN^{2-}, m = III, IV, V, and VI, and n = 0,1) were performed and compared to **4^{Ad}-U(CO$_2$)**. However, preliminary data analysis shows the UL$_3$ edge energy for $[((^{Ad}ArO)_3tacn)U(CO_2)]$ is virtually identical to that measured for the uranium IV complex $[((^{Ad}ArO)_3tacn)U(N_3)]$. This observation confirms the +IV oxidation state in $[((^{Ad}ArO)_3tacn)U^{IV}(\eta^1\text{–}CO_2^{\bullet -})]$, implying that the coordinated carbon dioxide ligand is in fact reduced by one electron. Future computational studies will attempt to shed light on the peculiar electronic structure and spectroscopic features, such as the complexes relatively low $\nu_{as}(CO_2)$ redshift.

IV. Conclusions

Herein we have reported the synthesis, characterization, and reactivity of both transition metal and actinide elements supported by tripodal ligand scaffolds. The carbon-anchored ligands TIME normally form polynuclear species while the nitrogen-anchored ligands TIMEN can coordinate to transition metals in a 1:1, κ^3 fashion, rendering the coordinated metal centers in well-protected pockets.

Particularly promising are the aryl-substituted TIMEN^{Ar} ligands that give rise to coordinatively and electronically unsaturated transition metal complexes. The cobalt complexes of these ligands show interesting reactivities towards dioxygen and organic azides. Our laboratory has also shown that tripodal aryloxide-functionalized triazacyclononane ligands can effectively encapsulate the highly reactive uranium center. Unique binding and small-molecule activation at low-valent uranium centers, resulting in potentially effective agents for functionalization of otherwise inert molecules, was observed. Contrasting the reactivity between the *t*-butyl and adamantyl ligand derivatives has shown that there is an extreme difference in the reactivity of small molecules such as carbon dioxide. By developing a family of well-protected uranium complexes and examining the reactivity with small molecules, general trends can be established for these systems, including a correlation between the oxidation state and geometrical parameters. Herein, several examples of tripodal ligand systems were described, including both the synthesis and metalation of these important molecules. All of the resulting complexes described are distinctive as they represent a set of isostructural complexes possessing a range of metals, oxidation states, and differing electronic and magnetic behaviors. This presents a benefit for the understanding of fundamental transition metal and actinide chemistry – uranium in particular. These ligands open the door for exploration of the nature of covalency and the role of f-orbitals in actinide-ligand bonding can be advanced.

References

1. (a) Trofimenko, S. *Prog. Inorg. Chem.* **1986**, *34*, 115–210. (b) Trofimenko, S. *Chem. Rev.* **1993**, *93*, 943–980. (c) Trofimenko, S. *"Scorpionates, The Coordination Chemistry of Polypyrazolylborate Ligands"*; Imperial College Press: London, **1999**.
2. Schrock, R. R. *Acc. Chem. Res.* **1997**, *30*, 9–16.
3. Borovik, A. S. *Acc. Chem. Res.* **2005**, *38*, 54–61.
4. Betley, T. A.; Peters, J. C. *Inorg. Chem.* **2003**, *42*, 5074–5084.
5. Slugovc, C.; Padilla-Martinez, I.; Sirol, S.; Carmona, E. *Coord. Chem. Rev.* **2001**, *213*, 129–157.
6. Diaz-Requejo, M. M.; Belderrain, T. R.; Trofimenko, S.; Perez, P. J. *J. Am. Chem. Soc.* **2001**, *123*, 3167–3168.
7. Diaz-Requejo, M. M; Belderrain, T. R.; Perez, P. J. *Chem. Commun.* **2000**, 1853–1854.
8. Morilla, M. E.; Diaz-Requejo, M. M.; Belderrain, T. R.; Nicasio, M. C.; Trofimenko, S.; Perez, P. J. *Chem. Commun.* **2002**, 2998–2999.
9. Egan, J. W.; Haggerty, B. S.; Rheingold, A. L.; Sendlinger, S. C.; Theopold, K. H. *J. Am. Chem. Soc.* **1990**, *112*, 2445–2446.
10. Parkin, G. *Chem. Commun.* **2000**, 1971–1985.
11. Pearson, R. J. *J. Am. Chem. Soc.* **1963**, *85*, 3533–3539.
12. Cummins, C. C.; Lee, J.; Schrock, R. R.; Davis, W. D. *Angew. Chem. Int. Ed. Engl.* **1992**, *31*, 1501–1503.
13. Verkade, J. G. *Acc. Chem. Res.* **1993**, *26*, 483–489.

14. (a) Yandulov, D. V.; Schrock, R. R. *J. Am. Chem. Soc.* **2002**, *124*, 6252–6253. (b) Yandulov, D. V.; Schrock, R. R. *Science* **2003**, *301*, 76–78.
15. Yandulov, D. V.; Schrock, R. R.; Rheingold, A. L.; Ceccarelli, C.; Davis, W. M. *Inorg. Chem.* **2003**, *42*, 796–813.
16. (a) Fujita, K.; Schenker, R.; Gu, W. W.; Brunold, T. C.; Cramer, S. P.; Riordan, C. G. *Inorg. Chem.* **2004**, *43*, 3324–3326. (b) Tard, C.; Liu, X. M.; Ibrahim, S. K.; Bruschi, M.; De Gioia, L.; Davies, S. C.; Yang, X.; Wang, L. S.; Sawers, G.; Pickett, C. J. *Nature* **2005**, *433*, 610–613.
17. (a) Mayer, H. A.; Kaska, W. C. *Chem. Rev.* **1994**, *94*, 1239–1272. (b) Muth, A.; Walter, O.; Huttner, G.; Asam, A.; Zsolnai, L.; Emmerich, C. J *Organomet. Chem.* **1994**, *468*, 149–163. (c) Winterhalter, U.; Zsolnai, L.; Kircher, P.; Heinze, K.; Huttner, G. *Eur. J. Inorg. Chem.* **2001**, 89–103. (d) Peters, J. C.; Feldman, J. D.; Tilley, T. D. *J. Am. Chem. Soc.* **1999**, *121*, 9871-9872. (e) Ott, J.; Venanzi, I. M.; Ghilardi, C. A.; Midollini, S.; Orlandini, A. *J. Organomet. Chem.* **1985**, *291*, 89–100. (f) Dawson, J. W.; Lane, B. C.; Mynott, R. J.; Venanzi, L. M. *Inorg. Chim. Acta* **1971**, *5*, 25–29.
18. Betley, T. A.; Peters, J. C. *J. Am. Chem. Soc.* **2003**, *125*, 10782–10783.
19. (a) Sacconi, L.; Bertini, I. *J. Am. Chem. Soc.* **1968**, *90*, 5443–5446. (b) Ghilardi, C. A.; Midollini, S.; Sacconi, L. *Inorg. Chem.* **1975**, *14*, 1790–1795.
20. (a) Herrmann, W. A. *Angew. Chem. Int. Edit.* **2002**, *41*, 1291–1309. (b) Bourissou, D.; Guerret, O.; Gabbai, F. P.; Bertrand, G. *Chem. Rev.* **2000**, *100*, 39–91. (c) Dias, H. V. R.; Jin, W. C. *Tetrahedron Lett.* **1994**, *35*, 1365–1366.
21. (a) Jenkins, D. M.; Di Bilio, A. J.; Allen, M. J.; Betley, T. A.; Peters, J. C. *J. Am. Chem. Soc.* **2002**, *124*, 15336–15350. (b) Jenkins, D. M.; Peters, J. C. *J. Am. Chem. Soc.* **2003**, *125*, 11162–11163.
22. Nakai, H.; Tang, Y. J.; Gantzel, P.; Meyer, K. *Chem. Commun.* **2003**, 24–25.
23. Frankell, R.; Birg, C.; Kernbach, U.; Habereder, T.; Noth, H.; Fehlhammer, W. P. *Angew. Chem. Int. Edit.* **2001**, *40*, 1907–1910.
24. Chaudhuri, P.; Wieghardt, K. In: *"Progress in Inorganic Chemistry"*; vol. 50; John Wiley & Sons: Hoboken, New Jersey, **2001**, pp. 151–216.
25. Nakai, H.; Hu, X. L.; Zakharov, L. N.; Rheingold, A. L.; Meyer, K. *Inorg. Chem.* **2004**, *43*, 855–857.
26. Wang, H. M. J.; Lin, I. J. B. *Organometallics* **1998**, *17*, 972–975.
27. Arduengo, A. J.; Dias, H. V. R.; Calabrese, J. C.; Davidson, F. *Organometallics* **1993**, *12*, 3405–3409.
28. (a) Grundemann, S.; Kovacevic, A.; Albrecht, M.; Faller, J. W.; Crabtree, R. H. *Chem. Commun.* **2001**, 2274–2275. (b) Grundemann, S.; Kovacevic, A.; Albrecht, M.; Faller, J. W.; Crabtree, R. H. *J. Am. Chem. Soc.* **2002**, *124*, 10473–10481.
29. Hu, X.; Castro-Rodriguez, I.; Meyer, K. *Organometallics* **2003**, *22*, 3016–3018.
30. (a) Boehme, C.; Frenking, G. *Organometallics* **1998**, *17*, 5801–5809. (b) Deubel, D. V. *Organometallics* **2002**, *21*, 4303–4305. (c) Abernethy, C. D.; Codd, G. M.; Spicer, M. D.; Taylor, M. K. *J. Am. Chem. Soc.* **2003**, *125*, 1128–1129.
31. Hu, X.; Meyer, K. *J. Am. Chem. Soc.* **2004**, *126*, 16322–16323.
32. Nemcsok, D.; Wichmann, K.; Frenking, G. *Organometallics* **2004**, *23*, 3640–3646.
33. (a) Glueck, D. S.; Hollander, F. J.; Bergman, R. G. *J. Am. Chem. Soc.* **1989**, *111*, 2719–2721. (b) Glueck, D. S.; Wu, J. X.; Hollander, F. J.; Bergman, R. G. *J. Am. Chem. Soc.* **1991**, *113*, 2041–2054. (c) Mindiola, D. J.; Hillhouse, G. L. *J. Am. Chem. Soc.* **2001**, *123*, 4623–4624.
34. Vandersluys, W. G.; Burns, C. J.; Huffman, J. C.; Sattelberger, A. P. *J. Am. Chem. Soc.* **1988**, *110*, 5924–5925.
35. (a) Clark, D. L.; Sattelberger, A. P.; Andersen, R. A. *Inorg. Syn.* **1997**, *31*, 307–315. (b) Stewart, J. L.; Andersen, R. A. *Polyhedron* **1998**, *17*, 953–958.
36. Thyagarajan, S.; Shay, D. T.; Incarvito, C. D.; Rheingold, A. L.; Theopold, K. H. *J. Am. Chem. Soc.* **2003**, *125*, 4440–4441.
37. Castro-Rodriguez, I.; Olsen, K.; Gantzel, P.; Meyer, K. *J. Am. Chem. Soc.* **2003**, *125*, 4565–4571.

38. Pauling, L. *"The Nature of the Chemical Bond"*; 3rd edn. Cornell University Press: Ithaca, NY, **1960**.
39. Huheey, J. E.; Keiter, E. A.; Keiter, R. L. *"Inorganic Chemistry: Principles of Structure and Reactivity"*; 4th edn. HarperCollins College Publishers, **1993**.
40. Brennan, J. G.; Andersen, R. A.; Robbins, J. L. *J. Am. Chem. Soc.* **1986**, *108*, 335–336.
41. Parry, J.; Carmona, E.; Coles, S.; Hursthouse, M. *J. Am. Chem. Soc.* **1995**, *117*, 2649–2650.
42. Evans, W. J.; Kozimor, S. A.; Nyce, G. W.; Ziller, J. W. *J. Am. Chem. Soc.* **2003**, *125*, 13831–13835.
43. Castro-Rodriguez, I.; Meyer, K. *J. Am. Chem. Soc.* **2005**, *127*, 11242–11243.
44. Castro-Rodriguez, I.; Nakai, H.; Zakharov, L. N.; Rheingold, A. L.; Meyer, K. *Science* **2004**, *305*, 1757–1759.
45. Brennan, J. G.; Andersen, R. A.; Zalkin, A. *Inorg. Chem.* **1986**, *25*, 1756–1760.
46. Brennan, J. G.; Andersen, R. A.; Zalkin, A. *Inorg. Chem.* **1986**, *25*, 1761–1765.
47. Gibson, D. H. *Chem. Rev.* **1996**, *96*, 2063–2095.
48. Aresta, M.; Nobile, C. F. *J. Chem. Soc. Dalton* **1977**, 708–711.
49. Aresta, M.; Nobile, C. F.; Albano, V. G.; Forni, E.; Manassero, M. *J. Chem. Soc. Chem. Comm.* **1975**, 636–637.
50. Dohring, A.; Jolly, P. W.; Kruger, C.; Romao, M. J. *Z. Naturforsch. (B)* **1985**, *40*, 484–488.
51. Tanaka, H.; Nagao, H.; Peng, S. M.; Tanaka, K. *Organometallics* **1992**, *11*, 1450–1451.
52. Lee, H. J.; Lloyd, M. D.; Harlos, K.; Clifton, I. J.; Baldwin, J. E.; Schofield, C. J. *J. Molec. Biol.* **2001**, *308*, 937–948.
53. Castro-Rodriguez, I.; Meyer, K. *Chem. Commun.* **2007**, 1353–1368.

β-CYCLODEXTRIN-LINKED Ru COMPLEXES FOR OXIDATIONS AND REDUCTIONS

W.-D. WOGGON, ALAIN SCHLATTER and HAO WANG

Department of Chemistry, University of Basel, St. Johanns-Ring 19, CH-4056 Basel, Switzerland

I. Introduction

Oxidation states of ruthenium ranging from +VIII to –II render ruthenium complexes a unique scaffold for both oxidations and reductions. We review here some of our results in both areas employing an enzyme-like design, i.e., suitable ruthenium complexes are covalently attached to β-cyclodextrins (β-CDs) which combines the site of reactivity with a binding pocket for lipophilic substrates.

ADVANCES IN INORGANIC CHEMISTRY
VOLUME 60 ISSN 0898-8838 / DOI: 10.1016/S0898-8838(08)00002-0

II. Mimicking the Enzymatic Cleavage of Carotenoids

A. The Central Cleavage of β-Carotene

The central cleavage of β-carotene **1** is most likely the major pathway by which mammals produce the required retinoids (*1*), in particular, retinal **2**, which is essential for vision and is subsequently oxidized to retinoic acid **3** and reduced to retinol **4**. An alternative excentric cleavage of **1** has been reported involving scission of the double bond at C7′–C8′ producing β-8′-apocarotenal **5** (*2a,2b*) which is subsequently oxidized to **2** (Fig. 1) (*2c*). The significance of carotene metabolites such as **2**, **3** and **4** to embryonic development and other vital processes such as skin and membrane protection is a major concern of medicinal chemistry.

The enzyme catalyzing the formation of retinal **2** by means of central cleavage of β-carotene **1** has been known to exist in many tissues for quite some time. Only recently, however, the active protein was identified in chicken intestinal mucosa (*3*) following an improvement of a novel isolation and purification protocol and was cloned in *Escherichia coli* and BHK cells (*4,5*). Iron was identified as the only metal ion associated with the (overexpressed) protein in a 1:1 stoichiometry and since a chromophore is absent in the protein heme coordination and/or iron complexation by tyrosine can be excluded. The structure of the catalytic center remains to be elucidated by X-ray crystallography but from the information available it was predicted that the active site contains a mononuclear iron complex presumably consisting of histidine residues. This suggestion has been confirmed by

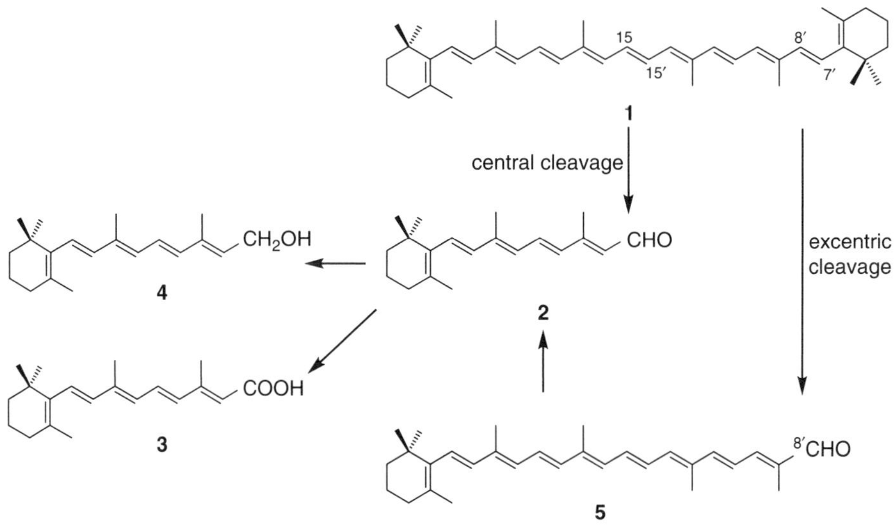

Fig. 1. Central and excentric cleavage of β-carotene **1**.

FIG. 2. Active site of apocarotenoid-15,15-oxygenase, adapted from Schulz *et al.* (*2c*).

investigation of the related enzyme apocarotenoid-15,15′-oxygenase (ACO) from Synechocystis sp. PCC 6803. The protein was expressed in *E. coli* and has been recently crystallized (*2c*). β-Carotene itself is not a substrate but rather all-E apocarotenal **5**. The X-ray structure of the protein displays a nearly perfect octahedral coordination of four histidines and one water molecule to Fe(II) (Fig. 2); similar to the active site suggested for the enzyme which catalyzes the central cleavage.

Comparison within the family of carotenoid oxygenases show that the four histidines are strictly conserved as well as their close environment. Presumably, all members of this family, including the protein catalyzing central cleavage of β-carotene, share a common chain fold, possess similar active centers and follow a similar reaction mechanism, *vide infra*.

The enzymatic reaction mechanism was determined by incubating α-carotene **6**, a non-symmetric substrate of the enzyme, under a $^{17}O_2$ atmosphere in $H_2^{18}O$ followed by isolation and characterization of derivatives of the cleavage product **2** (*6*). Accordingly, the enzyme cleaving the central double bond of **1** was found to be a non-heme iron monooxygenase (β-carotene 15,15′-monooxygenase) and not dioxygenase as termed earlier (Fig. 3). From the chemical point of view this enzymatic reaction is very unusual for various reasons: (i) the reaction

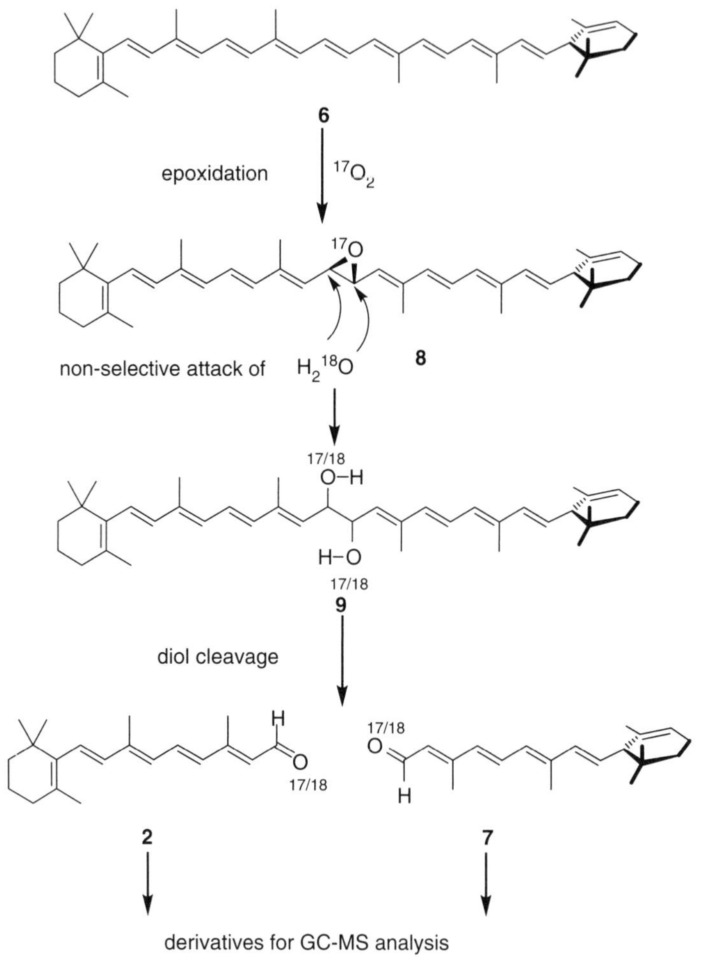

FIG. 3. The reaction mechanism of the central cleavage of carotenoids, catalyzed by β-carotene 15,15′-monooxygenase (*6*).

proceeds completely regiospecific oxidizing only one E-configured, conjugated double bond out of nine in the linear part of the molecule; (ii) the proposed mechanism implies a multistep sequence including epoxidation, see **8**, an epoxide hydrolase reaction and C–C bond cleavage of the diol **9**. For each of these steps enzymatic reactions are known to exist. Epoxidation is known to be catalyzed by cytochromes P450 (*7*).

As well as non-heme oxygenases (*8*), epoxide hydrolases of bacterial origin have been developed so far as to become preparatively useful (*9*) and P450 catalyzed diol cleavage occurs during the side chain cleavage of cholesterol (*10*) and during the biosynthesis of biotin (*11*). Alternatively and in agreement with the labeling experiments the cleavage of **6** and **1**, respectively, could proceed via a Crigee-type

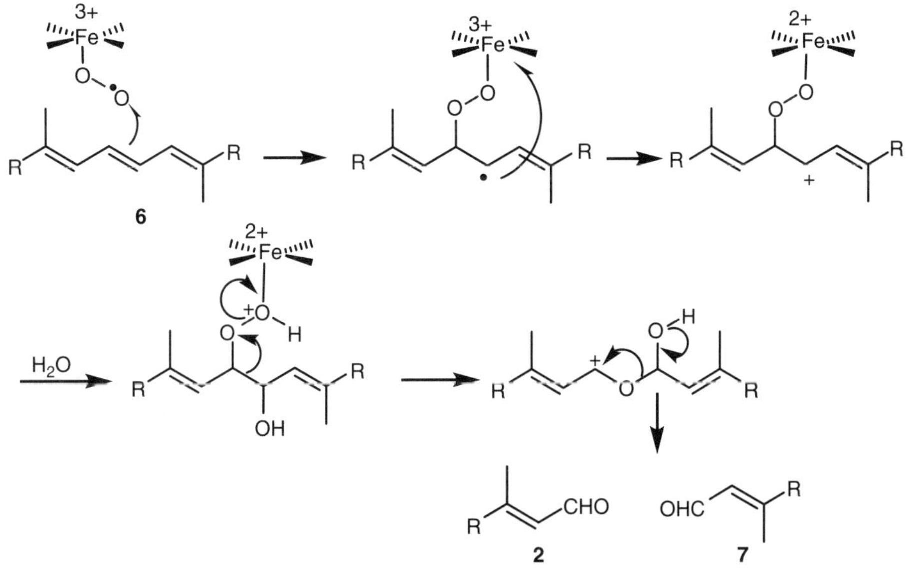

FIG. 4. Crigee-type mechanism to cleave carotenoid double bonds.

mechanism as shown in Fig. 4. A sequence of these events, however, catalyzed by a single enzyme is quite unique in particular for a non-heme monooxygenase.

Therefore, it is highly challenging to an organic chemist to design a synthetic catalyst which in principle can act the same way as the natural one does, to produce retinal **2** by selectively cleaving just the central C(15)–C(15′) double bond of β-carotene **1**. Moreover, the mimic would definitely help to understand a plausible reaction pathway of this very special class of enzymes.

B. Design and Synthesis of Enzyme Mimics of β-Carotene 15,15′-monooxygenase

To mimic the regioselective enzyme we envisaged to synthesize receptor **10** having the following fundamental criteria, namely (i) the association constant, K_a, of **1**–**10** is in orders of magnitude greater than that for retinal **2** to rule out any sort of product inhibition; (ii) introduction of a reactive metal complex which is capable of cleaving E-configured double bonds; and (iii) the use of a co-oxidant that is not reactive to β-carotene **1** in the absence of the metal complex.

After initial molecular modeling studies using the MOLOC program (*12*), the supramolecular construct **10** consisting of two β-CDs moieties linked by a porphyrin spacer was designed for the binding of **1**. Each of the CDs was shown to be capable of binding one of the cyclohexenoid end-groups of β-carotene **1**, leaving the porphyrin linker to span over the long polyene chain, that also shows, at least in static view, that the

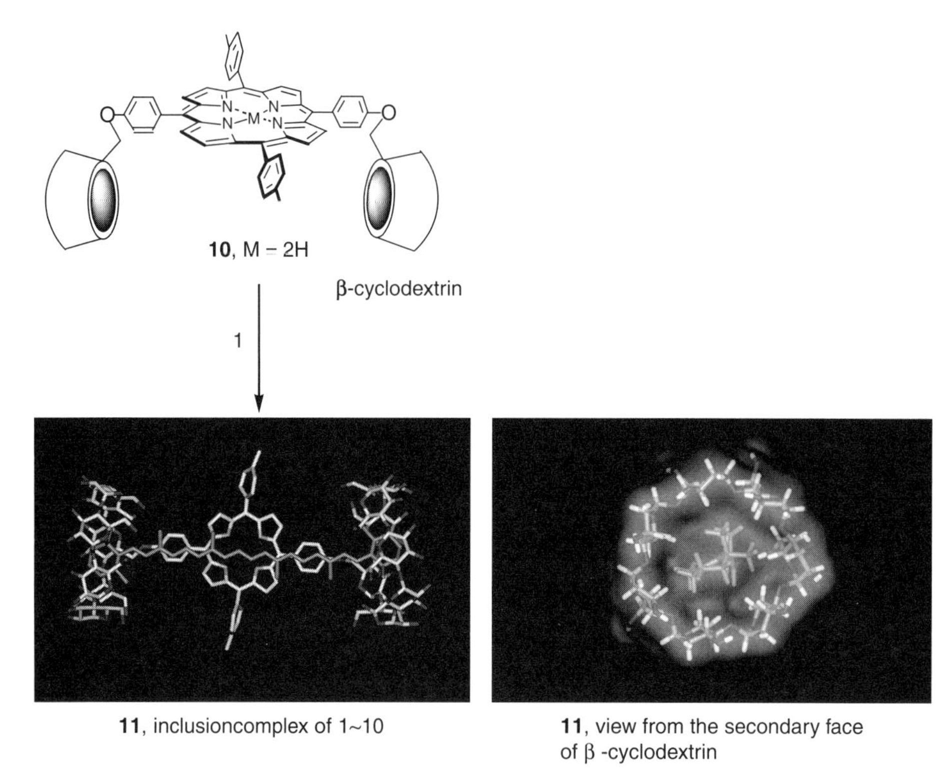

FIG. 5. Inclusion complex of β-carotene **1** and the receptor **10**.

15,15′-double bond would of **1** be placed under the reactive metal center (Fig. 5). Although the enzyme is known to be a non-heme monoxygenase the porphyrin has been chosen as a metal ligand due to structural and photophysical features, *vide infra*. In the absence of **1**, different conformations of **10** are possible due to the rotation of the single bonds around the ether linkages; in the presence of **1**, however, an induced fit should be observed yielding the inclusion complex **11**. Free-base porphyrins, such as **10** or its corresponding Zn complex **12** display a characteristic fluorescence at around 600–650 nm and the ability of carotenoids to quench this fluorescence was envisioned as a sensitive probe for the binding interaction of the two entities in an aqueous medium. It can be reasonably postulated that a CD dimer such as **10** and **12** should be capable of providing a K_a for **1** in the region of 10^5–$10^7\,M^{-1}$. Further, one can expect $K_{a(1-12)} > K_{a(1-10)}$ due to conformational differences of the respective porphyrin macrocycles.

The syntheses of the receptors **10** and **12** were carried out by the treatment of bisphenol porphyrin **13** and its Zn complex **14**, respectively, with portionwise addition of a large excess of β-cyclodextrin-6-*O*-monotosylate (CD-Tos) **15** using caesium carbonate as base in N,N dimethyl formamide (DMF) (Fig. 6) (*13*).

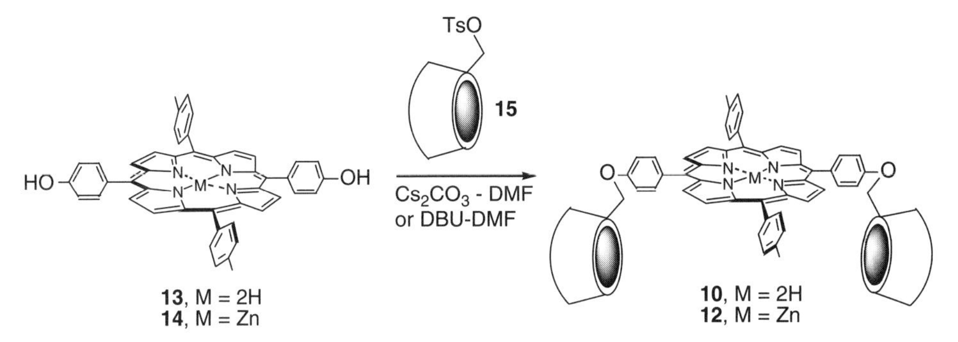

FIG. 6. Syntheses of receptors **10/12**.

The fluorescence quenching experiments (Fig. 7) revealed a binding constant $K_{a\ (\mathbf{1-12})} = 8.3 \times 10^6\,M^{-1}$. Due to its saddle-shaped conformation, the metal-free porphyrin **10** displays a smaller binding constant $K_{a\ (\mathbf{1-10})} = 2.4 \times 10^6\,M^{-1}$ (*13*). This satisfied the first of our strategic criteria for mimicking the biological system, as the binding constant for retinal **2** to β-CDs is smaller by three orders of magnitude and hence, no product inhibition should be expected after the oxidation of central double bond of β-carotene **1**.

For the choice of a metalloporphyrin capable of cleaving E-configured, conjugated double bonds, we chose a ruthenium porphyrin because preliminary experiments with (E, E)-1, 4-diphenyl-1,3-butadiene **17** and the complex **18** in the presence of *tert*-butyl hydroperoxide (TBHP) looked promising giving aldehydes **19** and **20** in good yield (Fig. 8) (*14*). The advantage of co-oxidants such as TBHP (or cumene hydroperoxide) is the inertness towards olefins in the absence of metal complexes. Accordingly β-carotene **1** showed no degradation within 24 h when treated with excess of TBHP alone.

A possible mechanism for this transformation, similar to the proposed enzymatic cleavage of carotenoids (Fig. 3), involves O=Ru=O porphyrin **21** catalyzed epoxidation of **17** to **22**, followed by nucleophilic attack of TBHP and ring opening with assistance of **23**. Subsequent fragmentation yields the aldehydes (Fig. 8).

Finally the synthesis of the bis-cyclodextrin Ru porphyrin receptor **24** was pursued in analogy to the preparation of **12** as described earlier, and the stage was set to attempt carotenoid cleavage. A biphasic system was established in which carotenoids are extracted from a 9:1 mixture of hexane and chloroform into water phase containing **24** (10 mol%) and TBHP. The reaction products, released from the receptor, were then extracted into the organic phase. Aliquots of which were subjected to HPLC conditions using C30 reverse phase column and detecting the signals at the respective wavelength of the formed apocarotenals and retinal, respectively, using a diode array UV-Vis detector. Quantification was done by means of external calibration curves. Various carotenoids **1** and **25–27** were investigated

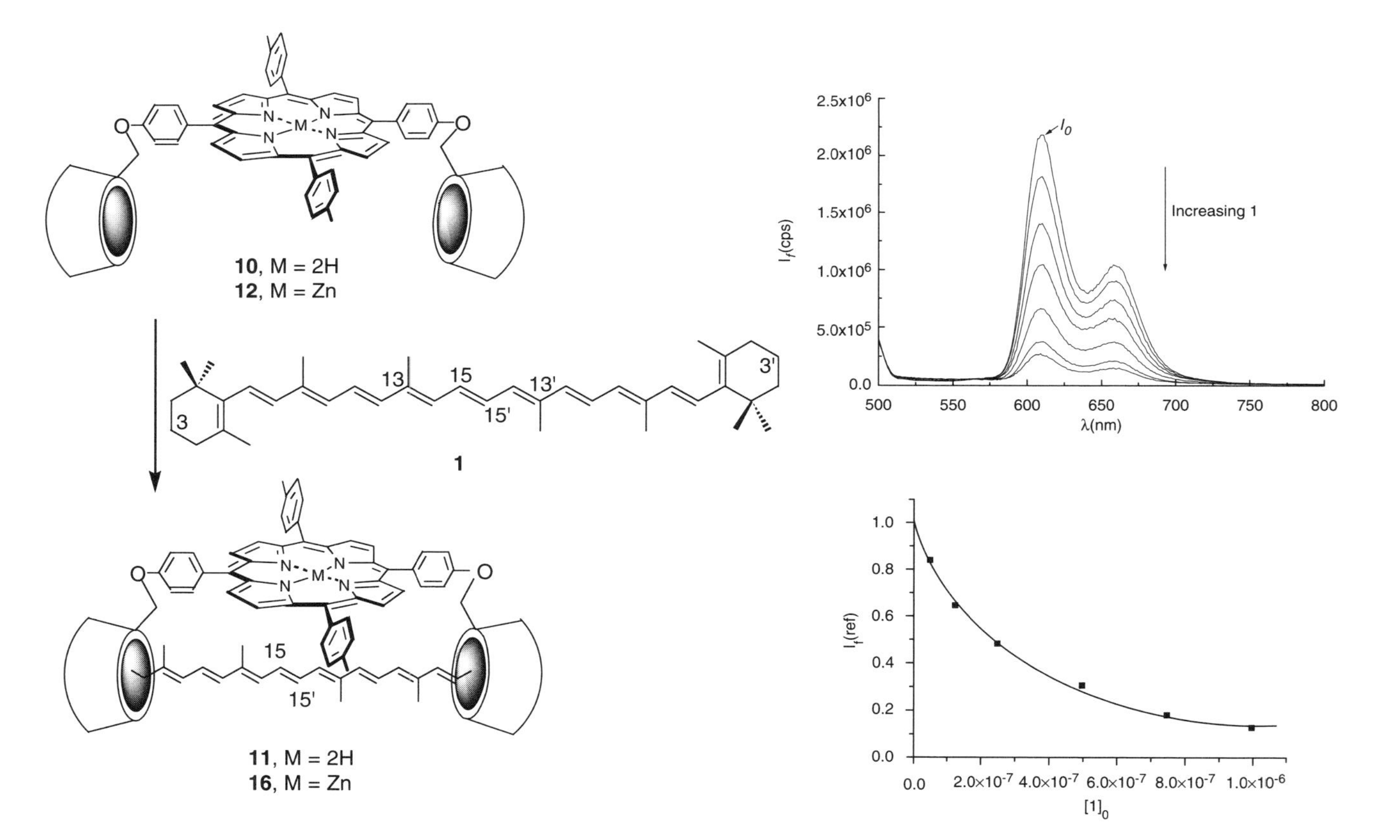

FIG. 7. Quenching of the fluorescence of the receptors **10** and **12** on binding the substrate **1**, $\lambda_{exc} = 420$ nm, $\lambda_{em} = 654$ nm. Graph top, overlayed fluorescence emission spectra of host **12** with increasing added concentrations of β-carotene **1**; graph bottom, relative fluorescence intensities of aqueous solutions of **12** as a function of added β-carotene concentration. K_a was determined using a non-linear least-squares fitting program (*13*).

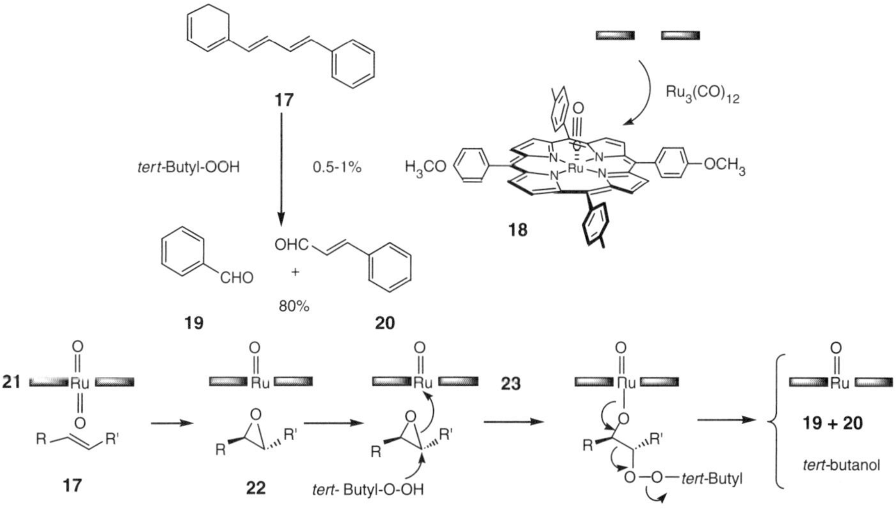

FIG. 8. Ru porphyrin/TBHP-catalyzed double bond cleavage.

and exclusive cleavage of the central double bond was observed for **27** a substrate analogue of β-carotene 15, 15′-monooxygenase (Fig. 9) (*15*); the other carotenoids gave up to 25% bond scission at excentric double bonds (*16*).

These results suggests that stronger hydrophobic interactions between the aromatic end-group of **27** and the β-CD cavity are favoring the 1:1 inclusion complex with the central double bond just under the reactive O=Ru=O center. In contrast, **1**, **25** and **26** slide within the CD cavity exposing three double bonds rather than one to the reactive Ru=O. Determination of the binding constant of **27** to the receptor **10** supports this interpretation, i.e., $K_{a\,(\mathbf{27-10})} = 5.0 \times 10^6\,M^{-1}$ which is about two times larger than $K_{a\,(\mathbf{1-10})} = 2.4 \times 10^6\,M^{-1}$.

Thus the supramolecular enzyme model **29** (Fig. 10), binding carotenoids purely by hydrophobic interactions, catalyzes the cleavage of these polyolefins like the enzyme β-carotene 15,15′-monooxygenase. The regioselectivity of the enzyme model, however, depends on specific interactions of the end-groups of carotenoids, reflected in different K_a values, with the hydrophobic interior surface of the β-CDs. In contrast, β-carotene 15,15′-monooxygenase exclusively catalyzes the bond scission of the central double bond of all natural carotenoids investigated so far. Even artificial substrates such as **27** and **30**, lacking one methyl group in the polyene, are cleaved at the central double bond (Fig. 11) (*17*). However, we observed two exceptions from the rule (*18*). The synthetic carotenoids **31** and **32**, originally designed as inhibitors, turned out to be substrates being cleaved at the C14′–C15′ double bond (Fig. 11). These results hint to hydrophobic interactions of the aromatic rings, placed in the center of the carotenoids, to aromatic residues of the ligand sphere of iron, most

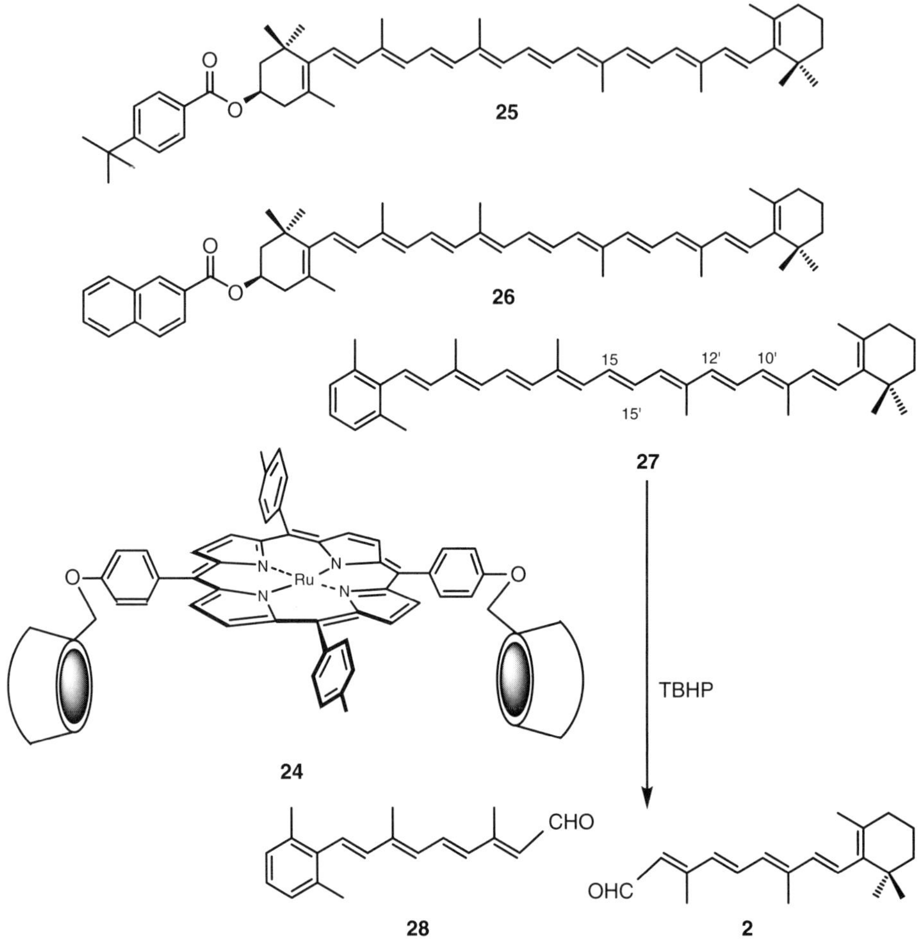

Fig. 9. Cleavage of the substrate analogue **27** by the enzyme mimic **24**+TBHP.

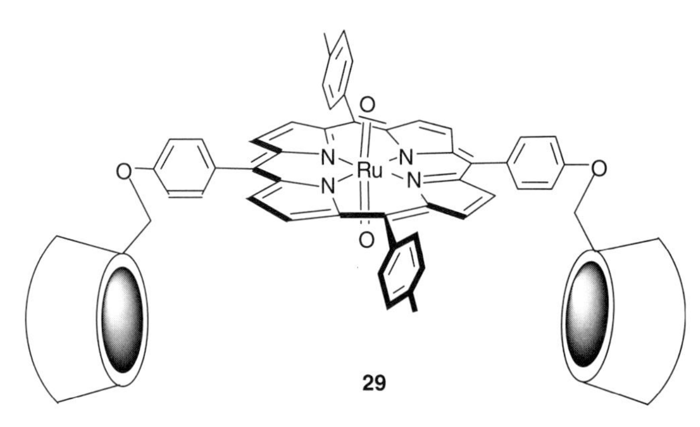

Fig. 10. Structure of the supramolecular catalyst generated *in situ* from **24** and TBHP.

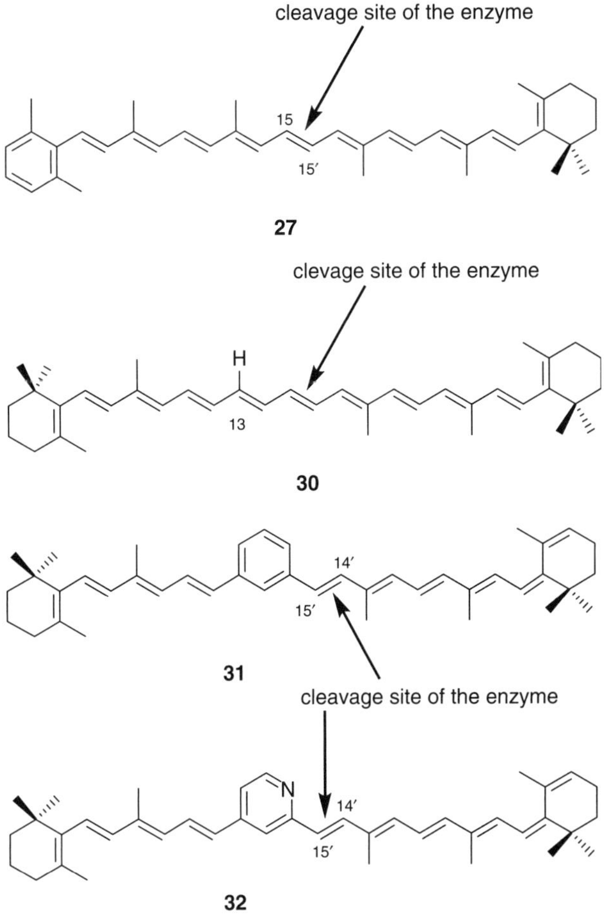

FIG. 11. Regioselectivity of β-carotene 15,15′-monooxygenase action on non-natural substrates.

likely to the histidines. As a consequence the substrate is "moved" within the active site by about 1.5 Å exposing the double bond adjacent to the usual one. Accordingly, hydrophobic interactions are significant to the regioselectivity of oxidation in both the enzyme and the enzyme mimic.

C. The Excentric Cleavage of Carotenoids

β-Carotene can also be cleaved enzymatically at double bonds other than C15–C15′ producing apocarotenals such as **5** that can be further degraded to retinal **2** (Fig. 12) (*2a,2b*).

To mimic this reaction our first approach aimed to construct **33**, a tube-like bis-β-cyclodextrin linked to a ruthenium porphyrin which

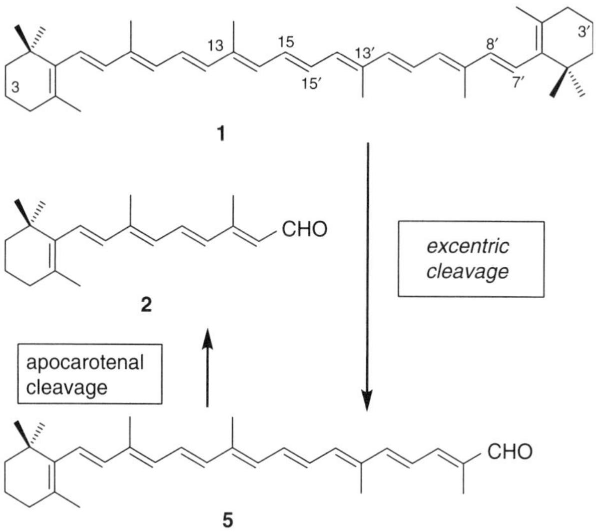

FIG. 12. Excentric cleavage of β-carotene.

was expected to bind **1** yielding the inclusion complex **34** which exposes the double bonds at C12′–C8′ to the ruthenium porphyrin (Fig. 13).

The synthesis of the bis-β-cyclodextrin **35** is outlined in Fig. 14. Interestingly, however, the X-ray structure of **36**, the de-tosylated **35**, revealed that β-carotene would not be incorporated into both CD units due to the unfavorable orientation of the diamide linker which blocks the entrance to the second β-CD. In agreement with this result are experiments with the target Ru complex **37** which displayed central cleavage on β-carotene **1** yielding retinal **2** as the major product (Fig. 15) (*19*).

Two other catalysts **38** and **39** were synthesized and employed for excentric cleavage. Best results were obtained with **39** for which attack at excentric double bonds of a diphenyl carotenoid **40** was observed, albeit with low selectivity (Fig. 16).

D. SUMMARY

We have prepared β-CD-linked ruthenium porphyrins which catalyze the central cleavage of carotenoids. This supramolecular construct is a catalytically competent mimic of the enzyme β-carotene 15,15′-monooxygenase, an enzyme we have isolated, overexpressed and investigated the reaction mechanisms. Several modifications of the binding site for lipohilic substrates such as carotenoids were investigated in order to accomplish enzyme mimics for the excentric cleavage of carotenoids. These approaches were met with limited

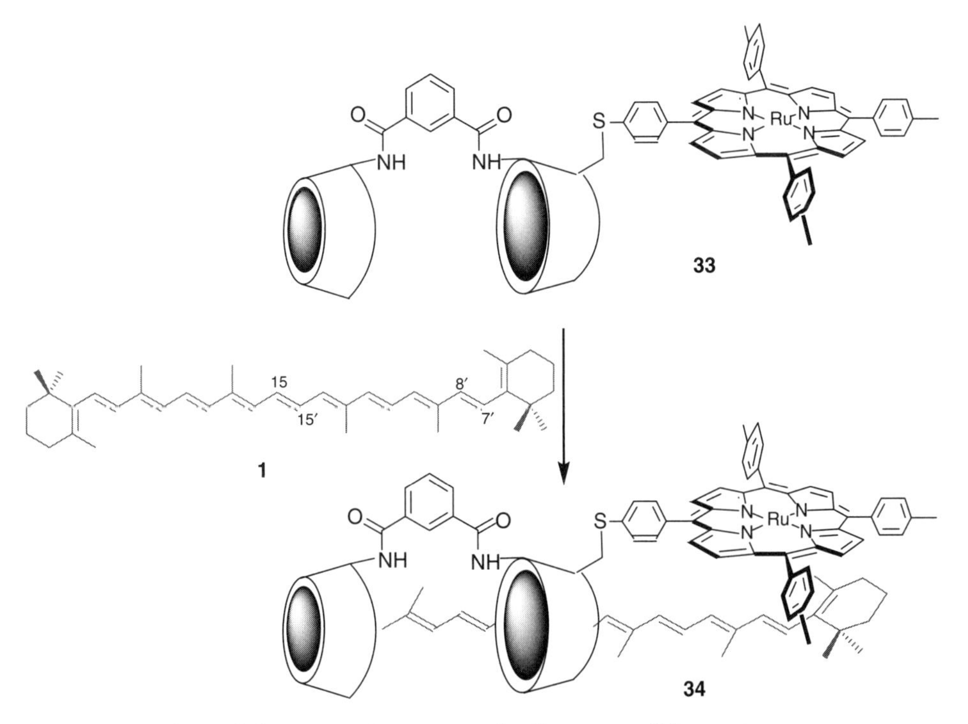

FIG. 13. A model system for excentric cleavage of β-carotene.

success, we could indeed prepare a complex which preferentially attacks excentric double bonds, the selectivity, however, to cleave double bonds at C-14′, 12′, 10′ and 8′ was low. After all, this is not a surprise in view of the possible control of orientation of carotenoids in a rather flexible binding site.

III. Hydrogen Transfer Reactions Catalyzed by Ruthenium Complexes Linked to β-Cyclodextrin

A. Ruthenium Catalyzed Asymmetric Transfer Hydrogenation (ATH) to Ketones

The asymmetric reduction of prochiral ketones to their corresponding enantiomerically enriched alcohols is one of the most important molecular transformations in synthetic chemistry (*20,21*). The products are versatile intermediates for the synthesis of pharmaceuticals, biologically active compounds and fine chemicals (*22,23*). The racemic reversible reduction of carbonyls to carbinols with superstoichiometric amounts of aluminium alkoxides in alcohols was independently discovered by Meerwein, Ponndorf and Verley (MPV) in 1925 (*24–26*). Only in the early 1990s, first successful versions of catalytic

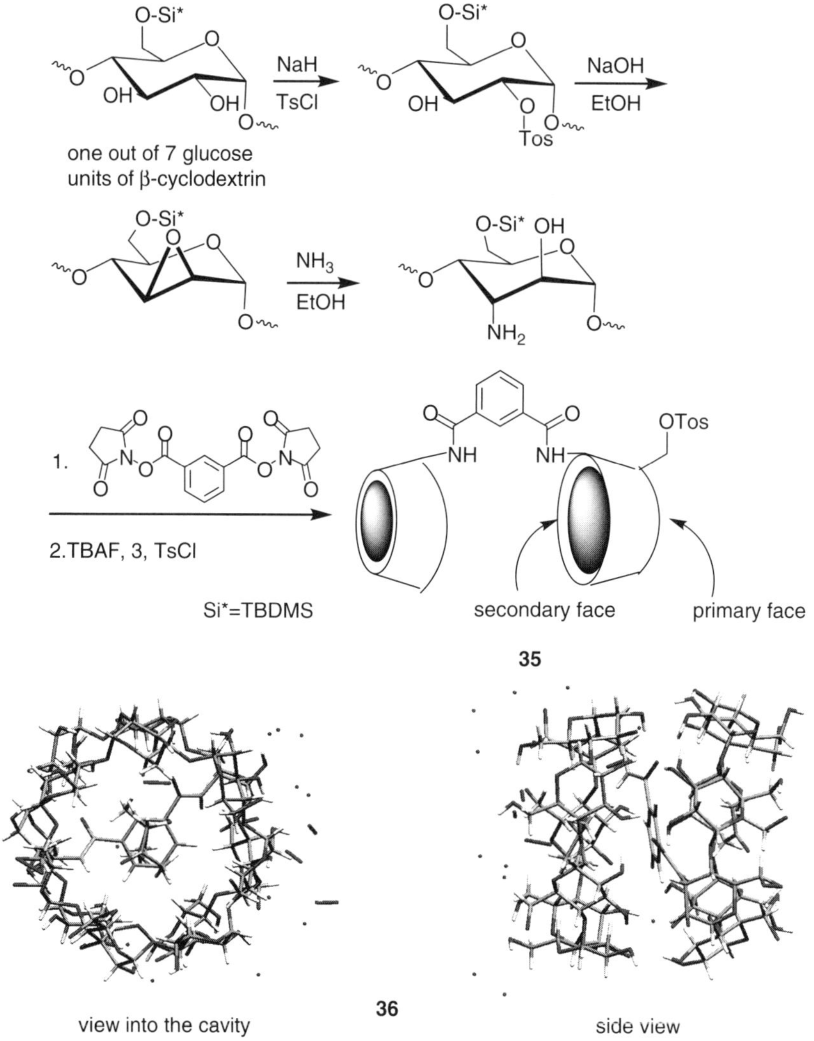

FIG. 14. Synthesis and structure of a β-carotene receptor.

asymmetric MPV reactions have been reported using C_2 symmetric chiral ligands. Pioneering efforts were made by the group of Pfaltz using Ir(I) dihydrooxazole complexes (*27*) and the group of Genêt using chiral diphosphine Ru(II) catalysts (*28*). Evans showed that lanthanide complexes with amino alcohol derived ligands (*29*) reduce aromatic ketones with enantioselectivities up to 97%.

In 1996, Noyori and co-workers discovered that Ru(II) $^{6}\eta$-arene complexes containing either a chiral 1,2-amino alcohol such as in **41** or a chiral N-monotosylated 1,2-diamine ligand, see **42**, serve as excellent catalysts. It was a breakthrough for catalytic ATH reactions to ketones in terms of enantioselectivities, catalyst loading and

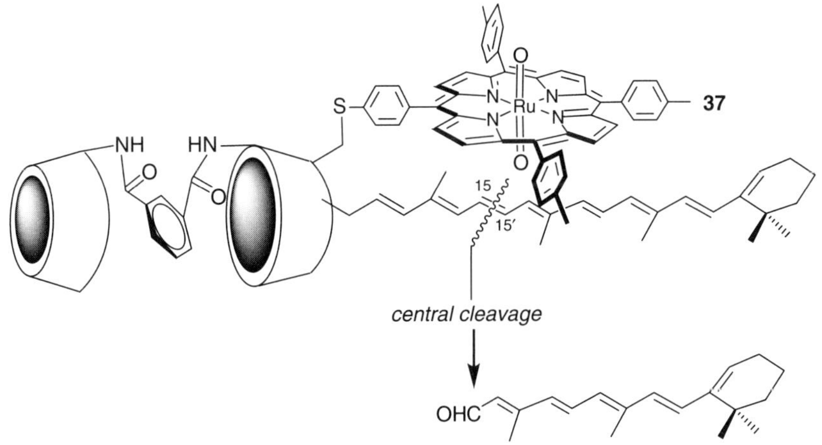

FIG. 15. Cleavage of β-carotene with the bis-β-cyclodextrin Ru complex **37**.

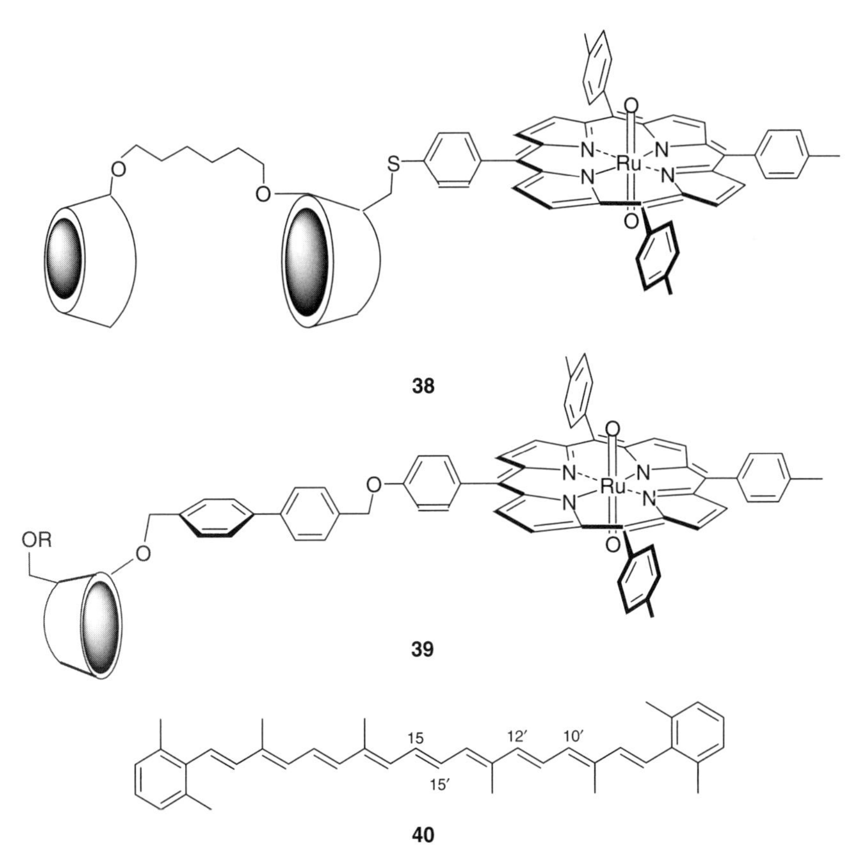

FIG. 16. Two further catalysts **38** and **39** which were employed to accomplish excentric cleavage of the substrate **40**.

substrate scope (*30*). For example, acetophenone can be reduced quantitatively to its corresponding (S)-alcohol with up to 98% ee in the presence of an azeotropic mixture of HCOOH/NEt_3 (Fig. 17). This ATH system is strictly chemoselective, i.e., double bonds, halogens, nitro groups and esters are not affected under such conditions. Several efforts have been made in the past decade to improve the catalytic properties by designing and employing new chiral ligands. Nevertheless, due to excellent reactivity, selectivity and low costs the system described by Noyori remains the standard to which all new procedures are related. Experimental and theoretical studies revealed that ATH occurs via metal–ligand bifunctional catalysis (*31*), in contrast to pathways involving metal coordinated alkoxides (*32*). The mechanism proceeds through a concerted transfer of both a hydride and a proton from the catalyst to the carbonyl passing via a pericyclic-like transition state (Fig. 18).

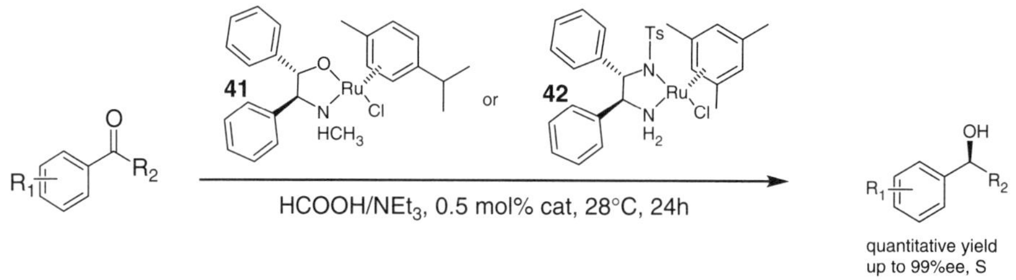

FIG. 17. Noyori's ATH system using ruthenium complexes of chiral 1,2-amino alcohols **41** and monotosylated 1,2-diamines **42**, respectively.

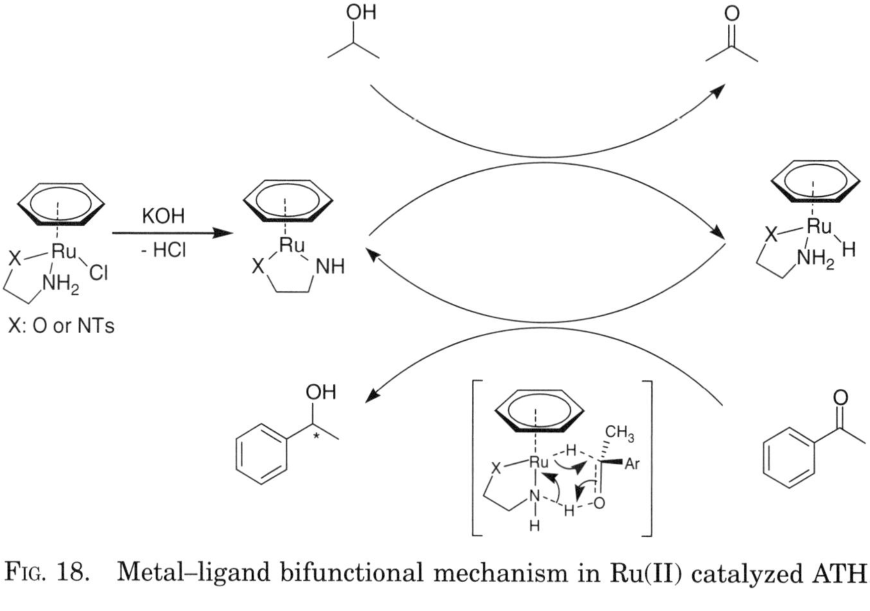

FIG. 18. Metal–ligand bifunctional mechanism in Ru(II) catalyzed ATH.

Accordingly, ATH is a versatile method for the enantioselective reduction of prochiral aromatic ketones. Aliphatic unconjugated ketones, however, are only reduced in poor to moderate enantioselectivities up to date (*33–36*).

We reasoned that this limitation is not due to the reactivity of the ketones but rather a problem related to enantioface-differentiation at prochiral sp^2 systems if the substituents are not distinct enough in size. Hence, we believed Ru $^6\eta$-arene complexes linked to CDs would be suitable since a sterically more demanding or simply longer substituent (L) would bind to the cavity of β-CD such that the difference between (**L**) and (**S**) becomes artificially greater and the orientation of the substrate's keto group relative to the Ru-H might be triggered by the binding interaction of (**L**) to the β-CD cavity (Fig. 19).

CDs are cyclic oligosaccharides comprised of α-1,4-linked glucopyranose units (*37–40*). The following properties make CDs attractive components in organic chemistry and supramolecular catalysis/enzyme mimics in particular: (i) CDs are water soluble; (ii) their hydrophobic cavity can host a variety of lipophilic guest molecules; (iii) their structure is well defined due to intramolecular H bonds; (iv) the hydroxyl groups allow functionalization; and (v) CDs are chiral and therefore applicable to enantioselective reactions. Exploiting these features, in particular those of β-CD the cyclic heptasaccharide, Breslow (*41*) started his pioneering work in the late 1970s mimicking enzymatic reactions such as esterases. Several β-CD-based enzyme models followed in the next decades (*42,43*). Regarding catalytic reductions, Sugimoto compared the reduction of aryl trifluoromethyl ketones with either NAH (nicotinamide without the dinucleotide part) or sodium borohydride in the presence of β-CD. Enantioselectivities exceeding 10% ee could not be observed at conversions ranging from 20% to 99% (*44*). A rare example exceeding 50% ee has been described by Rao. Three out of 17 substrates showed ee's higher than 60% in the asymmetric reduction of azido arylketones using sodium borohydride as the hydrogen source and stoichiometric amounts of β-CD, but no aliphatic substrates were reported (*45*).

We believed that an enzyme-like system containing a substrate binding site (β-CD) and an attached site of reactivity, a chiral Ru

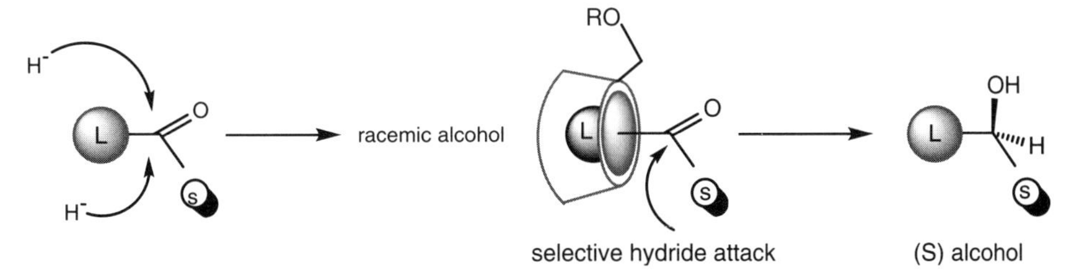

FIG. 19. CD complexation enhances the difference between substituents at C=O facilitating enantioface-selective hydride addition.

complex, would be suitable to accomplish high enantioselectivity in the hydrogen transfer reaction to aromatic and in particular to aliphatic ketones, see for example Fig. 20. We report here part of our ongoing studies regarding new β-CD-based Ru(II) catalysts with Ru ligands such as chiral amino alcohols, monosulfonated 1,2-diamines and α-pycolyl amines as well as mechanistic insights into this transformation.

B. ATH with Ruthenium Complexes of Amino Alcohol Complexes Linked to the Primary Face of β-Cyclodextrin

B.1. Reduction of aromatic ketones

In general, β-cyclodextrin-attached ligands to bind ruthenium were prepared as shown for **43** in Fig. 21 (*46*). Substitution of commercially

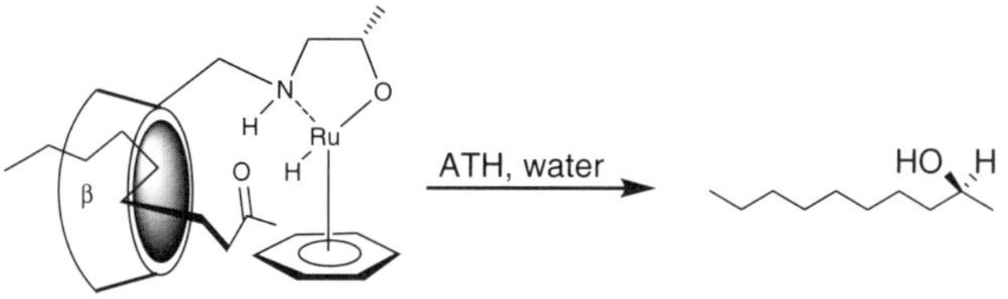

FIG. 20. Design of new β-CD-based Ru(II) catalysts for ATH to challenging aliphatic ketones.

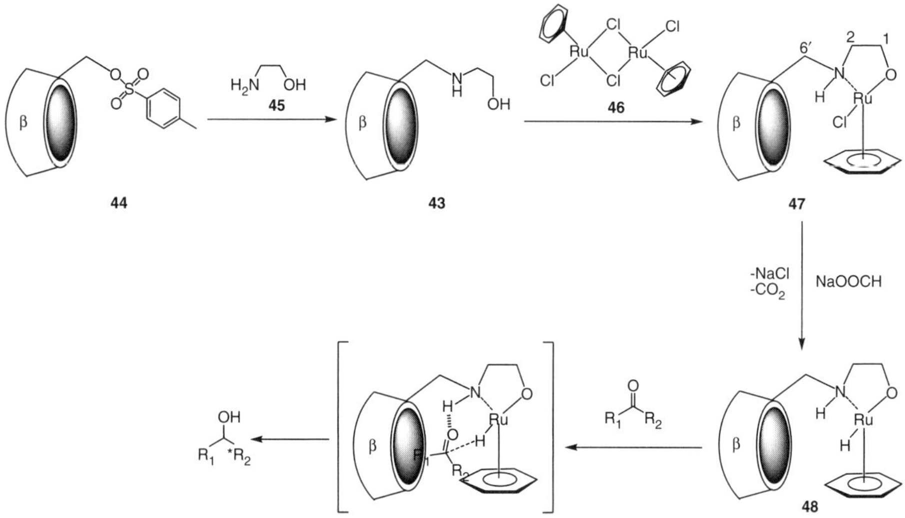

FIG. 21. General synthesis of β-CD-linked ruthenium complexes; asymmetric transfer hydrogenation is described as a metal–ligand bifunctional mechanism according to (*31*).

available β-CD tosylate **44** with amino ethanol **45** furnished the crystalline β-CD derivative **43** in a single step. Treatment of **43** with the Ru(II) dimer **46** gave quantitatively Ru complexes such as **47**; this reaction can be controlled by ^{1}H-NMR spectroscopy since protons at C1, C2 and C6′ of **47** display downfield shifts by >0.5 ppm relative to **43**. For catalytic reactions, the Ru complexes such as **48** were formed *in situ* and subsequent ketone reductions were pursued in the presence of excess sodium formate at room temperature during 12 h. Enantioselective reductions employing **48** and various substituted acetophenones revealed up to 47% ee in favor of the R alcohol. Since it is known from work by other research groups (*47*) that asymmetric reduction catalyzed by Ru/amino alcohol complexes, lacking a CD unit, is largely dependent on the chirality at the carbinol carbon atom (C1), we prepared a series of ligands **43** and **49–62** (Fig. 21) to improve enantioselectivity, reaction rate and catalyst loading. The corresponding ruthenium complexes were first screened with *p*-Cl acetophenone **63** (Fig. 22).

Twelve out of 15 ligands show good to excellent yields. The only unreactive one is the indanol amin ligand **54** with only 3% conversion. Regarding enantioselectivities the best results were obtained with the acyclic ligands **49** and **51**. It was therefore decided to investigate aliphatic ketones.

B.2. Reduction of aliphatic ketones using the ruthenium complex of ***49***

The ruthenium complex of **49** was chosen due to its easy accessibility and because preliminary experiments had shown that simple non-prochiral aliphatic ketones such as 4-methyl-cyclohexanone are quantitatively reduced. This positive outcome encouraged us to test various prochiral aliphatic ketones **64–75**. The results using standard conditions are summarized in Fig. 23. Most substrates could be reduced in good yields and the enantioselectivities of six alcoholic products are higher than 85%, 2-decanone **67** and geranylacetone **68** showed even ee's of 95%, demonstrating that the concept works well not only for aromatic ketones.

Some ketones **70, 72, 73** and **75** gave low yields most likely due to low solubility in aqueous solution and small binding constants of the inclusion complexes. It is important to note that double bonds were not reduced in compounds **68–70** and styrene could be stirred at 50°C for three days in the presence of the catalyst without alkane formation, i.e., the reaction proceeds completely chemoselective. This is also valid for the α- and β-ketoesters **76–79**, though for these substrates the enantioselectivities are low (Fig. 23).

The results demonstrate clearly the capacity of the catalytic system to reduce aliphatic ketones enantioselectively which has not been accomplished so far with other ruthenium catalysts under hydrogen transfer conditions (*46*).

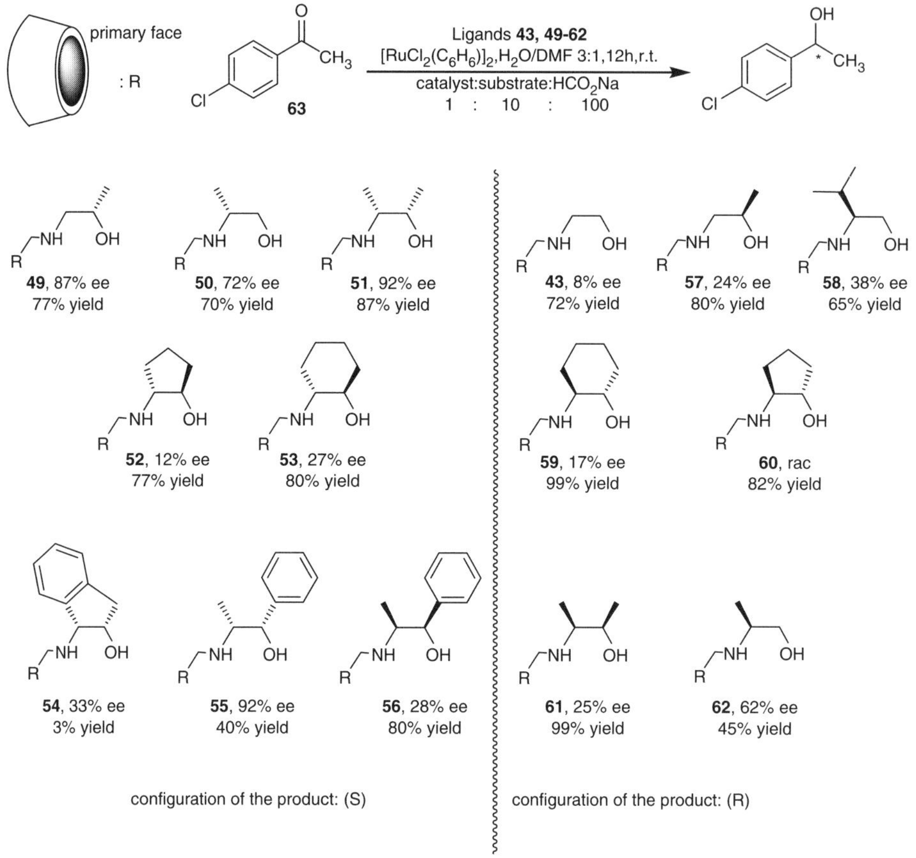

primary face

: R

63

Ligands **43, 49-62**
$[RuCl_2(C_6H_6)]_2$, H_2O/DMF 3:1, 12h, r.t.
catalyst:substrate:HCO_2Na
1 : 10 : 100

49, 87% ee
77% yield

50, 72% ee
70% yield

51, 92% ee
87% yield

43, 8% ee
72% yield

57, 24% ee
80% yield

58, 38% ee
65% yield

52, 12% ee
77% yield

53, 27% ee
80% yield

59, 17% ee
99% yield

60, rac
82% yield

54, 33% ee
3% yield

55, 92% ee
40% yield

56, 28% ee
80% yield

61, 25% ee
99% yield

62, 62% ee
45% yield

configuration of the product: (S)

configuration of the product: (R)

FIG. 22. Reduction of **63** with various ruthenium amino alcohol complexes attached to the primary face of β-CD; yield and ee of the resulting alcohol is given below the ruthenium ligand.

B.3. Reduction of ***63*** *with ruthenium complexes of monotosylated 1,2 diamines linked to the primary face of β-CD*

To improve the rate of reduction the amino alcohol ligand of the ruthenium complexes was exchanged for monotosylated 1,2-diamine ligands. For exploratory experiments *N*-tosylethane-1,2-diamine was prepared by monotosylation of ethane-1,2-diamine and attached to the primary face of β-CD yielding **80**. With β-CD as the only chiral unit the ruthenium complex of **80** could reduce aromatic and aliphatic standard ketones **63** and **69** in 91% yield, 25% ee (S) and 68% yield, 58% ee, respectively, within only 4 h under standard conditions (Fig. 24).

Reactivity studies revealed that the ruthenium complexes of monotosylated 1,2-diamines linked to β-CD such as **80** react about three times faster than their amino alcohol counterparts. Several

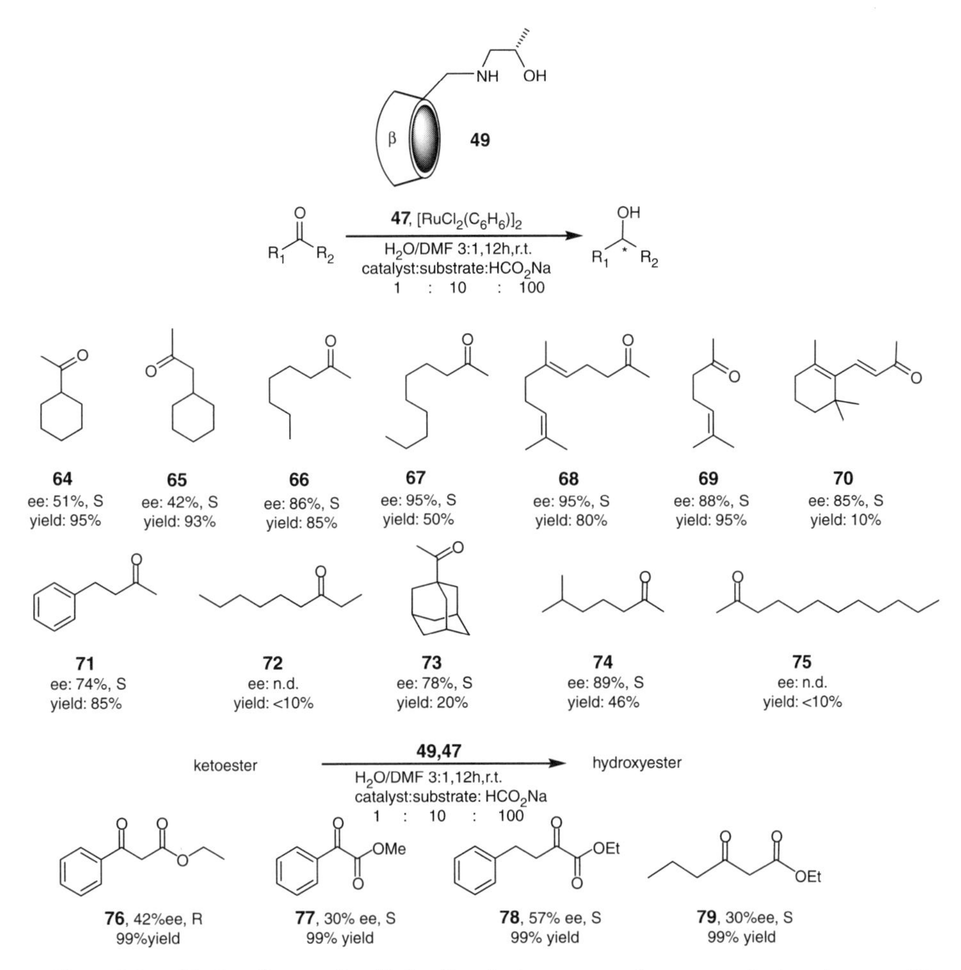

FIG. 23. Reduction of aliphatic ketones and some ketoesters with ruthenium complexes of **49**; yields and ee values refer to isolated pure products.

efforts have been made to lower the catalyst loading using **80**+**47** and standard reduction conditions. Respectable 56% yield was obtained with 5 mol% catalyst whereas it could be increased to 67% performing the reduction in a twice as concentrated solvent system. A 34% yield was afforded decreasing the catalyst loading to 2% which corresponds to a turnover number of 17. Nevertheless, the rate acceleration using the ligand system **80** encouraged us to introduce chirality in the monotosylated diamino binding site for ruthenium to increase enantioselectivity. Since the required chiral diamines were not commercially available, ligand systems were synthesized with >98% ee as shown, for example, for **81** using the chiral tosylaziridine **82** as a key intermediate in the sequence (Fig. 25) (*48*). Starting from corresponding amino acids **83**–**86** were prepared using the same strategy.

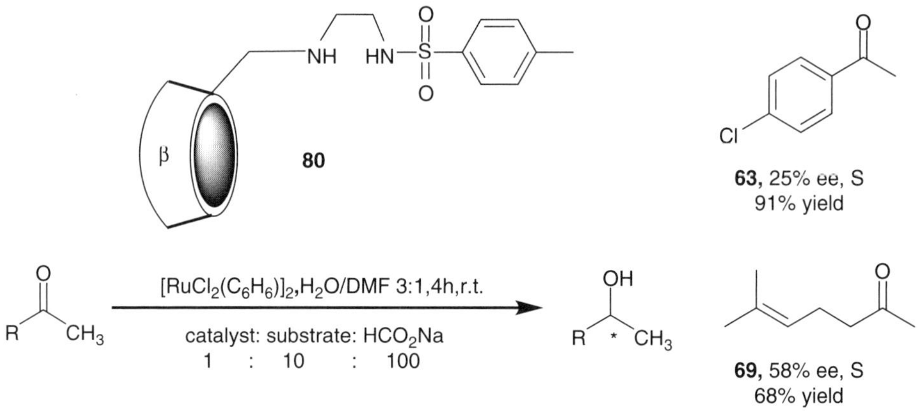

FIG. 24. ATH with **80**.

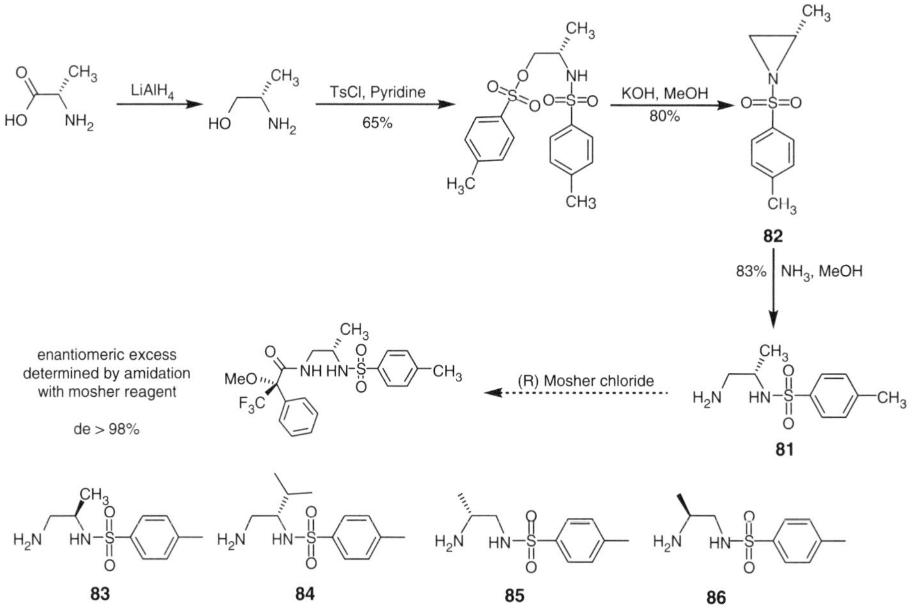

FIG. 25. Synthesis of chiral monotosylated 1,2-diamines.

Compounds **81** and **83–86** were linked to β-CD, the ruthenium η-complexes prepared *in situ* and reductions carried out with the standard substrate **63** (Fig. 26). Comparing the results of ruthenium complexes with ligands **87–91** reveals that any substituent adjacent to the tosyl group leads to modest to good ee values but reduces the reactivity of the catalyst considerably, see **87–89** (Fig. 26), improvement of the yields between 33% and 53% was only achieved at elevated temperatures (50°C). In contrast, ruthenium complexes with ligands

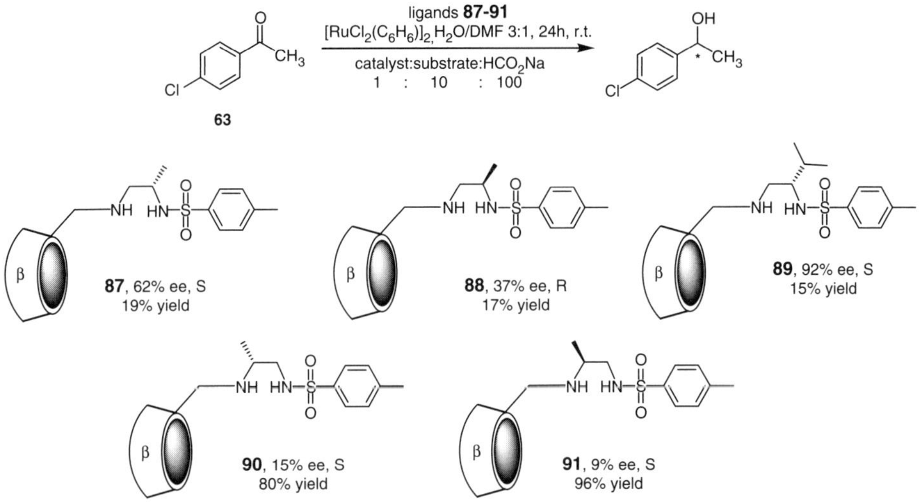

ligands **87-91**
$[RuCl_2(C_6H_6)]_2$, H_2O/DMF 3:1, 24h, r.t.
catalyst:substrate:HCO_2Na
1 : 10 : 100
63

87, 62% ee, S
19% yield

88, 37% ee, R
17% yield

89, 92% ee, S
15% yield

90, 15% ee, S
80% yield

91, 9% ee, S
96% yield

FIG. 26. ATH with ruthenium complexes of chiral monotosylated 1,2-diamino ligands.

90 and **91** display the reverse behavior, i.e., good to excellent yields but very low ee values. In general, the dependence of ee values from the substitution pattern obtained with these ligands corresponds to those observed with amino alcohols, see Fig. 21, but the presence of both the tosyl group and substituents at the ruthenium binding site obviously poses a conformational problem with respect to binding and/or orientation of the substrate in the CD cavity. Further, the rate acceleration observed with ligand **80** diminishes when substituents are attached to the diamine linker.

C. ATH WITH RUTHENIUM COMPLEXES OF α-PICOLYL AMINES LINKED β-CYCLODEXTRINS

Replacing the OH-, respectively the $NHSO_2R$ functionality of the ligand site of ruthenium by a pyridine ring leads to α-pycolyl amines. Using commercially available non-chiral pyridin-2-ylmethanamine the ligand **92** was prepared in 48% yield. Exploratory ATH experiments with the ruthenium complex of **92** revealed very poor activity at room temperature but excellent yields at 70°C (Fig. 27). Consequently synthetic chiral diamines (*48*) were attached to β-CD and the ruthenium complexes of ligands **93–96** used at 70°C to reduce standard substrate **63**. The result given in Fig. 26 displaying excellent yields but only moderate ee values, the absolute configurations of the products correspond to reactions with ruthenium complexes of amino alcohols and monotosylated diamine.

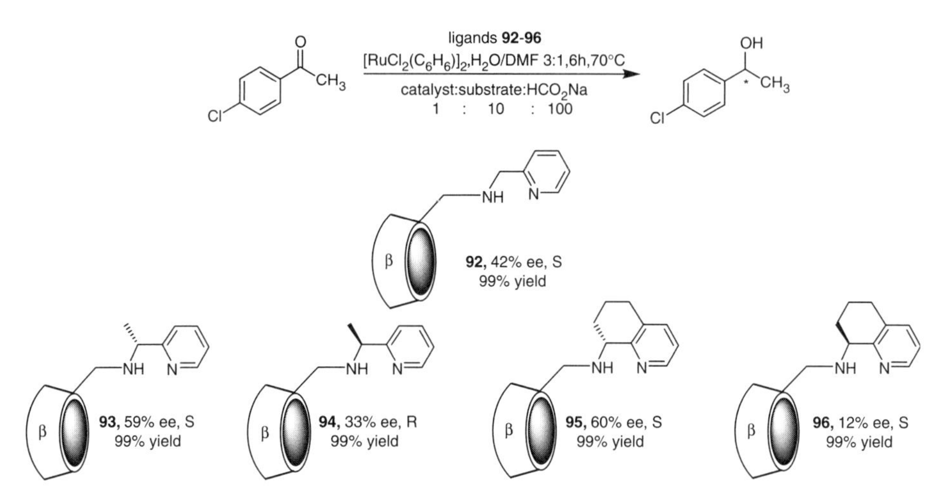

FIG. 27. ATH with α-pycolyl amines as ruthenium ligands.

D. STRUCTURES OF INCLUSION COMPLEXES, PRE CATALYSTS AND MECHANISTIC INVESTIGATIONS

The structures of β-CD-modified ligands were routinely elucidated by ^{1}H-NMR and ESI mass spectroscopy. Formation of Ru chloro complex **97** from **49** and **47** was demonstrated to be quantitative by ^{1}H-NMR (Fig. 28). Significant downfield shifts (0.36–0.56 ppm) in **97** of several protons relative to free ligand **49** were observed.

All aromatic protons of the arene moiety in **97** appeared as a singlet at 5.34 ppm. The fact that only one set signals was observed in the ^{1}H- and ^{13}C-NMR of **97** suggests the presence of one diastereoisomer with defined chirality at ruthenium, and since the complex **98** is formed under retention of configuration (*47*) the catalytically competent **98** induces rather high enantioselectivity during hydrogen transfer reactions. Absolute proof for this interpretation, however, is missing because neither **97** nor **98** gave crystals sufficient for X-ray analysis.

Binding constants of the standard substrate *p*-Cl-acetophenone **63** ($K_a = 1040\,M^{-1}$) and the corresponding product e.g., S-**99** ($K_a = 400\,M^{-1}$) with **49** were determined by ^{1}H-NMR titrations in D_2O/DMF-D_7. Increasing the guest concentration significant downfield shifts were observed for H at C3 and H at C5 of the β-CD unit which are pointing inside the cavity (Fig. 29). The alcohol binds less than half as good as the starting material. This result is essential for the catalytic process since product inhibition is not a serious problem. The larger affinity of **63** to β-CD can be attributed to hydrogen bonding from the carbonyl oxygen to the N–H group of the linker; this interaction is also significant to catalysis performed by ruthenium hydride complex **98** (see Figs. 28 and 18).

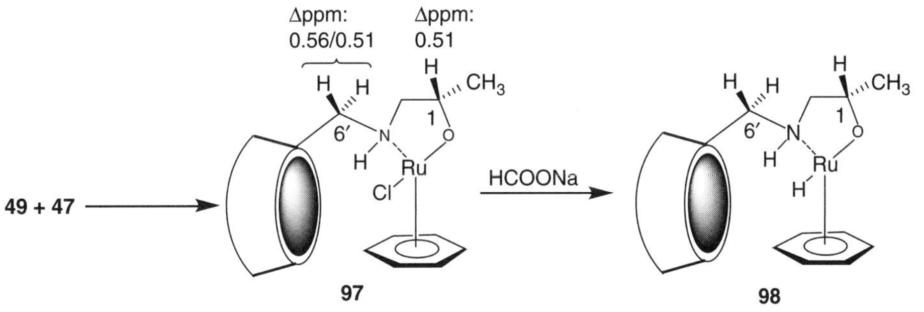

FIG. 28. NMR characterization of **97**, Δppm refer to **49**.

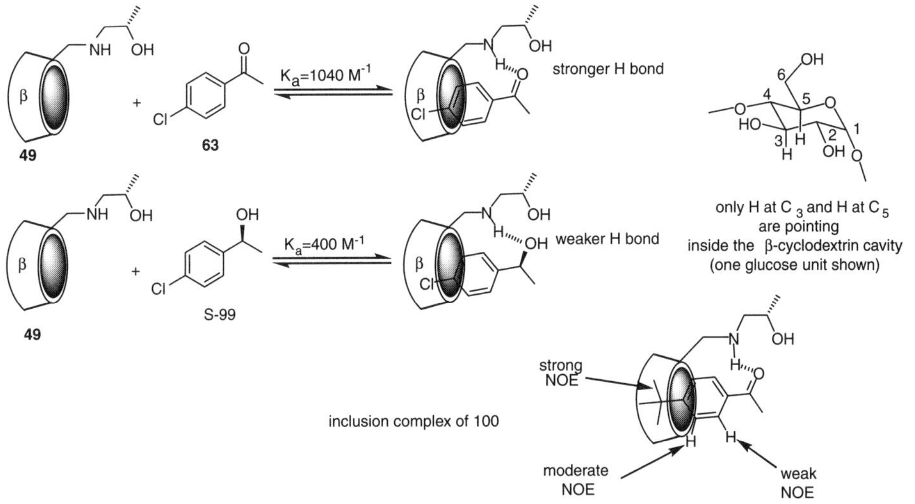

FIG. 29. Binding constants of **63**, **99** and interactions of **100** with the β-CD derivative **49**.

ROESY spectra of ligand **49** and *p*-*tert*-butylacetophenone **100** provided evidence for the binding of the aromatic ring within the cavity of β-CD because strong cross peaks of H at C3 and H at C5 of β-CD with the *tert*-butyl group of the ketone could be observed. Further, moderate cross peaks were observed for *meta*- and *ortho*-protons of the aromatic ring, indicating that the carbonyl compound enters the cavity with the hydrophobic *tert*-butyl rest ahead (Fig. 29).

To support the proposed hydride transfer as shown for the substrate geranylacetone **68** in Fig. 30 proline was linked to the primary face of β-CD yielding the tertiary amine **101** which was complexed with ruthenium and employed to ATH under standard conditions. No product was formed at all which is in good agreement with observations by other groups (*31,49,50*). The presence of a hydrogen

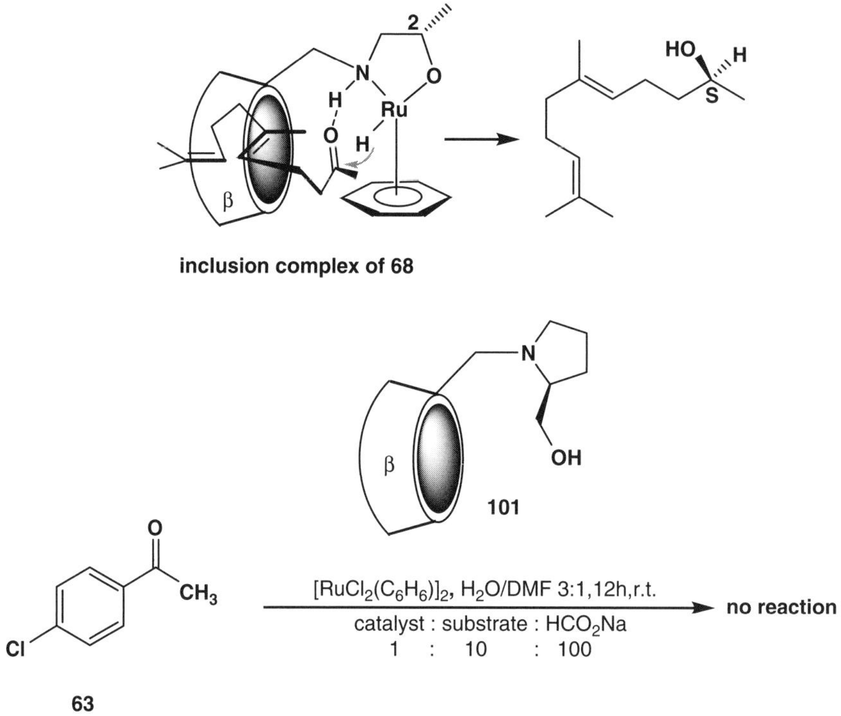

FIG. 30. The significance of he N–H group to reactivity of the ruthenium complex.

bond donor such as NH is essential to activate the carbonyl group of the substrate. This implies that the mechanism of our novel transformation resembles the metal–ligand bifunctional mechanism proposed by Noyori (*31*).

E. SUMMARY

β-CD was derivatized at the primary face with various synthetic ligands suitable to form ruthenium arene complexes. There is convincing spectroscopic evidence that the ruthenium complexes are formed and that lipophilic substrates bind into the cavity of β-CD.

One of the systems was found to be very efficient catalyzing enantioface-selective hydrogen transfer reactions to aromatic and in particular to aliphatic ketones with up to 95% ee. Regarding the latter reaction these are unprecedented ee values. The reaction mechanism of these transformations is best described as a metal–ligand bifunctional catalysis passing through a pericyclic-like transition state.

References

1. Britton, G.; Liaaen-Jensen, S.; Pfander, H. In: *"Carotenoids"*; Eds. Britton, G.; Liaaen-Jensen, S.; Pfander, S.; Birkhäuser: Basel, **1995**, pp. 1A–13.
2. (a) Wang, X.-D.; Tang, G.; Fox, J. G.; Krinsky, N. I.; Russell, R. M. *Arch. Biochem. Biophys.* **1991**, *258*, p. 8. (b) Yeum, K.-J.; Dos Anjos Ferreira, A. L.; Smith, D.; Krinsky, N. I.; Russell, R. M. *"Free Radicals in Biology and Medicine"*; **2000**, *29*, 105. (c) Kloer, D. P.; Ruch, S.; Al-Babili, S.; Beyer, P.; Schulz, G. E. *Science* **2005**, *308*, p. 267.
3. Wirtz, G. Ph.D. Dissertation, University of Basel, Basel, **1998**.
4. Wyss, A.; Wirtz, G.; Woggon, W.-D.; Brugger, R.; Wyss, M.; Friedlein, A.; Bachmann, H.; Hunziker, W. *Biochem. Biophys. Res. Commun.* **2000**, *271*, p. 334.
5. Wyss, A.; Wirtz, G.; Woggon, W.-D.; Brugger, R.; Wyss, M.; Friedlein, A.; Bachmann, H.; Hunziker, K. *Biochem. J.* **2001**, *354*, p. 521.
6. Leuenberger, M. G.; Engeloch-Jarret, C.; Woggon, W.-D. *Angew. Chem. Intl. Ed.* **2001**, *40*, p. 2614.
7. (a) Woggon, W.-D. *Acc. Chem. Res.* **2005**, *38*, 127–136. (b) Woggon, W.-D. In: *"Top. Curr. Chem. 184, Bioorganic Chemistry, Models and Applications"*; Ed. Schmidtchen, F.P.; Springer: Berlin, **1996**, p. 39.
8. Ono, T.; Nakazono, K.; Kosaka, H. *Biochem. Biophys. Acta* **1982**, *709*, p. 84.
9. Drauz, K.; Waldmann, H. *"Enzyme Catalysis in Organic Synthesis*; 2nd edn., vol. 2; Wiley-VCH Verlag GmbH: Weinheim, Germany, **2002**.
10. Lieberman, S.; Lin, Y. Y. *J. Steroid Biochem. Mol. Biol.* **2001**, *78*, p. 1.
11. Stok, J. E.; de Voss, J. J. *Arch. Biochem. Biophys.* **2000**, *384*, p. 351.
12. The MOLOC program was provided by F. Hoffmann-La Roche Inc., Basel.
13. French, R. R.; Wirz, J.; Woggon, W.-D. *Helv. Chim. Acta* **1998**, *81*, p. 1521.
14. French, R. R.; Holzer, P.; Leuenberger, M. G.; Woggon, W.-D. *Angew. Chem. Intl. Ed.* **2000**, *39*, p. 1267.
15. French, R. R.; Holzer, P.; Leuenberger, M. G.; Nold, M. C.; Woggon, W.-D. *J. Inorg. Biochem.* **2002**, *88*, p. 295.
16. Holzer, P. Ph.D. Dissertation, University of Basel, Basel, 2002.
17. Wirtz, G. M.; Bornemann, C.; Giger, A.; Müller, R. K.; Schneider, H.; Schlotterbeck, G.; Schiefer, G.; Woggon, W.-D. *Helv. Chem. Acta* **2001**, *84*, p. 2301.
18. Leuenberger, M. G. Ph.D. Dissertation, University of Basel, Basel, 2002.
19. Wang, H. Ph.D. Dissertation, University of Basel, Basel, 2006.
20. Noyori, K. *"Asymmetric Catalysis in Organic Chemistry"*; Wiley: New York, **1994**.
21. Ohkuma, T.; Noyori, R. In: *"Comprehensive Asymmetric Catalysis I"*; Eds. Pfaltz, A.; Jacobsen, E.N.; Yamamoto, H.; Springer: Berlin, **1999**, pp. 199–246.
22. Brands, K. M. J. *J. Am. Chem. Soc.* **2003**, *125*, p. 2129.
23. Tombo, G. M. R.; Bellus, D. *Angew. Chem. Int. Ed.* **1991**, *30*, p. 1193.
24. Meerwein, H.; Schmidt, R. *Liebigs Ann. Chem.* **1925**, *444*, p. 221.
25. Verley, A. *Bull. Soc. Chim. Fr.* **1925**, *37*, p. 871.
26. Ponndorf, W. Z. *Angew. Chem.* **1926**, *39*, p. 138.
27. Müller, D.; Umbricht, G.; Weber, B.; Pfaltz, A. *Helv. Chim. Acta* **1991**, *74*, p. 232.
28. Genêt, J.-P.; Ratovelomanana-Vidal, V.; Pinel, C. *Synlett* **1993**, p. 478.
29. Evans, D. A.; Nelson, S. G.; Gagné, M. R.; Muci, A. R. *J. Am. Chem. Soc.* **1993**, *115*, p. 9800.
30. Fujii, A.; Hashiguchi, S.; Uematsu, N.; Ikariya, T.; Noyori, R. *J. Am. Chem. Soc.* **1996**, *118*, p. 2521.
31. Yamakawa, M.; Ito, H.; Noyori, R. *J. Am. Chem. Soc.* **2000**, *122*, p. 1466.
32. Zassinovich, G.; Mestroni, G. *Chem. Rev.* **1992**, *92*, p. 1051.
33. Corey, E. J.; Bakshi, R. K.; Shibata, S.; Chen, C.-P.; Singh, V. K. *J. Am. Chem. Soc.* **1987**, *109*, p. 7925.

34. Ohkuma, T.; Sandoval, C. A.; Srinivasan, R.; Lin, Q.; Wei, Y.; Muñiz, K.; Noyori, R. *J. Am. Chem. Soc.* **2005**, *127*, p. 8288.
35. Reetz, M. T.; Li, X. *J. Am. Chem. Soc.* **2006**, *128*, p. 1044.
36. Sterk, D.; Stephan, M.; Mohar, B. *Org. Lett.* **2006**, *8*, p. 5935.
37. Szejtli, J. *"Cyclodextrins and their Inclusion Complexes"*; Wiley, Akademiai Kiado: Budapest, **1982**.
38. Atwood, J. L.; Davies, J. E. D.; MacNicol, D. D.; Vögtle, F. (Eds.) *"Supramolecular Chemistry"*; Pergamon: Oxford, **1996**.
39. Lehn, J.-M. *"Supramolecular Chemistry"*; VCH: Weinheim, **1995**.
40. Szejtli, J. *Chem. Rev.* **1998**, *98*, p. 1803.
41. Breslow, R.; Trainor, G.; Ueno, A. *J. Am. Chem. Soc.* **1983**, *105*, p. 2739.
42. Breslow, R.; Czarnik, A. W.; Lauer, M.; Leppkes, R.; Winkler, J.; Zimmermann, S. *J. Am. Chem. Soc.* **1985**, *108*, p. 1969.
43. Milovic, N. M.; Badjic, J. D.; Kostic, N. *J. Am. Chem. Soc.* **2004**, *126*, p. 696.
44. Baba, N.; Matsumura, Y.; Sugimoto, T. *Tetrahedron Lett.* **1978**, *44*, p. 4281.
45. Reddy, M. A.; Bhanumathi, N.; Rao, K. R. *Chem. Commun.* **2001**, p. 1974.
46. Schlatter, A.; Woggon, W.-D. *Angew. Chem. Int. Ed.* **2004**, *43*, p. 6731.
47. (a) Takehara, J.; Hashiguchi, S.; Fujii, A.; Inoue, S.; Ikariya, T.; Noyori, R. *Chem. Commun.* **1996**, *233*. (b) Noyori, R. *Adv. Synth. Catal.* **2003**, *1–2*, p. 345.
48. Schlatter, A. Ph.D. Dissertation, University of Basel, Basel, 2007.
49. Alonso, D.; Brandt, P.; Nordin, S. J. M.; Andersson, P. *J. Am. Chem. Soc.* **1999**, *121*, p. 9580.
50. Noyori, R.; Hashiguchi, S. *Acc. Chem. Res.* **1997**, *30*, p. 97.

CATALYTIC DISMUTATION VS. REVERSIBLE BINDING OF SUPEROXIDE

IVANA IVANOVIĆ-BURMAZOVIĆ

Department of Chemistry and Pharmacy, University of Erlangen-Nürnberg, Egerlandstrasse 1, 91058 Erlangen, Germany

I. Introduction

Superoxide ($O_2^{\bullet -}$) is the reactive radical anion formed following a one-electron reduction of dioxygen during numerous oxidation reactions under normal conditions in both living and non-living systems (*1*). It is a reducing agent in the anionic form (Eq. (1)), and an oxidant in the protonated form (pK_a (HO_2) = 4.69) (Eq. (2)) (*2*).

$$O_2 + e^- \rightarrow O_2^{\bullet -} \quad E^\circ = -0.16\,V \tag{1}$$

$$O_2^{\bullet -} + e^- + H^+ \rightarrow HO_2^- \quad E^\circ = +0.9\,V \tag{2}$$

Thus, superoxide can react with almost all redox-active metal centers (Scheme 1). In general, going through similar redox reaction steps metal complexes can interact with superoxide either as catalysts for its dismutation (superoxide dismutase (SOD) mimetics), or in a stoichiometric manner (Scheme 1).

These interactions are of a general chemical, as well as biological importance, since they occur in living organisms within the enzymatic

ADVANCES IN INORGANIC CHEMISTRY
VOLUME 60 ISSN 0898-8838 / DOI: 10.1016/S0898-8838(08)00003-2

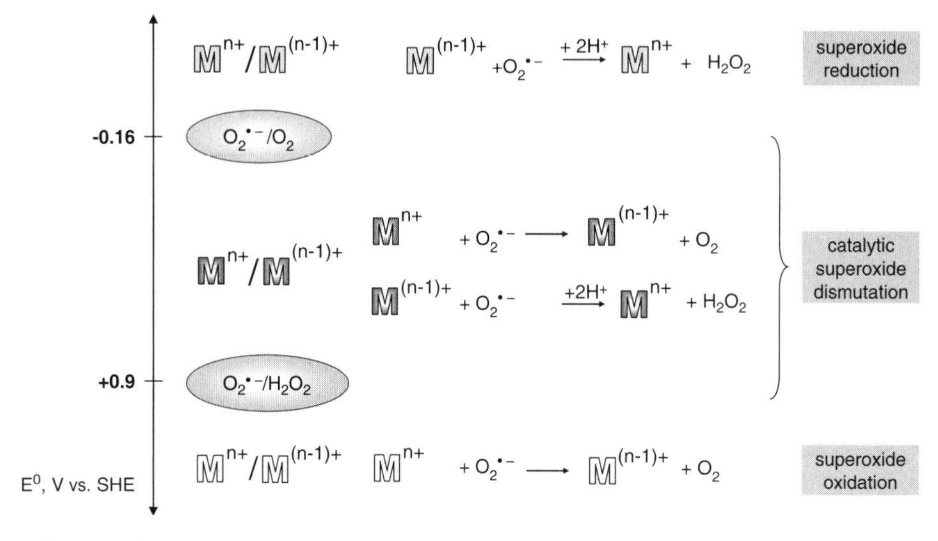

SCHEME 1.

(SODs and superoxide reductases (SORs)) (*1,3*) catalytic cycles and as undesired processes which lead to the pathophysiological conditions (see later). At the same time these reactions can be used in chemical catalysis. Namely, oxidative binding of superoxide to a metal center (M) generates a $M^{(n-1)+}$-superoxo adduct, which undergoes inner-sphere electron transfer resulting in a M^{n+}-peroxo species (Eq. (3)). Such species, depending on its electronic properties, are reactive towards different electron-rich and/or electron-poor organic substrates of synthetic and biological importance (*4*). Metal-peroxo complexes are invoked not only in the SODs and SORs catalytic mechanisms (*5*), but also as key intermediates in the mechanisms of metalloenzyme-catalyzed dioxygen (cytochrome P450, NO synthase, cytochrome oxidase) (*6*) and hydrogen peroxide activation (catalases and peroxidases) (*7*).

$$M^{(n-1)+} + O_2^{\bullet -} \longrightarrow M^{(n-1)+} - O_2^{-} \xrightarrow{e^-} M^{n+} - O_2^{2-} \quad (3)$$

The metal-peroxo species are considered to have a side-on structure (bidentate coordination of the peroxide ligand) and to be very unstable in protic medium (*8*). Under physiological conditions, after the first protonation and formation of a hydroperoxo intermediate (Scheme 2), the second protonation of this intermediate can proceed in two distinctly different pathways. In one case the second protonation results in the release of hydrogen peroxide from the metal center, leaving the metal oxidation state unchanged (Scheme 2). This is a crucial step in the catalytic cycles of SODs and SORs, especially in the catalytic mechanism of manganese SODs, which exist in the

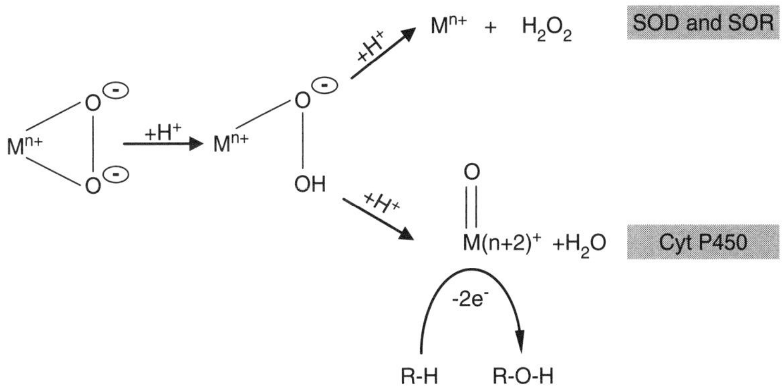

M^{n+} (side-on O_2^{2-}) $\xrightarrow{+H^+}$ M^{n+}–O(OH)$^-$ $\xrightarrow{+H^+}$ M^{n+} + H_2O_2 (SOD and SOR)

M^{n+}–O(OH)$^-$ $\xrightarrow{+H^+}$ $M(n+2)^+=O$ + H_2O (Cyt P450); R-H $\xrightarrow{-2e^-}$ R-O-H

SCHEME 2.

hydrophobic mitochondrial matrix. If protonation is not efficient, the enzyme remains in an inhibited peroxo form (*9*). In another possible pathway, the second protonation assists in the O–O bond cleavage and formation of a high-valent metal-oxo species (as in the case of catalase, peroxidase and cytochrome P450) (*6,7*).

We are interested in the reactions between superoxide and metal centers in non-heme and heme complexes, as the essential processes behind versatile manifestations and utilization of superoxide chemistry. We are also interested in characterization of the product species ($M^{(n-1)+}$-superoxo or M^{n+}-peroxo), since this enables understanding and tuning of their reactivity towards different organic substrates of industrial and biological interest. Such investigations help in perceiving the possible interactions in which superoxide can be involved under physiological and pathological conditions, and possible processes that can be utilized in chemical catalysis. The novelties regarding the catalytic superoxide dismutation by seven-coordinate manganese and iron complexes, and regarding the activation of superoxide by iron-heme complexes, which does not lead to $O_2^{\bullet -}$ catalytic disproportionation, are summarized in this chapter.

II. Catalytic Superoxide Dismutation by Seven-Coordinate Manganese and Iron Complexes as SOD Mimetics

A. State-of-the-Art

A.1. Life with superoxide

While dioxygen is essential to life on this planet, and its high redox potential is exploited as the driving force in the efficient metabolism of energy stores, the high reactivity of dioxygen, while affording this benefit, is also potentially hazardous for the very same reasons.

Reactions with oxygen must be carefully controlled during metabolism in order to avoid unwanted side-reactions, especially the production of the one-electron reduced form, superoxide. For this reason, O_2-activating centers generally utilize two redox-active centers to circumvent this one-electron reduction (*10*). Nevertheless, small quantities of superoxide are produced in ordinary circumstances, as well as in the presence of ionizing radiation (*11*). As a reactive species, superoxide is potentially hazardous to all cellular macromolecules, and it can give rise to other potentially hazardous reactive species. Some of the superoxide damaging effects are summarized in Scheme 3 (*11,12*). Nature's defense against the threats caused by the production of superoxide comes in the form of SOD enzymes, which catalyze the disproportionation of superoxide into dioxygen and hydrogen peroxide (Eq. (4)), and SOR enzymes, which catalyze the reduction of superoxide into hydrogen peroxide (Scheme 1). In the SOR mechanism the oxidized enzyme is then returned to the reduced state by reducing agents other than superoxide.

$$2O_2^{\bullet -} + 2H^+ \xrightarrow{SOD} O_2 + H_2O_2 \quad (4)$$

Four distinct forms of SODs are found in nature, which fall into three families. The Cu/Zn SODs occur primarily in cytoplasm of eukaryotes and chloroplasts, but have also been found in a few species of bacteria, nickel-containing SODs are known in some prokaryotes,

Superoxide toxicity

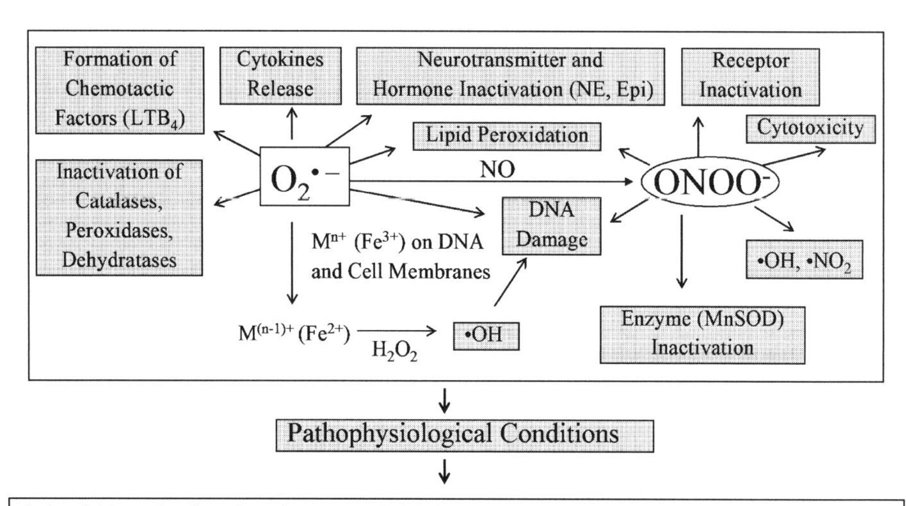

SCHEME 3.

while the structurally related Mn- and FeSODs are found in mitochondrial matrix, prokaryotes and a few families of plants, respectively (*13*). These enzymes are extremely efficient, with rates of dismutation approaching the diffusion limit (*12d*).

Under normal circumstances, these enzymes are responsible for keeping the formation of superoxide under control. However, in disease states the overproduction of $O_2^{\bullet -}$ exceeds the ability of the endogenous SOD system to remove or eliminate it, resulting in superoxide anion mediated damage. Many disease states have been linked to $O_2^{\bullet -}$ and its daughter radical species, and some of them are summarized in Scheme 3 (*12d,12f,14*). Since each of these pathophysiological conditions are associated with the $O_2^{\bullet -}$ production, an agent that removes $O_2^{\bullet -}$ would enable treatment of a wide range of diseases. For this purpose, the SOD enzymes (natural, recombinant and modified) have been used in preclinical animal studies and clinical trials and have been shown to possess efficacy in the disease states connected with the overproduction of $O_2^{\bullet -}$ (*12d,12f*). However, there are major drawbacks associated with the use of the enzymes as potential medicaments: immunological problems, lack of oral activity, short half-lives *in vivo*, costs of production, solution instability, reduced efficacy due to the large size and inability to access the target tissues (e.g., native enzymes do not penetrate the blood-brain barrier) (*12d,15*). To overcome the problems associated with enzymes, several investigations have been directed to design non-proteinaceous, synthetic, low molecular weight mimetics of the SOD enzymes (synzymes), for use as human pharmaceuticals in the treatment and prevention of diseases caused by overproduction of superoxide (*12d*).

A.2. SOD mimetics and related mechanisms

Among many different complexes that have been synthesized in attempt to mimic the structure and/or functionality of SODs (*16–22*), the most active SOD mimetics known to date are seven-coordinate Mn(II) complexes with macrocyclic ligands derived from C-substituted pentaazacyclopentadecane *[15]aneN*$_5$ and its pyridine derivative (Scheme 4) (*12d,16a,23–25*). Some of them possess SOD activity that exceeds the one of native mitochondrial MnSOD, and are the first SOD mimetics which entered clinical trials (*12d,16a,23,26–28*). A few Fe(III) complexes with the same type of ligands have also been studied and they are one of the best iron-based SOD catalysts (*18*). It should be stressed that the decomposition of superoxide catalyzed by these complexes has been quantified by direct stopped-flow method, in the presence of a substantial superoxide excess over catalyst, as a reliable method for determining true SOD activity (*29*).

The intriguing question is how the seven-coordinate geometry around the metal center favors its remarkable catalytic activity, knowing that in the native MnSOD and FeSOD enzymes the active

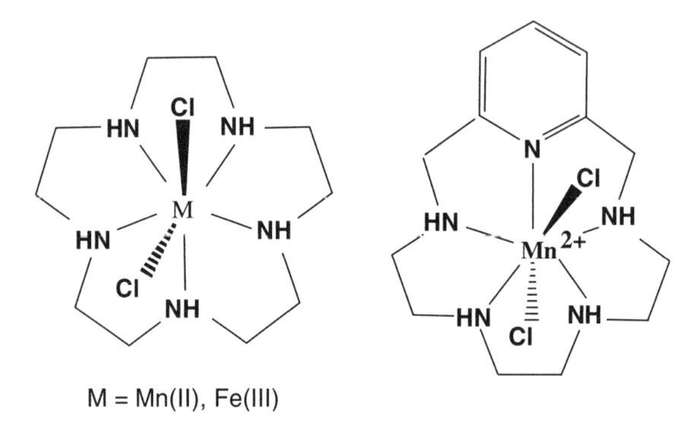

SCHEME 4.

metal center has a five-coordinate geometry (*3a,14f,30*). All SOD catalysts, whether native or synthetic, act in a "ping-pong" fashion, cycling between oxidized and reduced forms during each half-cycle of the $O_2^{\bullet -}$ disproportionation reaction (Scheme 1). Nature created the most efficient mechanism to enable fast catalytic turnover by choosing inner-sphere proton-coupled electron transfer, as an efficient path which does not require charge separation in the transition state, and by requiring the minimal perturbations to form catalytic intermediates. Superoxide binds without displacement of the other ligands in an associative manner, requiring minimal geometrical changes, and the metal center cycles between five- and six-coordinate forms in each successive half-cycle (*3a*).

The postulated mechanism for the catalytic action of the seven-coordinate Mn(II)(*[15]aneN$_5$*) type mimetics is more complicated (*22,24,31,32*). The oxidation of Mn(II) has been found to be the rate-determining step. The reactions follow two alternate but distinct pathways for this step: a pH dependent outer-sphere reaction and a pH independent inner-sphere reaction (Scheme 5). Each of these pathways involves large conformational changes of the ligand, through folding of the macrocycle to form a pseudo-octahedral geometry. Therefore, it has been postulated that only complexes with significant conformational flexibility are able to catalytically disproportionate superoxide (*22,25*). The folding model has been used to explain the differences in SOD activity caused by C-substituents on the macrocycle. A good correlation exists between the catalytic rate constants and the ability of the macrocycle to fold into the pseudo-octahedral geometry (*25,31,32*). Consequently, the lack of SOD activity in the case of the complex with analogue ligand system-containing imine groups (Scheme 6) can be interpreted in terms of its conformational rigidity. It was claimed that for the pH independent pathway, which contains inner-sphere binding of superoxide to a vacant coordination site on the manganese center, the rate of formation of a vacant axial coordination site is the rate-determining step, whereas

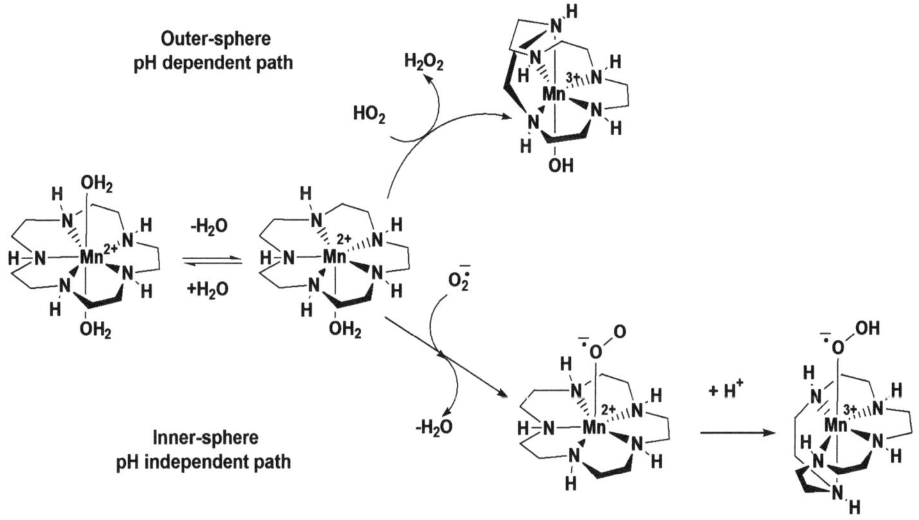

Scheme 5.

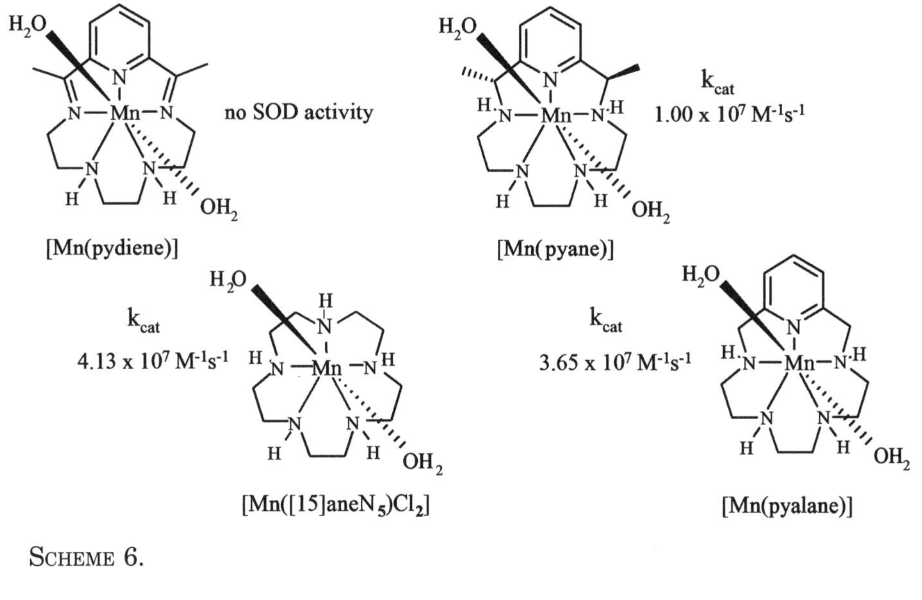

Scheme 6.

for the pH dependent process the proton-coupled electron transfer step was found to be rate-limiting (*22,25,31*).

B. Water Exchange on Seven-Coordinate Mn(II) Complexes: Insight in the Mechanism of Mn(II) SOD Mimetics

In the case of the Mn(II) SOD mimetics (see earlier), the loss of one water ligand from the seven-coordinate structure is the first step in the reaction cascade (Scheme 5) to form the corresponding six-coordinate intermediate, which can enter the SOD catalytic cycle. It was assumed that the water release and formation of a six-coordinate intermediate, requiring conformational rearrangements, is the rate-limiting step, and the catalytic rate constants for the inner-sphere pathway (k_{IS}) were compared with the water-exchange rate constant for the $[Mn(H_2O)_6]^{2+}$ ion (*22,24,31*). Water exchange on $[Mn(H_2O)_6]^{2+}$ follows an interchange associative (I_a) mechanism based on $\Delta V^{\neq}$ with a seven-coordinate intermediate and was measured to be $2.1 \times 10^7\ s^{-1}$ (*33*). The exchange of the aqua ligands of seven-coordinate complexes (Eq. (5)) however should follow a dissociative pathway with a six-coordinate intermediate and it was unknown so far how the exchange rate constant of these complexes is influenced by the structure of the ligands.

$$[MnL(H_2O)_2] + H_2O^* \rightleftharpoons [MnL(H_2O)(H_2O)^*] + H_2O \qquad (5)$$

We performed a detailed study of the acid–base properties of some pentaaza macrocyclic seven-coordinate manganese(II) complexes (Scheme 6) in aqueous solution and their water-exchange rate constants and activation parameters measured by temperature and pressure dependent ^{17}O-NMR techniques, which was performed to get more insight in the mechanism and to understand better the details of the influence of the ligand structure on the lability i.e., reactivity of the complexes (*34*). Three of them, without the imine groups in the macrocyclic ligand, are proven SOD mimetics (*27,35*). As mentioned earlier, it was reported that the imine groups-containing [Mn(*pydiene*)Cl_2] complex was SOD inactive, most probably due to its low conformational flexibility and consequently low ability to form a six-coordinate pseudo-octahedral intermediate (*22*). We have included this complex in our studies in order to compare its water-exchange parameters with those of the more flexible and SOD active [Mn(*[15]aneN$_5$*)Cl_2], [Mn(*pyane*)Cl_2] and [Mn(*pyalane*)Cl_2] complexes. Additionally we have synthesized and structurally characterized a new [Mn(L)Cl_2] complex (Fig. 1), which is the acyclic analog of [Mn(*pydiene*)Cl_2] (Scheme 6) and consequently has higher conformational flexibility (*34*).

B.1. Solution behavior

All studied complexes in aqueous solution are present as diaqua $[Mn(L)(H_2O)_2]^{2+}$ species, whereas two chloro ligands are replaced by water molecules in axial positions (*35,36*). For studying the water-exchange processes it was necessary first to determine the acid–base properties of the complexes by potentiometric titration in order to define the pH range where the diaqua form of the studied complexes is the predominant one. The potentiometric titrations of

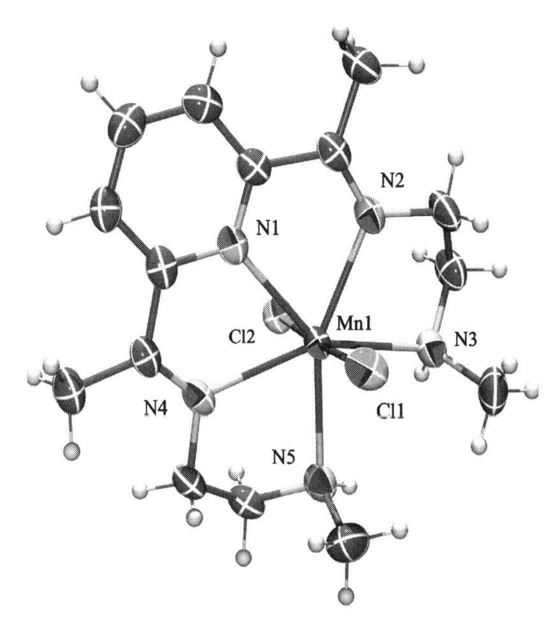

FIG. 1. ORTEP drawing for $[Mn(L)Cl_2]$, showing the labeling scheme and the 50% probability ellipsoids for the non-hydrogen atoms (*34*).

$[Mn(\textit{[15]aneN}_5)(H_2O)_2]^{2+}$ and $[Mn(\textit{pyalane})(H_2O)_2]^{2+}$ were already reported by Riley *et al.* (*35*), and we could confirm their results (*34*). *[15]aneN*$_5$, *pyane* and *pyalane* exhibit three protonation equilibria, which corresponds to the protonations of three secondary amine nitrogens. The pyridine ring-containing *pyane* and *pyalane* ligands are less basic than *[15]aneN*$_5$, but the binding constants for $[Mn(\textit{pyane})(H_2O)_2]$ and $[Mn(\textit{pyalane})(H_2O)_2]$ are somewhat higher than in the case of $[Mn(\textit{[15]aneN}_5)(H_2O)_2]$. All three complexes are stable between a pH around 5 and 10, and have a stable form with the monoprotonated ligand at one of the secondary amine nitrogen atoms. The binding constants for the imine groups-containing $[Mn(\textit{pydiene})(H_2O)_2]$ and $[Mn(L)(H_2O)_2]$ complexes are just slightly lower than in the case of $[Mn(\textit{[15]aneN}_5)Cl_2]$. However, the ligand protonation immediately leads to the complex decomposition, and in the basic region the base catalyzed water addition to the imine bonds occurs. A similar methanol and water addition to imine bonds was also observed in the literature (*37*). Therefore, due to the acid and base catalyzed hydrolytic process these two complexes are stable only in a quite narrow pH range. Such solution behavior can be a reason for the lack of catalytic SOD activity of the imine complexes. In any case, $[Mn(\textit{pydiene})(H_2O)_2]$ was found to be stable at $pH = 6.1$ in the temperature range between 5°C and 85°C, whereas stability of $[Mn(L)(H_2O)_2]$ was not high enough for performing the water-exchange measurements.

B.2. Water exchange

Rate constants and the corresponding activation parameters for the water-exchange reactions (Eq. (1)) of the studied complexes, resulted from the temperature and pressure dependent measurements, are summarized in Table I. The values of activation entropies and activation volumes are positive for all studied complexes suggesting a dissociative nature (I_d) of the water-exchange mechanism. The exchange rate constants, k_{ex}, however are influenced by the acceptor abilities of the spectator ligands. In terms of a dissociative mode of the water-exchange process, the extended π-acceptor system of the $[Mn(pydiene)(H_2O)_2]^{2+}$ complex leads to a less labile Mn–OH_2 bond and to a slower reaction. From another side, the lack of pyridine ring in the *[15]aneN*$_5$ ligand results in its decreased π-acceptor ability and consequently leads to the labilization of the coordinated water molecule (Table I). In contrast, pH has no influence on the exchange reaction rates of the studied complexes. This suggests that in the applied pH ranges there are no changes in the manganese(II) coordination sphere. Even in the case of $[Mn(pydiene)(H_2O)_2]^{2+}$, where at pH higher than 7.6 the predominant species possesses the hydrolyzed form of the *pydiene* ligand, the exchange rate constants are consistent within the experimental error. This can be related to the fact that the new electron-withdrawing hydroxyl group compensates a decrease in π-acceptor ability of the pentadentate ligand upon addition of OH^- to the double bond.

In the literature there are only few studies on the water-exchange processes of the manganese(II) species in general (*33,38–41*), and the only seven-coordinate Mn(II) complexes studied are $[Mn(EDTA)(H_2O)]^{2-}$ and its derivatives (*38,39,42,43*). Such studies are essential for understanding the mechanism of the manganese-containing SOD mimetics. The volume of activation for the water-exchange reaction

TABLE I

ACTIVATION PARAMETERS AND RATE CONSTANTS FOR THE WATER-EXCHANGE REACTION OF $[Mn(pydiene)(H_2O)_2]^{2+}$, $[Mn(pyane)(H_2O)_2]^{2+}$ AND $[Mn(pyalane)(H_2O)_2]^{2+}$ AND LITERATURE VALUES FOR $[Mn(H_2O)_6]^{2+}$ (*34*)

Complex	pH	k_{ex} (298 K) ($10^7\ s^{-1}$)	$\Delta H^{\neq}$ (kJ mol^{-1})	$\Delta S^{\neq}$ (J K^{-1} mol^{-1})	$\Delta V^{\neq}$ (cm^3 mol^{-1})	Literature
$[Mn(H_2O)_6]^{2+}$	–	2.1 ± 0.1	32.9 ± 1.3	$+5.7\pm5.0$	-5.4 ± 0.1	(*33*)
$[Mn(pydiene)(H_2O)_2]^{2+}$	6.1	2.0 ± 0.1	39.8 ± 1.3	$+28.6\pm4.4$	$+3.4\pm0.1$	(*34*)
$[Mn(pydiene)(H_2O)_2]^{2+}$	7.6	1.7 ± 0.1	35.9 ± 1.7	$+13.6\pm5.8$	–	(*34*)
$[Mn(pydiene)(H_2O)_2]^{2+}$	8.7	1.7 ± 0.1	40.3 ± 1.0	$+28.8\pm3.5$	–	(*34*)
$[Mn(pydiene)(H_2O)_2]^{2+}$	9.5	1.6 ± 0.1	39.5 ± 1.1	$+25.7\pm3.8$	$+5.3\pm0.2$	(*34*)
$[Mn(pyane)(H_2O)_2]^{2+}$	7.7	5.3 ± 0.3	37.2 ± 0.4	$+27.6\pm1.5$	$+5.2\pm0.7$	(*34*)
$[Mn(pyane)(H_2O)_2]^{2+}$	10.9	5.8 ± 0.3	33.3 ± 0.6	$+15.2\pm1.9$	$+3.4\pm0.7$	(*34*)
$[Mn(pyalane)(H_2O)_2]^{2+}$	8.0	4.7 ± 0.3	39.1 ± 0.6	$+33.0\pm2.1$	$+3.2\pm0.1$	(*34*)
$[Mn([15]aneN_5)(H_2O)_2]^{2+}$	8.0	≥10				(*34*)

was reported only for six-coordinate $[Mn(H_2O)_6]^{2+}$ (*33*) and $[Mn_2^{II}(ENOTA)(H_2O)_2]$ (ENOTA = triazacyclononane-based ligand) (*41*). For seven-coordinate 3d metal complexes a dissociative mode of substitution reactions in general (although there are some exceptions (*44*)), and water exchange in particular, with a six-coordinate intermediate is to be expected. The difference between six- and seven-coordinate Mn(II) species is clearly represented by the negative and positive values of the activation volumes for six-coordinate $[Mn(H_2O)_6]^{2+}$ ($\Delta V^{\neq} = -5.4\,cm^3\,mol^{-1}$) (*33*) and $[Mn_2^{II}(ENOTA)(H_2O)_2]$ ($\Delta V^{\neq} = -10.7\,cm^3\,mol^{-1}$) (*41*) and the studied seven-coordinate complexes (Table I), respectively, since the activation volume is a more sensitive parameter than $\Delta H^{\neq}$ and $\Delta S^{\neq}$, leading to a better understanding of the nature of the substitution processes. The exchange rate constant of $[Mn(EDTA)(H_2O)]^{2-}$ is reported to be $(4.4 \pm 0.3) \times 10^8\,s^{-1}$ (298 K) (*38*), which is much faster than in the case of the complexes studied in this chapter. This is due to the strong π-donor ability and high negative charge of the EDTA ligand and also due to the fact that in $[Mn(EDTA)(H_2O)]^{2-}$ the water molecule is in the sterically crowded equatorial plane with five donor atoms, whereas in our complexes water molecules are in the axial positions. In the case of six-coordinate species, where an associative mechanism for the water exchange would be expected, the increase in the π-acceptor ability of the spectator ligands has an opposite effect than in the case of seven-coordinate complexes, namely it accelerates the exchange rate. It can be clearly observed if we compare the water-exchange rate constants for $[Mn(H_2O)_6]^{2+}$, $[Mn(phen)(H_2O)_4]^{2+}$ and $[Mn(phen)_2(H_2O)_2]^{2+}$ which are found to be $(5.8 \pm 0.1) \times 10^6$ (calculated for 0°C from literature data) (*33*), $(13 \pm 2) \times 10^6\,s^{-1}$ and $(31 \pm 3) \times 10^6\,s^{-1}$ (both at 0°C), respectively (*40*).

B.3. Correlation with SOD activity

Since it has been reported that in the inner-sphere SOD catalytic pathway (Scheme 5) the water-exchange process is the rate-limiting one, the inner-sphere catalytic rate constants k_{IS} were correlated with the water-exchange rate constants on $[Mn(H_2O)_6]^{2+}$ (*22,31*). However, it seems that it is not possible to draw a direct correlation between these rate constants. Firstly, k_{IS} (which is pH independent) according to the observed rate law for dismutation of superoxide ($V = -d[O_2^{\bullet -}]/dt = [Mn][O_2^{\bullet -}]\{k_H^+[H^+]+k_{ind}\}$, $k_{ind} = 2k_{IS}$, $k_H^+ = 2k_{OS}/K_a$) has the unit $M^{-1}\,s^{-1}$, whereas the unit of the water-exchange rate constant is s^{-1}. Therefore the values for k_{IS}, which are in general for all reported complexes in the 0.15×10^7–$3.98 \times 10^7\,M^{-1}\,s^{-1}$ range and in particular for the $[Mn([15]aneN_5)Cl_2]$ and $[Mn(pyalane)Cl_2]$ complexes 0.91×10^7 and $1.01 \times 10^7\,M^{-1}\,s^{-1}$, respectively (*32,35*), are not directly comparable with $k_{ex} = 2.1 \times 10^7\,s^{-1}$ for $[Mn(H_2O)_6]^{2+}$. (The k_{IS} value for $[Mn(pyane)Cl_2]$ was not reported in the literature and

TABLE II

COMPARISON OF $K_{EX}/[H_2O]$ (*34*), K_{IS} (*32,35*) AND ΔE (*34*) VALUES

Complex	$k_{ex}/[H_2O]$ ($M^{-1}\,s^{-1}$)	k_{IS} ($M^{-1}\,s^{-1}$)	ΔE^a (kcal mol^{-1})
$[Mn(pydiene)(H2O)_2]^{2+}$	0.03×10^7	0	17.5
$[Mn(L)(H_2O)_2]^{2+}$	–	–	30.4
$[Mn(pyalane)(H_2O)_2]^{2+}$	0.08×10^7	1.0×10^7	18.4
$[Mn(pyane)(H_2O)_2]^{2+}$	0.09×10^7	–	16.2
$[Mn([15]aneN_5)(H_2O)_2]^{2+}$	$\geq 0.18 \times 10^7$	0.9×10^7	–

[a]B3LYP/LANL2DZp.

$[Mn(pydiene)Cl_2]$ is known to be SOD inactive.) One should divide k_{ex} by $[H_2O]$ (55.5 M), in order to obtain comparable values. In that way the values for k_{IS} would be usually much higher than the corresponding value of k_{ex}, suggesting somewhat different mechanism for the inner-sphere SOD catalytic pathway of seven-coordinate Mn(II) complexes. It would be more appropriate to compare k_{IS} with k_{ex} of the particular seven-coordinate complex and not with the value of k_{ex} for the six-coordinate $[Mn(H_2O)_6]^{2+}$, since we have shown that k_{ex} can vary by almost a factor of 10 (Table I). The corresponding $k_{ex}/[H_2O]$ values are given in Table II and they are almost one order of magnitude lower than corresponding k_{IS}, suggesting that the water release cannot be a rate-determining step (otherwise the inner-sphere SOD catalytic pathway should have consequently been slower than the experimentally observed one). Additionally, these two complexes have almost the same values of k_{IS}, however k_{ex} differs significantly. Such discrepancies could be explained in terms of an interchange (I_d) mechanism for the substitution of the water molecule by the incoming superoxide anion (as we have found for the water-exchange process) with formation of the outer-sphere precursor complex, rather than in terms of a limiting dissociative mechanism with a six-coordinate intermediate.

It is not surprising that the substitution processes on seven-coordinate 3d metal ions follow an interchange, rather than a limiting dissociative mechanism. In the case of seven-coordinate Fe(III) complex we even found an associative interchange mechanism for the substitution of solvent molecules as a result of the high π-acceptor ability of the fully conjugated pentadentate ligand system present in its equatorial plane (*44*).

B.4. *Six-coordinate intermediate and conformational flexibility*

DFT (density functional theory) calculations were performed for the $[Mn(pydiene)(H_2O)_2]^{2+}$, $[Mn(L)(H_2O)_2]^{2+}$, $[Mn(pyane)(H_2O)_2]^{2+}$ and $[Mn(pyalane)(H_2O)_2]^{2+}$ complexes, as well as for their corresponding six-coordinate structures $[Mn(ligand)(H_2O)]^{2+}$ in order to compare the

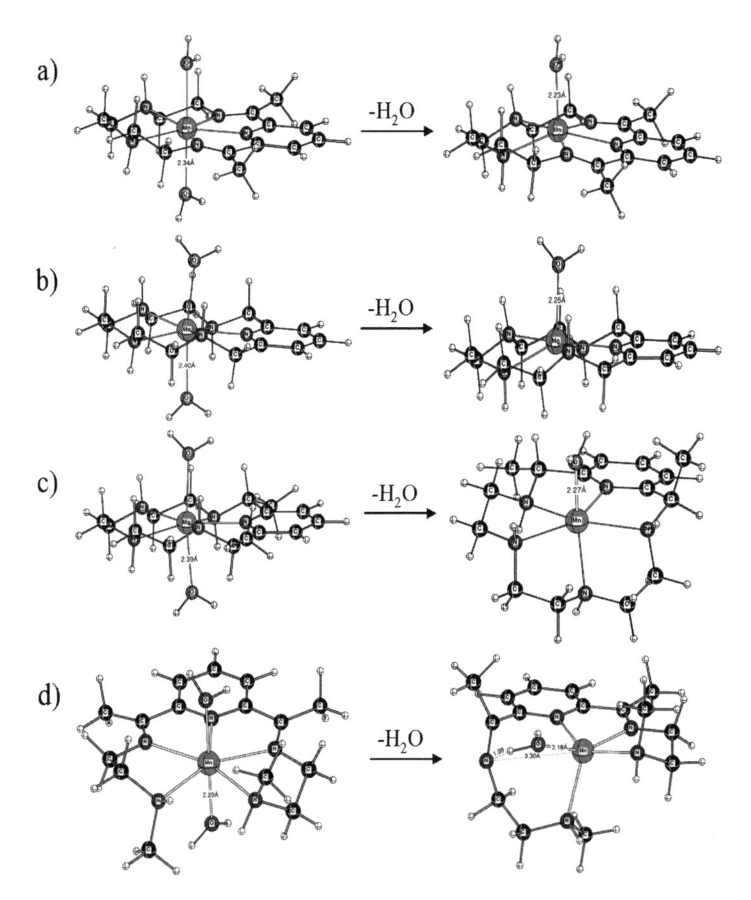

FIG. 2. With DFT methods (B3LYP/LANL2DZp) calculated structures for the seven-coordinate diaqua complexes (a) $[Mn(\textit{pydiene})(H_2O)_2]^{2+}$, (b) $[Mn(\textit{pyalane})(H_2O)_2]^{2+}$, (c) $[Mn(\textit{pyane})(H_2O)_2]^{2+}$ and (d) $[Mn(L)(H_2O)_2]^{2+}$ (on the left) and the corresponding six-coordinate species after loss of one water molecule (on the right) (*34*).

energy required for the water dissociation (ΔE) and the complex reactivity according to a dissociative mechanism with the six-coordinate species as intermediate structures. The ΔE values are given in Table II and the calculated structures are shown in Fig. 2. From Table II it can be seen that for the three complexes with cyclic ligands the energy differences are almost the same, whereas in the case of acyclic ligand ΔE is significantly different. Ligand *pyane* shows the most distinct folding, with the corresponding six-coordinate structure showing a pseudo-octahedral geometry. Surprisingly, *pyalane* is not able to fold and form a pseudo-octahedral coordination sphere. It shows a rigid behavior like the imine group-containing *pydiene* ligand. Interestingly, $[Mn(\textit{pyane})(H_2O)_2]^{2+}$ with the strongest folded ligand, has a lower k_{cat} value $(1 \times 10^7\,M^{-1}\,s^{-1})$ than

[Mn(*pyalane*)$(H_2O)_2]^{2+}$ ($3.65 \times 10^7\,M^{-1}\,s^{-1}$) (*27,32,35*), where the ligand is nearly unable to fold. The attempt to calculate the seven-coordinate structure of [Mn(*[15]aneN*$_5$)$(H_2O)_2$] resulted in a six-coordinate geometry around Mn(II), where only one water molecule is coordinated to the manganese center and the second non-coordinated water molecule, which is in a sort of second coordination sphere, forms hydrogen bonds with the *[15]aneN*$_5$ ligand. This can be correlated with the higher substitution lability of the [Mn(*[15]aneN*$_5$)$(H_2O)_2$] complex in aqueous solution, which is reflected in the faster water-exchange process as experimentally observed (Table I).

The acyclic L ligand, despite possessing two imine groups, is much more flexible than the studied macrocyclic ligands and interestingly, as a result of DFT calculations dissociation of one water molecule resulted in a five-coordinate square-pyramidal geometry around the Mn(II) center (Fig. 2b). This change from the seven- to five-coordination geometry upon water dissociation, viz. a break of the additional bond (Mn–N(imine)) is a reason for higher ΔE in the case of $[Mn(L)(H_2O)_2]^{2+}$ (Table II). Despite the fact that the $[Mn(L)Cl_2]$ complex is quite unstable in water, the increased flexibility of the acyclic ligand and possibility for the formation of the square-pyramidal structure with the vacant axial coordination site is quite promising for the further development of this class of ligands and corresponding SOD active complexes.

The interchange character of the water-exchange mechanism of the studied seven-coordinate complexes can be a reason why there is no clear correlation between their rates for the exchange process and the energies required for the dissociation of the coordinated water molecule. ΔE is also not possible to correlate with the catalytic rate constants published in the literature.

B.5. Conclusion

We have shown that although Mn(II) has a spherically symmetrical d^5 electronic configuration, coordination geometry and ligand electronic properties have an influence on the water-exchange process. In the case of seven-coordinate pentagonal-bipyramidal Mn(II) complexes, with water molecules in apical positions, the water exchange proceeds via an interchange dissociative (I_d) mechanism. k_{ex} are mainly controlled by the π-acceptor abilities of the ligands and decrease with an increase of the ligand π-acceptor ability, which also confirms a dissociative character of the water exchange. However, the corresponding water-exchanged rate constants strongly suggest that the water release and formation of a six-coordinate intermediate cannot be the rate-limiting step in the overall inner-sphere catalytic superoxide dismutation pathway, opposite to what has been postulated in the literature. This can be explained in terms of an interchange (I_d) mechanism for the substitution of the water molecule on the seven-coordinate Mn(II) center, where the incoming superoxide anion also plays a role in the overall substitution process. Additionally, it has been postulated that the complexes with ligand systems-containing imine groups do not have SOD activity (*22,25*) most probably due to the conformational rigidity and low ability to form a six-coordinate pseudo-octahedral intermediate upon release of one water molecule. However, we have also shown that conformational flexibility of the pentadentate ligand is not the key factor assisting in SOD activity. Further more, we have shown that the seven-coordinate complexes with acyclic imine-containing ligands have a possibility to form the square-pyramidal intermediate structure, with a vacant axial coordination site, upon release of one water molecule. This is quite promising for the further development of the complexes with acyclic pentadentate ligands as potential SOD mimetics.

C. Acyclic and Rigid Seven-Coordinate Complexes as SOD Mimetics

Since we have shown (see earlier) that the conformational flexibility of the pentadentate ligand is not a key requirement for the SOD activity of the seven-coordinate complexes, due to the fact that in an interchange substitution mechanism (operating in the case of these complexes) (*34*) efficient formation of a real six-coordinate (with pseudo-octahedral geometry) intermediate is not indispensable, we were interested in additional experimental approval of such mechanistic paradigm. For achieving this goal it was important to select an appropriate ligand system, with electronic properties which could provide SOD activity of the coordinated metal center and at the same time demonstrating high conformational rigidity.

C.1. Iron complexes

Our iron(III) seven-coordinate $[Fe(dapsox)(H_2O)_2]ClO_4$ (Scheme 7) (H_2dapsox = 2,6-diacetylpyridine-bis(semioxamazide)) (*45*) complex appeared to be an excellent candidate for these studies. The dapsox^{2-} anion is acyclic, completely planar and the extremely rigid pentadentate ligand, with specific electronic properties as a result of a fully conjugated system of double bonds over the entire ligand. This leads to the high stability of the $[Fe(dapsox)(H_2O)_2]ClO_4$ complex despite of the acyclic nature of dapsox^{2-}. Since free iron ions are more toxic than manganese ions, it is important that the chelate will form a very stable complex and prevent the release of iron ions. Despite this toxicity, complexes of Fe(III) are highly attractive as SOD mimetics due to their higher kinetic and thermodynamic stability than Mn(II) complexes (*12d*). In aqueous solution under physiological conditions, no evidence for the decomposition of the complex has been observed over a period of one year. Potentiometric titrations (Fig. 3) show that at pH ≈ 2, 50% of the complex is in the $[Fe(Hdapsox)(H_2O)_2]^{2+}$ form and 25% in the

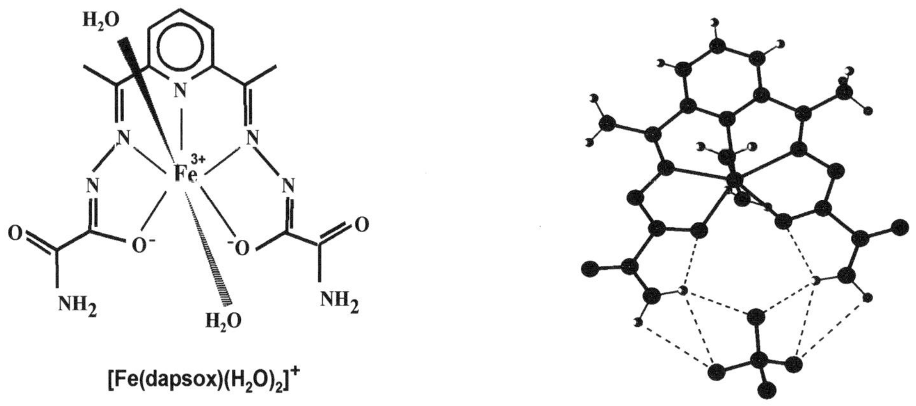

H_2O N N N Fe^{3+} N N O O^- ^-O O NH_2 H_2O NH_2

$[Fe(dapsox)(H_2O)_2]^+$

SCHEME 7.

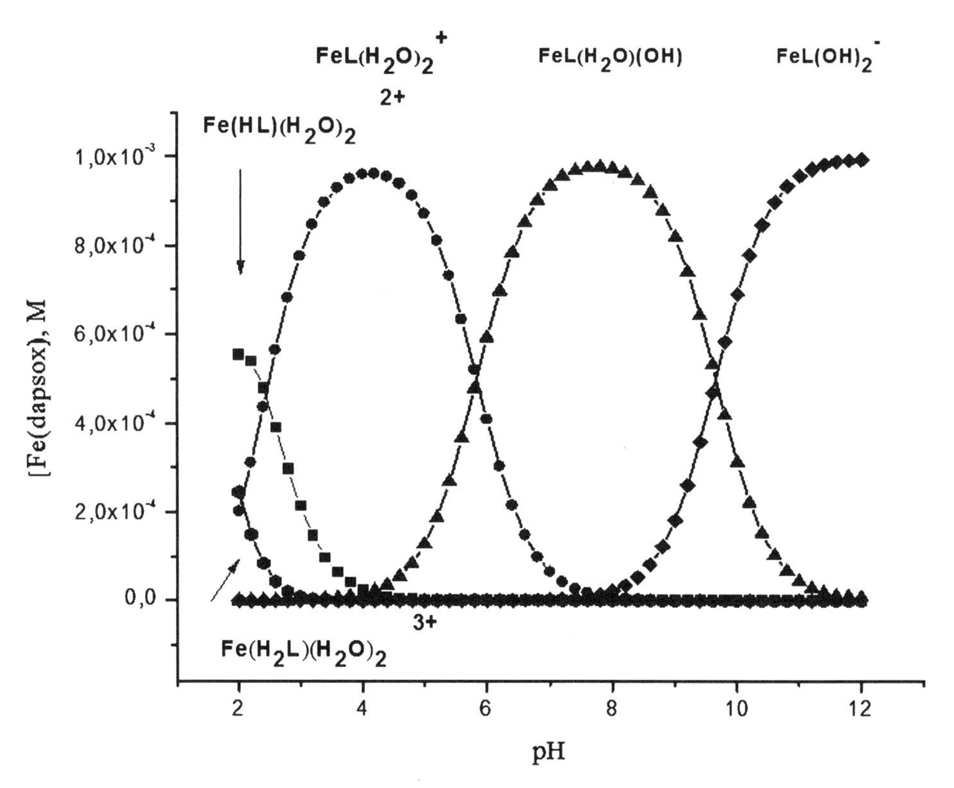

FIG. 3. Speciation of $[Fe(dapsox)(H_2O)_2]ClO_4$ in aqueous solution (*46*).

$[Fe(H_2dapsox)(H_2O)_2]^{3+}$ and $[Fe(dapsox)(H_2O)_2]^+$ forms, i.e., with a mono, diprotonated and deprotonated pentadentate chelate, respectively (*46*). However, there is no indication that the complex decomposes to release iron in solution under such conditions. A slow decomposition over a period of one month was observed in a 1 M acid solution.

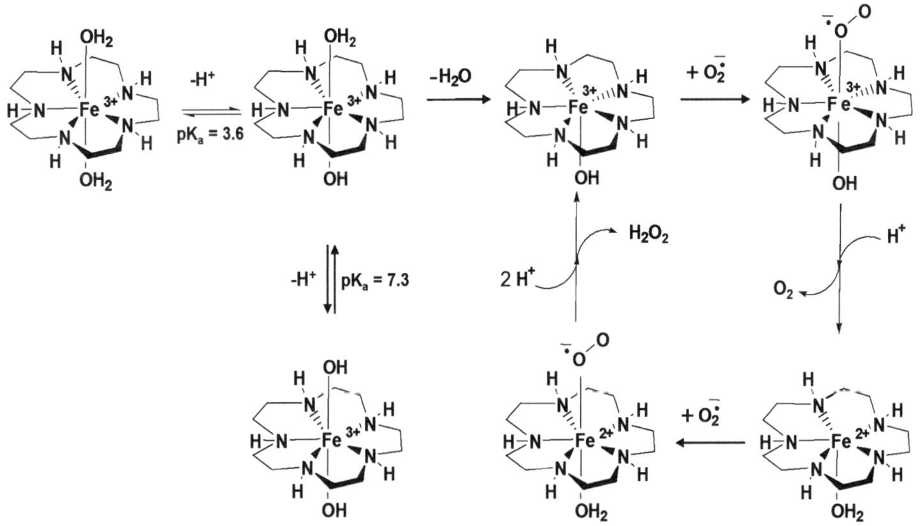

SCHEME 8.

Interestingly, for the seven-coordinate Fe(III) SOD mimetics with the macrocyclic *[15]aneN*$_5$ type of the ligands, a catalytic mechanism (different from that proposed for analogue Mn(II) complexes) in which the aqua-hydroxo form of the complex, $[Fe(ligand)(OH)(H_2O)]^{2+}$, is the catalytically active species (Scheme 8) has been proposed (*18*). A drawback of these complexes is the low pK_a values of the two coordinated water molecules, which results in the formation of inactive (inert) dihydroxo complexes at the physiological pH (*18*). Therefore, the idea is to design a chelate that will decrease the acidity of the iron center and so increase the concentration of the catalytically active aqua-hydroxo species at the physiological pH to promote an enhanced SOD activity. It is important to note that there are no reports on the SOD activity of Fe^{III} complexes of the *[15]aneN*$_5$ derivatives with a fused pyridine moiety present in the macrocyclic ligand.

The $dapsox^{2-}$ ligand, besides high complex stability, provides appropriate acid–base properties of the complex (Fig. 3), causes an increase in the pK_a values of the coordinated water molecules (pK_{a1} = 5.8 and pK_{a2} = 9.5) (*45b,46*), which are very close to the pK_a values of the native Fe^{III}-SOD enzyme ($\sim$5 and $\sim$9) (*3a*). Thus at the physiological pH almost 100% of the complex is in the catalytically active aqua-hydroxo form.

An interesting structural feature of $[Fe(dapsox)(H_2O)_2]^+$ is its pentagonal-bipyramidal (PBP) geometry with kinetically labile solvent molecules coordinated in the axial positions and the completely planar pentadentate chelate in the equatorial plane, which generally facilitates easy access of nucleophiles to the metal center without any steric hindrance above and below the pentadentate plane.

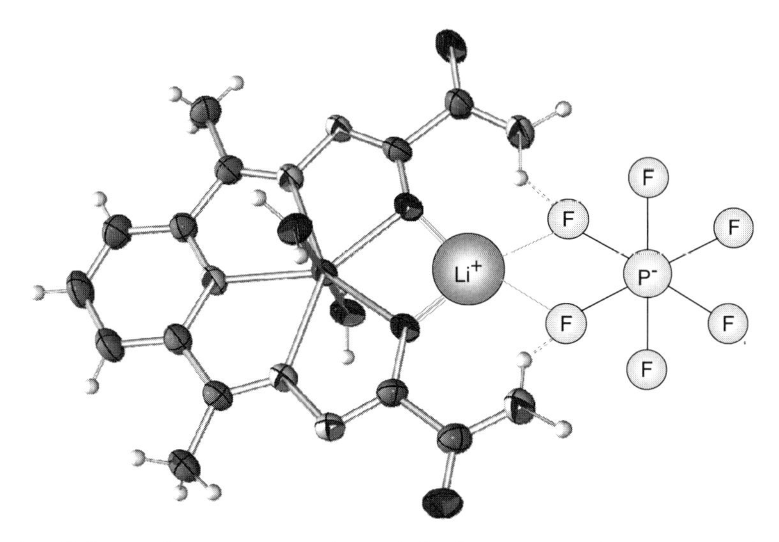

FIG. 4. Schematic representation of the host–guest interactions between the seven-coordinate $[Fe(dapsox)(solvent)_2]^+$ complex and lithium salt.

A tweezer-like structure of $dapsox^{2-}$ enables strong hydrogen bonding between amide $-NH_2$ groups on the ligand and anions of appropriate size. This type of interaction is not only present in the crystal structure (*45a*), but also can be observed in solution, especially in less polar solvents such as acetonitrile (*47*), and is an ion recognition effect which simulates properties of the enzymatic active site. Additionally, the negatively charged α-oxyazine oxygen atoms coordinated to the iron center bind the small lithium cation (electrostatic interactions) (Fig. 4). The more prominent interaction with Li^+ exists in the case of the Fe(II) form of the complex. The coordinated lithium cation close to the iron center decreases the electron density around the metal center and consequently increases its electrophility and reactivity towards anionic species (*47*). Simulation of an enzymatic active site, which could bind small metal cations and in that way enhance binding of superoxide, would be of importance.

The metal-centered redox potential is the most important criterion for the complex to be the SOD mimetic, since the catalytic disproportionation of $O_2^{\bullet -}$ requires redox reactions between complex and superoxide (Scheme 9) (*18*). The complex redox potential should fall between the redox potentials for the reduction and oxidation of $O_2^{\bullet -}$, viz. −0.16 and +0.89 V vs. NHE (normal hydrogen electrode), respectively (Scheme 1) (*2*).

Aqueous solutions of $[Fe(dapsox)(H_2O)_2]ClO_4$ in the pH range 1–12 exhibit a reversible redox wave for the Fe^{III}/Fe^{II} couple, and no complex decomposition or dimerization was observed (*46*). Furthermore, in the pH range 1–10 the metal-centered redox potential for $[Fe(dapsox)(H_2O)_2]ClO_4$ is in the range required for the possible SOD

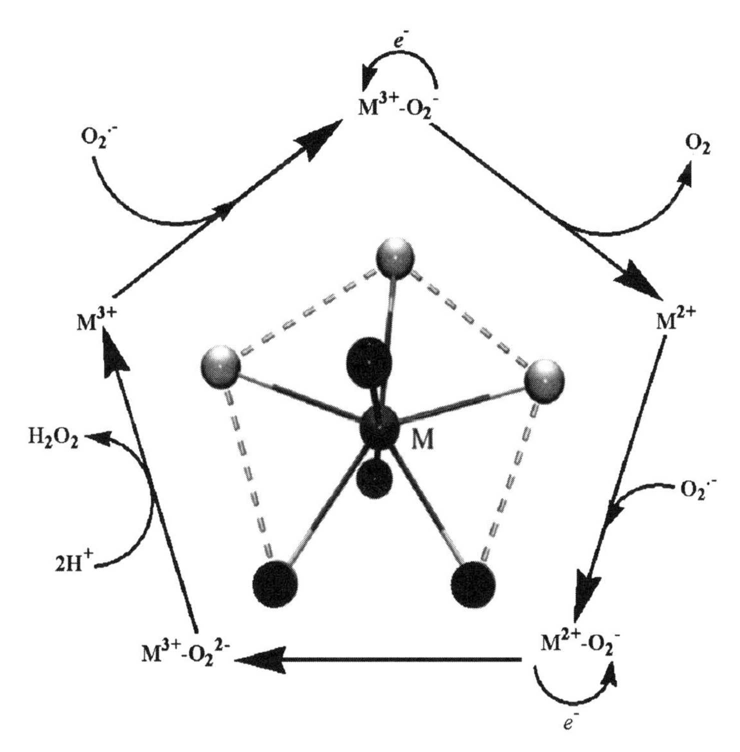

SCHEME 9.

activity (*46*). For the Fe(III) complexes with *[15]aneN5* type of chelates that are proven SOD catalysts, the redox potentials were not reported for the physiological pH. At this pH these complexes exist as an equilibrium mixture of the dihydroxo and aqua-hydroxo species (*18*). At pH ~3 they have the redox potential in the range of 0.35–0.45 V vs. NHE. In comparison, our complex at pH = 3 and 7.8 shows reversible redox behavior at 0.34 V and 0.05 V vs. NHE, respectively (*46*).

Protons are in general indispensable for the dismutation of superoxide (Eq. (4)). Also in the case of its dismutation catalyzed by a metal center, two protons are needed for the dissociation of the product (H_2O_2) from the metal center (Scheme 9). Therefore, a complex which can accept two protons upon reduction and release them upon oxidation is an excellent candidate for SOD activity. The studies on proton-coupled electron transfer in Fe- and Mn-SODs (*48*), demonstrated that the active site of MnSOD consists of more than one proton acceptor (Scheme 10). Since the assignment of species involved in proton transfer is extremely difficult in the case of enzymatic systems, relevant investigations on adequate model complexes could be of vast importance. H_2dapsox coordinates to Fe(II) in its neutral form, whereas in the case of Fe(III) it coordinates in the dapsox^{2-} form. Thus, oxidation and reduction of its iron complex is a proton-coupled electron transfer process (*46*), which as an energetically favorable

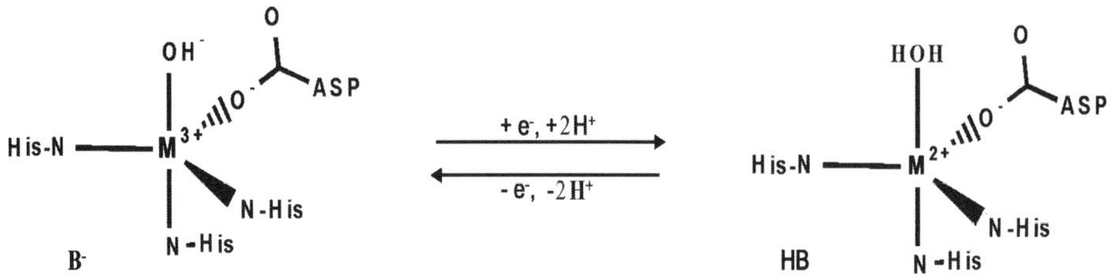

SCHEME 10.

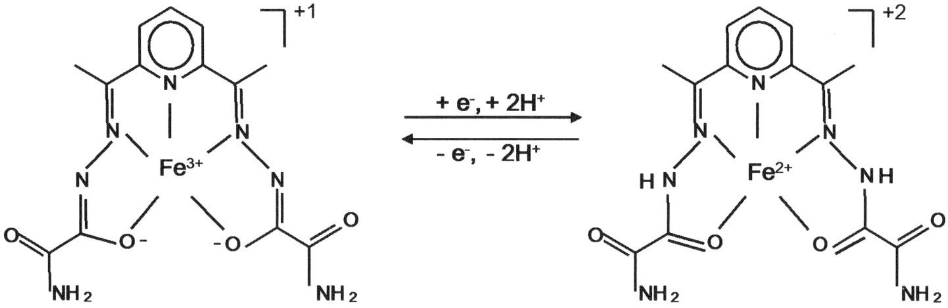

SCHEME 11.

redox process also operates in the case of natural Mn- and FeSODs. This point that makes our ligand system even more interesting due to the fact that for the other seven-coordinate SOD mimetics, the possibility of a proton transfer process in which the pentadentate ligand is involved, does not exist. Our electrochemical studies (*46*) reveal that $[Fe(dapsox)(H_2O)_2]ClO_4$ offers the possibility for a proton-coupled electron transfer in general, and "two-proton-one-electron-transfer" process in particular (Scheme 11).

A qualitative test for the interaction between superoxide and the reduced form of a potential SOD mimetic is an electrochemical experiment, where the reduced form of a complex and superoxide are generated *in situ* in aprotic solvent (most appropriate in DMSO = dimethyl sulfoxide). Aprotic solvent is needed to stabilize superoxide.

The cyclic voltammogram for $[Fe(dapsox)(H_2O)_2]ClO_4$ in DMSO (*49*) purged with nitrogen exhibits a reversible couple at -0.13 V vs. Ag/AgCl electrode (Fig. 5a), or -0.11 vs. NHE. The cyclic voltammogram in air-saturated DMSO in the scan range up to -0.4 V (Fig. 5b) shows again reversible redox wave for the Fe^{III}/Fe^{II} couple at slightly more negative potential, -0.18 V, since in the presence of oxygen it is more difficult to reduce Fe(III). When the scan proceeds towards more negative potentials (Fig. 5c), after the complex is reduced to the Fe(II) species, molecular oxygen is reduced to superoxide at -0.82 V. When the scan is then returned to 0.2 V, no corresponding anodic peak assigned to the oxidation of $O_2^{\bullet -}$ is found, in contrast to the reversible redox behavior for dioxygen in DMSO solutions (Fig. 5d). The intensity of

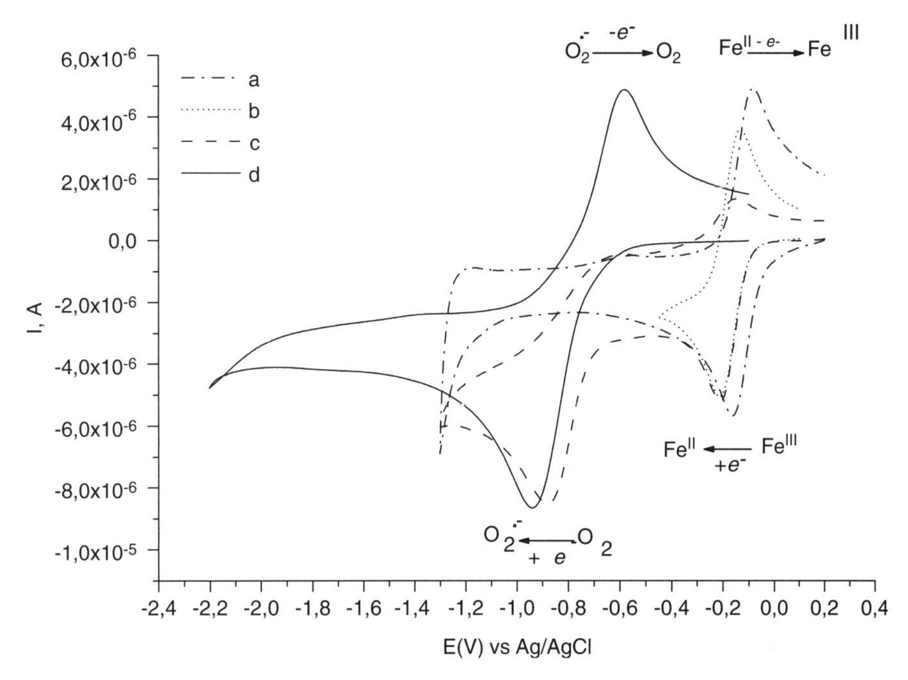

FIG. 5. Cyclic voltammograms of $[Fe(dapsox)(H_2O)_2]ClO_4$ in DMSO (a) purged with nitrogen, (b) purged with air, (c) purged with air and (d) pure DMSO purged with air (*49*).

the anodic peak corresponding to the oxidation of Fe(II) is also significantly decreased. This shows that the Fe(II) form of the complex (electrochemically generated) decomposes superoxide. In this particular experiment, complex concentration was lower than the concentration of the saturated oxygen solution in DMSO suggesting that the complex decomposes superoxide in a catalytically manner.

It should be mentioned that in aprotic media redox potential for the reduction of superoxide to peroxide, $E(O_2^{\bullet -}/O_2^{2-})$, is significantly catodically shifted, so that it is even more negative that the redox potential for the oxidation of superoxide to dioxygen. This is exactly the reason why the superoxide is stabilized in aprotic solvents, whereas peroxide is extremely unstable under such conditions. However, coordination of superoxide to the metal center induces effect similar to that caused by protonation, and the $O_2^{\bullet -}/O_2^{2-}$ redox potential shifts anodically. Thus, upon binding to a metal cation, superoxide can be reduced in aprotic media, as well.

The reactions with superoxide were studied in DMSO containing a controlled amount of water (.06%), which was in excess over the superoxide and complex concentrations (*49*). Water present in the DMSO solution plays an important role and enables the catalytic decomposition of $O_2^{\bullet -}$. Under absolute water-free conditions only a stoichiometric reaction between $O_2^{\bullet -}$ and the complex could be observed and the

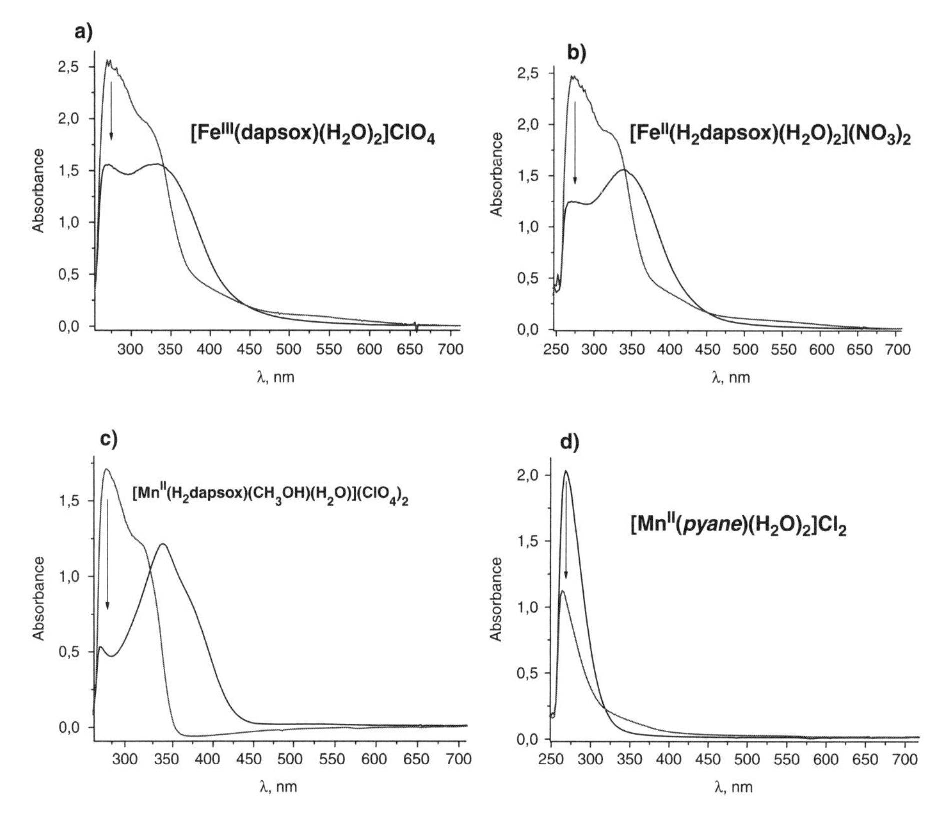

FIG. 6. UV-Vis spectra recorded before and after mixing 1 mM KO_2 with 5×10^{-5} M complex solutions in DMSO (a) $[Fe^{III}(dapsox)(H_2O)_2]ClO_4$, (b) $[Fe^{II}(H_2dapsox)(H_2O)_2](H_2O)(NO_3)_2$, (c) $[Mn^{II}(H_2dapsox)(H_2O)(CH_3OH)](ClO_4)_2$ and (d) [Mn(*pyane*)$(H_2O)_2]Cl_2$ (*49*).

catalytic process was suppressed. When the superoxide ($[KO_2] = 1$ mM) solution in DMSO is mixed with the catalytic amount of $[Fe(dapsox)(H_2O)_2]ClO_4$ (5×10^{-5} M), immediate decrease of the absorbance band at 270 nm in the UV/Vis spectrum is observed (Fig. 6a), suggesting a rapid superoxide decomposition. Upon addition of the acid into the solution, complete recovery of the $[Fe(dapsox)(H_2O)_2]ClO_4$ spectrum was obtained. The Fe(II) form of the complex (Fig. 7) (*49*) shows the same behavior, confirming the electrochemical observations and its ability to catalitically decompose superoxide. The products of superoxide disproportionation, O_2 and H_2O_2, were qualitatively detected in all four experiments (*50*).

The rapid process was quantified by following the corresponding absorbance decrease at 270 nm in a series of stopped-flow measurements, in which the catalytic concentration of the studied complexes was varied (*49*). Application of a microcuvette accessory (which reduced the dead time of the instrument down to 0.4 ms) enabled

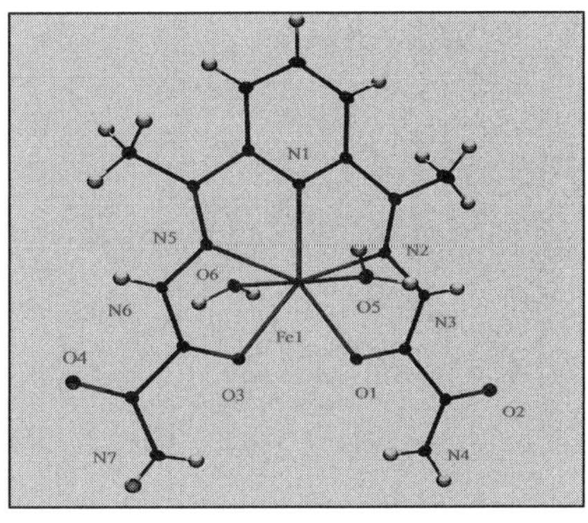

$[Fe^{II}(H_2dapsox)(H_2O)_2]^{2+}$

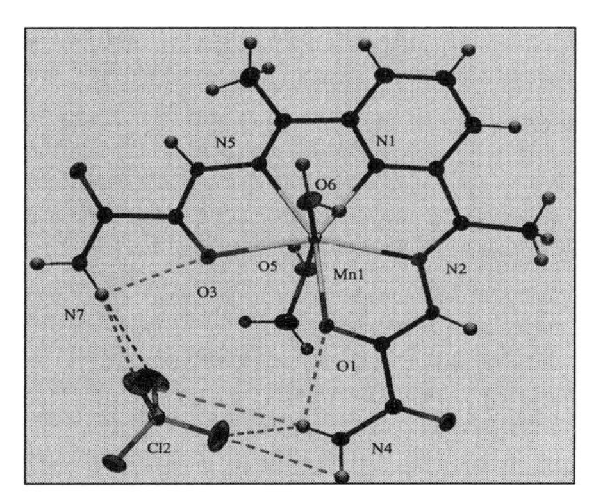

$[Mn^{II}(dapsox)(H_2O)(CH_3OH)]ClO_4$

FIG. 7. Crystal structures of $[Fe^{II}(H_2dapsox)(H_2O)_2]^+$ and $[Mn^{II}(H_2dapsox)(H_2O)(CH_3OH)](ClO_4)_2$ (*49*).

observation of the fast disappearance of the 270 nm absorption, which could best be fitted as a first-order process to obtain the characteristic k_{obs} (s^{-1}) value. When experiments were performed using the complex solutions with a higher amount of water, a larger absorbance change at 270 nm was observed for all complexes. However, it was not possible to quantify the corresponding rate constants since the higher water contents caused mixing problem at short timescale.

The obtained k_{obs} values are reported as a function of the complex concentration (Fig. 8), and a good linear correlation between k_{obs} and the complex concentration was observed for both oxidation forms of iron complexes. From the slope of the plot of k_{obs} vs. catalyst concentration the catalytic rate constants (k_{cat}) (*29*) were determined to be $(3.7 \pm 0.5) \times 10^6\,M^{-1}\,s^{-1}$ and $(3.9 \pm 0.5) \times 10^6\,M^{-1}\,s^{-1}$ for $[Fe^{III}(dapsox)(H_2O)_2]ClO_4$ and $[Fe^{II}(H_2dapsox)(H_2O)_2](NO_3)_2$, respectively (*49*). It is important to note that, it does not matter whether we start from the Fe(III) or Fe(II) form of the complex, identical spectral changes (Fig. 6a and 6b) and kinetic behavior (Fig. 8) for these two complexes is observed upon reaction with, which is consistent with the redox cycling of the complex during $O_2^{\bullet -}$ decomposition (Scheme 9).

These experiments show that the iron complex with acyclic and rigid pentadentate ligand can catalytically decompose superoxide cycling between the +3 and +2 oxidation state. This is not surprising, since both Fe(III) and Fe(II) can form stable seven-coordinate complexes (*51*), meaning that in the SOD catalytic cycle alternation between these two oxidation states does not require the changes in coordination geometry, and consequently does not require ligand rearrangement. However, manganese seven-coordinate structures are expected to be more demanding with respect to the ligand conformational flexibility, since Mn(III) is not able to form a stable seven-coordinate geometry (*51*).

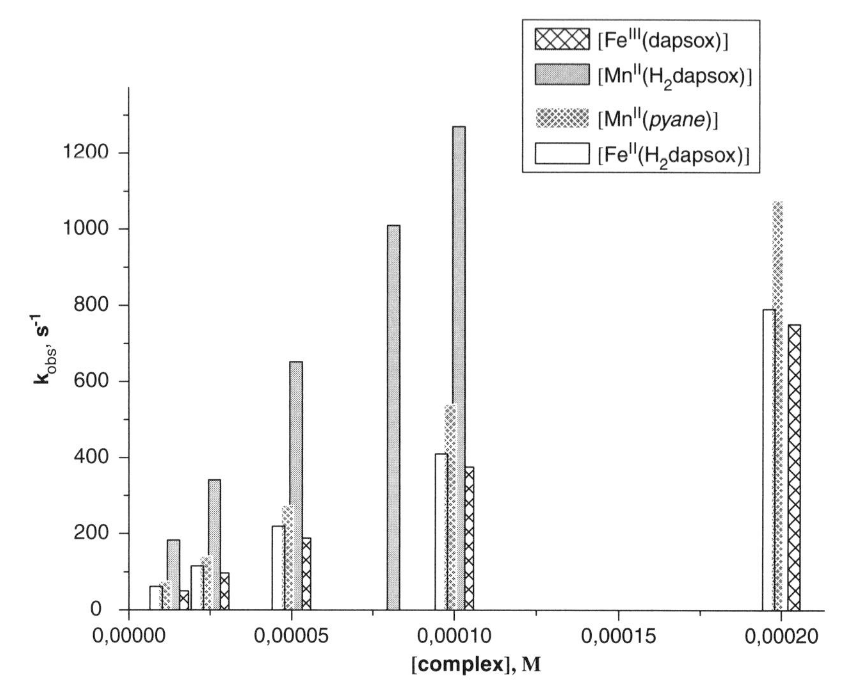

FIG. 8. Graphical comparison of k_{obs} for studied complexes.

C.2. Manganese complexes

As mentioned earlier, it was even more interesting to probe whether the rigid pentadentate ligand can support SOD activity of the Mn(II) seven-coordinate complex. Therefore we have synthesized the PBP $[Mn^{II}(H_2dapsox)(CH_3OH)(H_2O)](ClO_4)_2$ complex (*49*) (Fig. 7) with the same rigid ligand system as in the case of iron complexes. To compare its reactivity with the reactivity of a proven SOD catalyst under the selected experimental conditions, the interaction of [Mn(*pyane*) $(H_2O)_2]Cl_2$ with superoxide was parallel examined (*49*).

At first the redox properties of these two Mn(II) complexes were studied (*49*). Similar to the proven Mn(II) seven-coordinate SOD mimetics with *[15]aneN5* type of chelates (*35*), $[Mn^{II}(H_2dapsox)(CH_3OH)(H_2O)](ClO_4)_2$ is stable in the pH range 6–10.5 and in methanol exhibits a reversible redox potential at 0.8 V vs. NHE (*22,35*). The redox behavior in aqueous solutions for the macrocyclic manganese SOD mimetics was not reported. We measured the cyclic voltammograms for [Mn(*pyane*)$(H_2O)_2]Cl_2$ ($E_{ox} = 0.98$ V and $E_{red} = 0.35$ V) and $[Mn^{II}(H_2dapsox)(CH_3OH)(H_2O)](ClO_4)_2$ ($E_{ox} = 0.64$ V and $E_{red} = 0.20$ V) at pH = 7.8 (*49*), and both complexes show similar behavior with large peak separation.

Electrochemical measurements in air-saturated DMSO (Fig. 9) show (*49*), as in the case of iron complex that the analogue Mn(II) complex

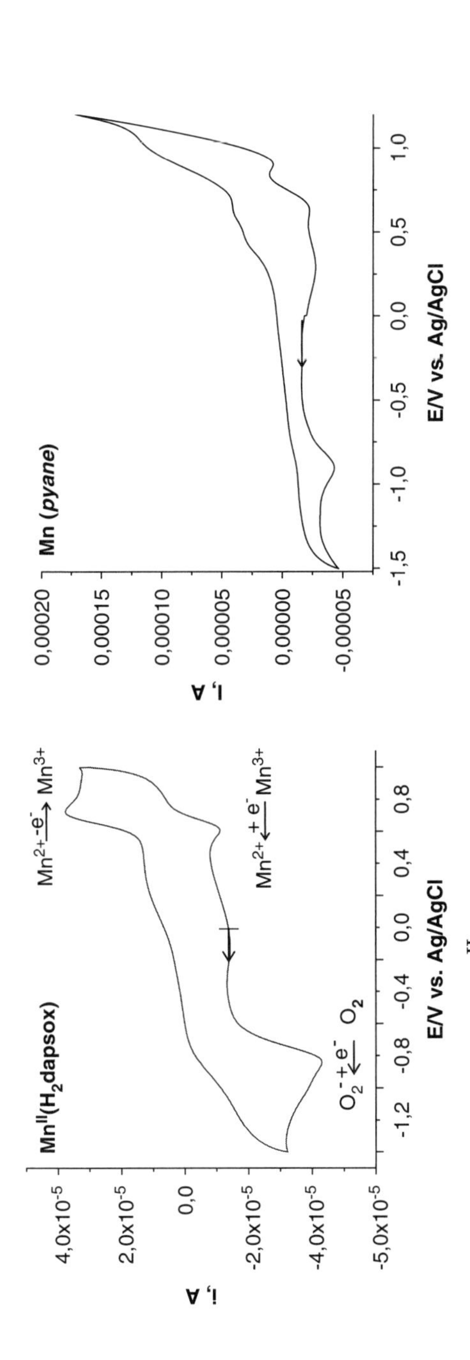

FIG. 9. Cyclic voltammograms of $[Mn^{II}(H_2dapsox)(H_2O)(CH_3OH)](ClO_4)_2$ and $[Mn(\textit{pycane})(H_2O)_2]Cl_2$ in DMSO purged with air (*49*).

can also catalytically decompose superoxide (disappearance of the anodic peak assigned to the oxidation of $O_2^{\bullet -}$ in the presence of the Mn(II) form of the complex), remaining unchanged (appearance of reversible redox wave for the Mn^{II}/Mn^{III} couple at 0.65 V). In comparison, the proven SOD mimetic $[Mn(\textit{pyane})(H_2O)_2]Cl_2$ upon the reaction with superoxide undergoes modification and in the scan range from 0 to 1.2 V and back to 0 V, three oxidation and reduction peaks appear (Fig. 9).

Immediately after mixing a superoxide solution with a complex solution in DMSO, rapid decomposition of $O_2^{\bullet -}$ (decrease in absorbance in the 240–330 nm range within the dead time of the stopped-flow instrument) was also observed in the case of both manganese complexes (Fig. 6). More prominent spectral changes follow the reaction with our rigid ligand. In the case of $[Mn(\textit{pyane})(H_2O)_2]Cl_2$, after fast superoxide decomposition, the complex starts to decompose slowly and results in the formation of a light brown colloid precipitate (presumably MnO_2) after ~3 h. Three hours after mixing with KO_2, acid was added to the solutions of $[Mn^{II}(H_2dapsox)(CH_3OH)(H_2O)](ClO_4)_2$, which resulted in the complete recovery of the staring complexes. This demonstrates that our acyclic complexes are more stable than the macrocyclic one under the applied experimental conditions, which is in agreement with the electrochemical observations (see earlier).

As for iron complexes the catalytic rate constants (k_{cat}) were determined by direct stopped-flow measurements of k_{obs} as a function of catalytic amount of manganese complexes (Fig. 8), and they are found to be $(1.2 \pm 0.3) \times 10^7$ and $(5.3 \pm 0.8) \times 10^6\ M^{-1}\ s^{-1}$ for $[Mn^{II}(H_2dapsox)(CH_3OH)(H_2O)](ClO_4)_2$ and $[Mn(\textit{pyane})(H_2O)_2]Cl_2$, respectively (*49*). The k_{cat} values shows that the both iron complexes and the macrocyclic manganese complex have almost the same catalytic activity, within the error limits, whereas $[Mn^{II}(H_2dapsox)(CH_3OH)(H_2O)](ClO_4)_2$ has approximately two times higher activity.

C.3. Reaction with superoxide in aqueous solutions

Stopped-flow measurements with superoxide in aqueous solution at physiological pH are not possible due to its fast self-dismutation under these conditions. Therefore, the indirect assays such as McCord–Fridovich, adrenalin and nitroblue tetrazolium (NBT) assays are widely used in the literature, not only for qualitative but also for quantitative detection of SOD activity of small molecular weight mimetics (*52*). Not going into details, we just want to stress that the indirect assays have very poor even qualitative reliability, since they can demonstrate the SOD activity of the complexes which does not react with superoxide at all. It has been reported in the literature that this is caused by the interference of hydrogen peroxide (*29*). We have observed that the direct reaction between complexes and indicator substances (cytochrome *c* and NBT used in our experiments) inevitably lead to over- or underestimation of the catalytic rate constants under the experimental conditions of assay. In that way the catalytic rate constants determined by using the McCord–Fridovich assay for $[Mn^{II}(H_2dapsox)(CH_3OH)(H_2O)](ClO_4)_2$ and $[Mn(\textit{pyane})(H_2O)_2]Cl_2$ were found to be at least one order of magnitude higher then those obtained by stopped-flow method. At the same time, for the iron complex it was even not possible to use this assay as a qualitative test, since in its reduced form the complex acted as a reductant of cytochrome *c*, and increased the rate of cytochrome *c* reduction when injected into solution, opposite of what should be observed. Both iron and manganese dapsox complexes reacted with NBT.

C.4. Conclusion

Although it has been postulated in the literature that only seven-coordinate complexes of macrocyclic ligands with prominent conformational flexibility could possess SOD activity (*24,25*), our seven-coordinate iron and manganese complexes with the acyclic and rigid H_2dapsox ligand demonstrate ability for catalytic decomposition of superoxide. Importantly, not only that the manganese complex with conformational unflexible pentadentate ligand possesses SOD activity, but it is even more active than the conformational flexible complex. Similar to what usually was found in the case of the macrocyclic pentadentate ligands (*18*), the manganese complex shows higher SOD activity than the corresponding iron complex. However, higher stability of the iron complex within a very wide pH range is its advantage in terms of a possible application.

The demonstrated SOD activity of the rigid seven-coordinate complexes is in agreement with our recent findings that the water release and formation of a six-coordinate intermediate, requiring conformational rearrangement of the ligand, is not the rate-limiting step in the overall inner-sphere catalytic SOD pathway of the proven macrocyclic SOD mimetics (*34*). Furthermore, it also shows that conformational flexibility of the pentadentate ligand is not the key factor assisting in SOD activity, and that the acyclic and rigid ligand systems can also be considered as structural motives for designing SOD mimetics. Their additional advantage can be the fact that their syntheses are more economic than the syntheses of macrocyclic ligands.

We have also shown that the indirect SOD assays, which are the mostly used methods for demonstrating complex SOD activity, are not very reliable and if, they can be applied only upon considering possible cross reactions between indicator substance and the studied complex in their different oxidation forms, in which they may occur within the SOD catalytic cycle. The direct stopped-flow method, where the high excess of superoxide over complex can be utilized, is a better probe for a complex SOD activity even though it requires DMSO medium. Importantly, as it was stressed by D. T. Sawyer *et al.*, even closer relation between the kinetic measurements in aprotic media than in bulk water can be drawn with the processes in mitochondria, which are the major source of superoxide in the aerobic organisms, since aprotic media "may be representative of a hydrophobic biological matrix" (*53*). Under less protic conditions, causing longer half-life of O_2^-, efficient superoxide decomposition is even more desirable.

III. Reversible Binding of Superoxide to Iron-Porphyrin Complex

A. State-of-the-Art

We are also interested in the activation of superoxide which does not lead to its catalytic disproportionation, namely, we are interested in stoichiometric reactions of different metal complexes, in the first place iron porphyrins, with superoxide. The biochemistry of dioxygen and its reduced forms (O_2^- and O_2^{2-}) is closely related to iron-porphyrin centers in different hemoproteins. Although superoxide is not a common substrate for hemoproteins it is formally bound to the iron center in oxygenated hemoglobin and in intermediate species involved in the catalytic cycle of cytochrome P450 and heme-copper assemblies (cytochrome *c* oxidases) (*6*). It is well established that in aprotic coordinating solvents (e.g., DMSO or MeCN) one equivalent of KO_2 reduces Fe(III) porphyrins to Fe(II), whereas an additional equivalent of KO_2 produces a species which has been formulated (based on ESR and vibrational spectral measurements and relatively broad pyrrole deuterium NMR signal) as Fe(III)-peroxo porphyrin adduct (Scheme 12) (*54,55*). Previously this species was described as Fe(II)-superoxo complex based on the unusual red shifting of the visible absorption spectrum and its ERP silent character (*56*). It is isoelectronic with a product of dioxygen binding to Fe(I) porphyrin complex and with an electrochemically reduced Fe(II) dioxygen adduct (*57*) (Scheme 13).

Fe(III)-peroxo porphyrin complexes comprising ligands of different electronic properties were extensively investigated by Valentine *et al.* (*4,58*) as synthetic analogues of intermediates that might occur during enzymatic reactions. All these complexes are high-spin species.

$$[Fe^{III}(porphyrin)]^+ \; O_2^- \longrightarrow [Fe^{II}(porphyrin)] + O_2$$

$$[Fe^{II}(porphyrin)]^+ \; O_2^- \longrightarrow [Fe^{III}(porphyrin)(O_2^{2-})]^-$$

Scheme 12.

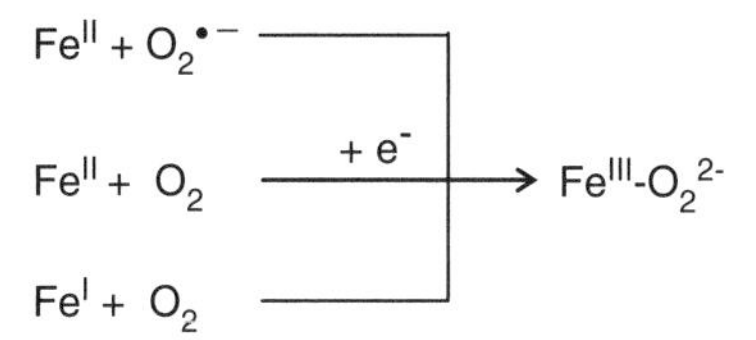

SCHEME 13.

The same authors studied the effect of the electron-withdrawing groups on the stability and type of reactivity of the corresponding Fe(III)-peroxo complexes and showed that electron-withdrawing groups attached to the porphyrin macrocycle stabilize peroxo species and decrease nucleophilic reactivity (*58*). Even the most stable Fe(III)-peroxo porphyrin complex $[Fe^{III}(F_{20}tpp)(O_2^{2-})]^-$, containing the very electron-poor F_{20}tpp porphyrin ligand (F_{20}tpp = 5,10,15,20-(pentafluorophenyl)porphyrin), is stable only under an inert atmosphere for several weeks (*58*). Interestingly, the same authors reported that addition of a potassium chelator to a solution of Fe(III)-peroxo species increases its stability (*54b*), suggesting the existence of an interaction between the potassium ion and the coordinated peroxo ligand.

It has been postulated on the basis of the crystal structure of $[Mn^{III}(tpp)(O_2^{2-})]^-$, that the peroxide ligand in Fe(III)-peroxo porphyrin species is coordinated in a side-on bidentate manner (Scheme 2) (*8*). Interestingly, based on the deuterium NMR studies the $[Mn^{III}(tpp)(O_2^{2-})]^-$ complex was characterized as the Mn(II)-superoxo species (*55*). This discrepancy has been explained by an alteration of the normal d orbital ordering, where the highest energy d orbital is a d_{yz}+O_2 π_{gz} hybrid not the $d_{x^2-y^2}$ (*8*). These examples show that definitive electronic and molecular structural assignments for iron and manganese porphyrin-superoxide reaction products are not straightforward tasks, especially in solutions. Also an accurate description of the superoxide/peroxide-binding mode in solution is not a trivial problem. As mentioned earlier, peroxide is considered to bind in a side-on fashion, whereas superoxide coordination is characterized by end-on binding mode. However, it seems that the end-on form of Fe(III)-peroxo species leads to nucleophilic attack and that axial *trans* coordination of DMSO results in the opening of the triangular peroxo chelate ring (Scheme 14), which increases the nucleophilicity of these species (*59*). Coordination of the axial ligand as a switch for chelate ring-opening of the peroxo ligand and a promoter of the nucleophilic reactivity was proposed to play a role in reactions of peroxo complexes derived from group VIII transition metals, as well as in enzymatic reactions (*59*). Interestingly, it has been demonstrated that side-on peroxo non-heme Fe(III) and Mn(III) species exhibit nucleophilic but not electrophilic reactivity (*60*). This again demonstrates that there are still unrevealed questions about interactions between superoxide and metal centers in general, and metalloporphyrins in particular, and the nature and reactivity of their products species.

B. Iron Complex of a Crown Ether-Porphyrin Conjugate Suitable for Studying Interactions with Superoxide

In the literature there are no quantitative studies on the kinetics and thermodynamics of stoichiometric superoxide reactions with metal centers in general, and metalloporphyrins in particular. More precisely, superoxide concentration and temperature dependent kinetic and thermodynamic measurements were never reported and consequently the rate constants, activation parameters or binding constants for this type of reactions (Scheme 15) are not known. (The catalytic rate constants for the superoxide disproportionation, i.e., dismutation, by metal complexes are known (see earlier), however in those measurements the concentration of a catalytic amount

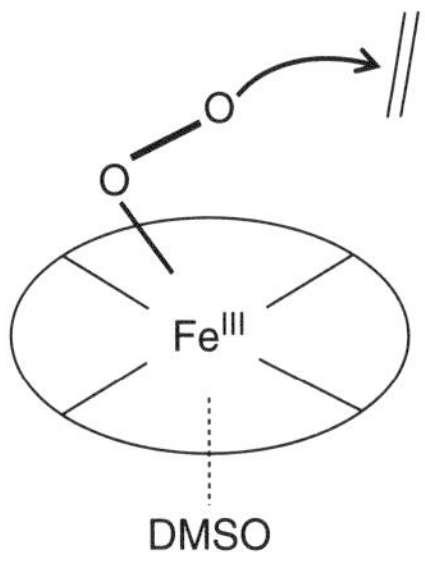

Scheme 14.

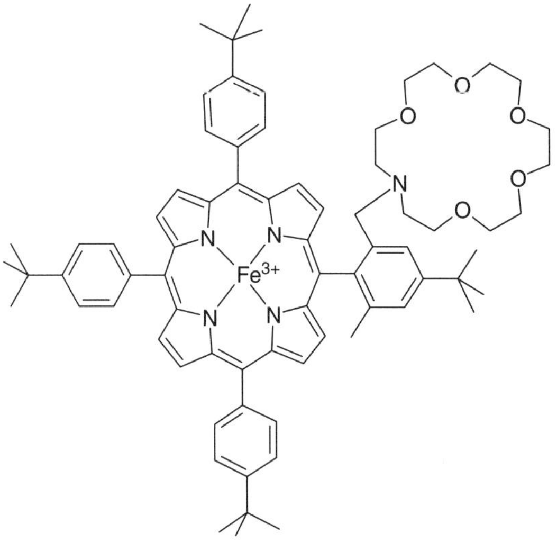

Scheme 15.

of complexes was varied and usually such constants were obtained from indirect SOD assays.) Therefore, we went beyond the qualitative descriptions of such processes and undertook detailed kinetic and thermodynamic study on the reactions of KO_2 with a novel iron complex (Scheme 15) of crown ether-porphyrin conjugate (H_2Por) in DMSO, in order to shed more light on this type of interactions (*61*).

H_2Por used in our studies is a tetraphenylporphyrin with *tert*-butyl groups in the *para* position of the porphyrin which enhance the solubility in organic solvents. Aza-18-crown-6 is attached through its amine group to a methyl group in *ortho*-position of one aryl ring. This close proximity of the crown ether gives the metallated form of the ligand its unique properties. The metalloporphyrin can function as a ditopic receptor by coordinating a potassium cation, which can then interact with a diatomic anionic ligand bound to the metal center. Such an interaction has already been confirmed by X-ray structure analysis of the corresponding zinc(II) and cobalt(III) complexes with one and two coordinated cyanide anions, respectively (*62*).

The Fe(III) complex of H_2Por, [Fe^{III}(Por)Cl], is a typical high-spin five-coordinate species with the chloro ligand pointing to the coordinated water molecule in the crown ether. The bis(cyano)iron(III) complex of H_2Por was also synthesized to study the influence of the potassium ion on the spin state of the iron center. NMR spectroscopy of our low-spin K[Fe^{III}Por$(CN)_2$] complex reveals the strong increase of π-acceptor ability of the cyanide ligands that comes from interaction with the positively charged potassium ion coordinated by the crown ether, which pulls electron density away from the ligands. These results show that our iron(III) porphyrin acts as a ditopic receptor for diatomic anions as it has been demonstrated in the crystal structures of the corresponding zinc and cobalt complexes (*62*). The strong electrostatic interaction, as the one between chelated K^+ and CN^-, is also to be expected between chelated K^+ and the peroxide anion (see later), and can enhance stability of the peroxo species. In DMSO solution our complex is present as the six-coordinate [Fe^{III}(Por)$(DMSO)_2$]$^+$ species.

The redox behavior of [Fe^{III}(Por)Cl] is in accordance with the redox behavior of [Fe^{III}(tpp)]$^+$ (*63*). The redox potential corresponding to the Fe(III)/Fe(II) couple of our complex is more positive than that of [Fe^{III}(tpp)]$^+$ (*63*), suggesting higher stability of the complex in the Fe(II) oxidation state. This was observed when [Fe^{II}(Por)] was chemically or electrochemically generated, since it is stable under air for a few hours, whereas [Fe^{II}(tpp)] oxidizes immediately under air. The significant stability of [Fe^{II}(Por)] enabled further kinetic and thermodynamic investigations.

B.1. Reactions with superoxide

The property of the [Fe^{III}(Por)Cl] complex to act as a ditopic receptor towards diatomic anions is very attractive for studying its reaction with superoxide. The reaction of Fe(III) porphyrins with superoxide is convenient to follow in DMSO as aprotic solvent, in which the solubility and stability of KO_2 (common source of O_2^-) are satisfactory. It is also known that the use of potassium chelators will enhance the solubility of KO_2. Moreover, the presence of potassium chelators can influence the properties of the product species that result from the reaction between Fe(III) porphyrin complexes and superoxide (*54b*).

Different from what has been found in the literature, upon addition of less than tenfold excess of KO_2 to the complex solution only formation of the hydroxo complex $[Fe^{III}(Por)OH]$ can be observed. This can be explained by the fact that KO_2 inevitably contains some KOH, which leads to the formation of the Fe(III) hydroxo complex. $[Fe^{III}(Por)OH]$ itself reacts with KO_2. Tenfold excess of KO_2 is needed for reduction of $[Fe^{III}(Por)(DMSO)_2]^+$ to the Fe(II) form of a complex. Reduction of the iron center by superoxide is confirmed by comparison of the resulting spectrum to the one obtained by electrochemical reduction of $[Fe^{III}(Por)(DMSO)_2]^+$ (Fig. 10).

When a larger than tenfold excess of superoxide is applied to the solution of $[Fe^{III}(Por)(DMSO)_2]^+$, the reduced $[Fe^{II}(Por)]$ complex is formed during the dead time of the stopped-flow instrument, and the second reaction step (Scheme 12), i.e., formation of the peroxo $[Fe^{III}(Por)(O_2^{2-})]^-$ complex is observed (Fig. 10). This second

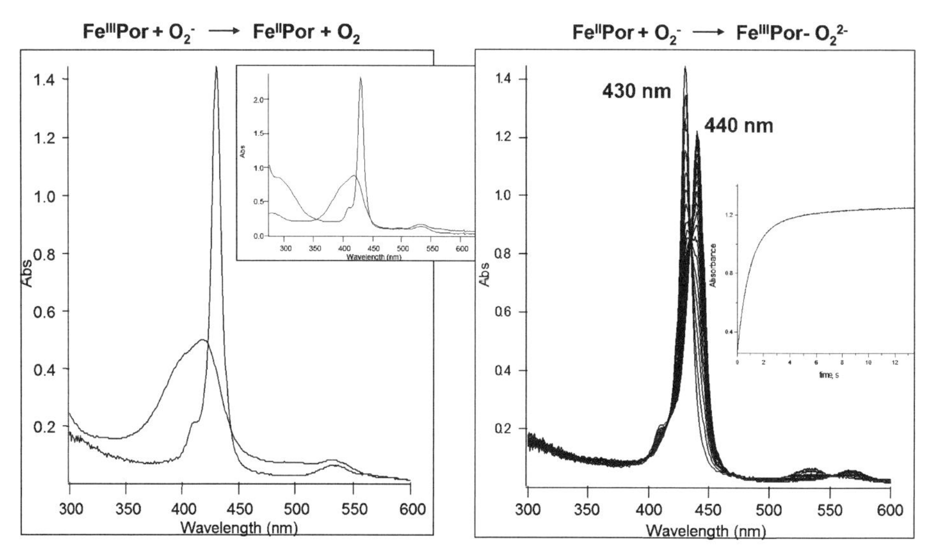

FIG. 10. UV/Vis spectra recorded in a tandem cuvette of the solution of 5×10^{-6} M $[Fe^{III}Por(DMSO)_2]^+$ and 5×10^{-5} M KO_2 in DMSO (a) before ($\lambda_{max} = 420$ nm) and (b) after ($\lambda_{max} = 430$ nm) mixing. Inset: UV/Vis spectra of 5×10^{-6} M $[Fe^{III}Por(DMSO)_2]^+$ (a) before ($\lambda_{max} = 420$ nm) and (b) after reduction ($\lambda_{max} = 430$ nm) by dithionite. Time-resolved spectra for the reaction between $[Fe^{II}(Por)]$ and KO_2 in DMSO (*61*).

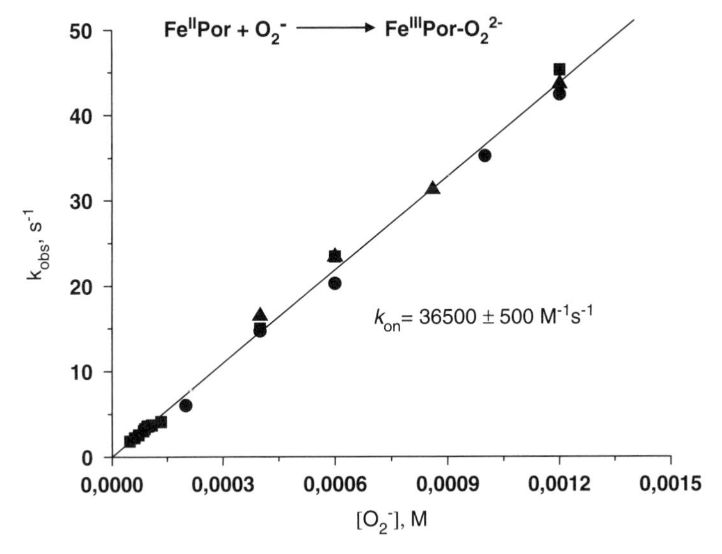

FIG. 11. Plots of k_{obs} vs. $[O_2^-]$ for the reaction of 5×10^{-6} M $[Fe^{II}(Por)]$ and KO_2 at 25°C in DMSO (*61*).

reaction step can also be followed starting with electrochemically generated $[Fe^{II}(Por)]$.

Although, the first reaction step, reduction of the Fe(III) to the Fe(II) porphyrin complex by KO_2, could not be studied in detail because of interference of the formation of the Fe(III) hydroxo species, the second reaction step, binding of superoxide to the Fe(II) species and formation of the Fe(III)-peroxo complex, could be studied in detail (*61*). To our knowledge, for the first time superoxide concentration (Fig. 11) and temperature dependent kinetic studies of reactions with superoxide have been performed by stopped-flow UV/Vis measurements, and as a result the second-order rate constant ($k_{on} = 36{,}500 \pm 500\,M^{-1}\,s^{-1}$) and corresponding activation parameters ($\Delta H^{\neq} = 61.2 \pm 0.9\,kJ\,mol^{-1}$ and $\Delta S^{\neq} = +48 \pm 3\,J\,K^{-1}\,mol^{-1}$) could be obtained (*61*). On the basis of the obtained results, we can conclude that strongly coordinated DMSO controls the reaction mechanism in the sense that its dissociation is the rate-determining step.

B.2. *Stability of Fe(III)-peroxo species and reversible binding of superoxide*

In contrast to the Fe(III)-peroxo porphyrin species reported in the literature (*58*), which are only stable under an inert atmosphere, our peroxo species is quite stable in the solution with excess of superoxide, even when exposed to air (*61*). The presence of nearby K^+ can account for this increased stability of K[Fe^{III}(Por)(O_2^{2-})]. The peroxo complex is indefinitely stable in a frozen DMSO solution, even upon exposure to air. In a closed system under air it is stable at room temperature for a week. When the solution is opened to the atmosphere, slow transformation (within a few hours) to the [Fe^{II}(Por)] species is observed, as monitored by UV/Vis spectroscopy. This reaction was found to be caused by absorption of moisture from the atmosphere, as no spectral changes are observed when the peroxo complex is purged with dry oxygen. The rate of decay of [Fe^{III}(Por)(O_2^{2-})]$^-$ (green solution, $\lambda_{max} = 440$ nm) is accelerated by passing wet air through the solution, which results in the formation of the brown [Fe^{II}(Por)] complex, as monitored by UV/Vis spectroscopy (Fig. 12). Further addition of superoxide to the brown [Fe^{II}(Porph)] complex results in reformation of the green peroxo complex. All these observations suggest that the binding of superoxide to the Fe(II) complex is reversible and that the decrease in concentration of O_2^-, caused by its decomposition in the presence of protons, shifts the equilibrium from the Fe(III)-peroxo to the Fe(II) complex (Scheme 16). Addition of a moderate excess of triflic acid to the peroxo complex causes partial formation of the [Fe^{II}(Porph)] complex by facilitating the decay of superoxide in the solution.

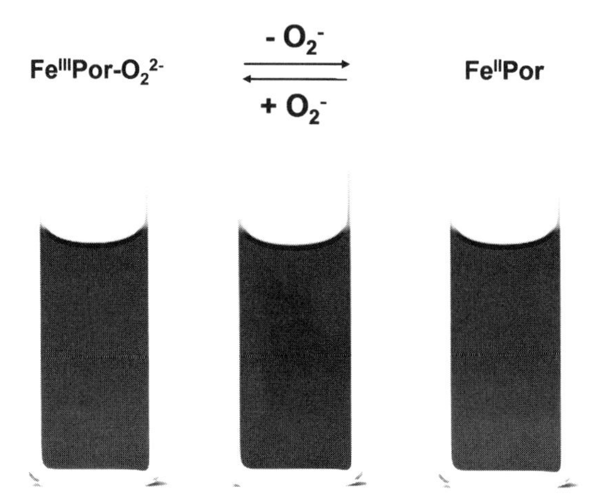

FIG. 12. Transformation of K[Fe^{III}Por(O_2^{2-})] to [Fe^{II}(Por)] by passing wet air through the DMSO solution.

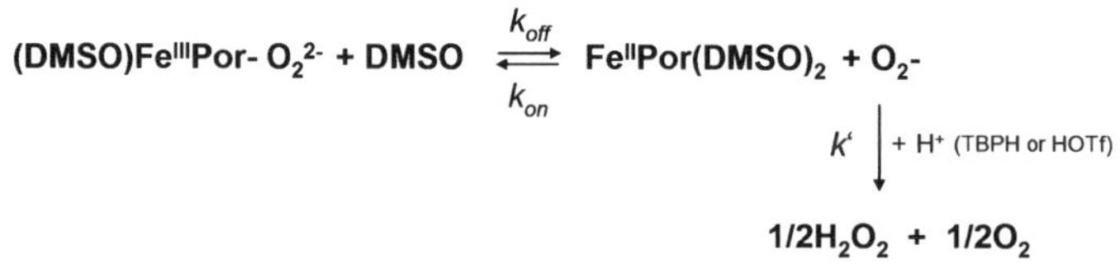

$$(DMSO)Fe^{III}Por\text{-}O_2^{2-} + DMSO \underset{k_{on}}{\overset{k_{off}}{\rightleftharpoons}} Fe^{II}Por(DMSO)_2 + O_2^-$$

$$O_2^- \xrightarrow[k']{+ H^+ \text{ (TBPH or HOTf)}} 1/2 H_2O_2 + 1/2 O_2$$

SCHEME 16.

When a large excess of acid is added, the $[Fe^{III}(Porph)(DMSO)_2]^+$ complex is formed.

As mentioned earlier (Scheme 2), Fe(III)-peroxo complexes usually decompose to high-valent iron-oxo species. To test for the presence of high-valent iron-oxo species as a decay product of our peroxo complex, the trap TBPH (2,4,6-tri(*tert*-butyl)phenol) was utilized. Typically, upon oxidation, TBPH forms an oxygen-centered radical which results in an increase of absorbance at 380, 400 and 630 nm (*64*). However, upon addition of TBPH to the solution of $[Fe^{III}(Por)(O_2^{2-})]^-$ no evidence for high-valent iron-oxo species was observed, but $[Fe^{II}(Por)]$ was produced instead (Fig. 13). In a subsequent study, TBPH was also found to react with superoxide; comparison of the spectral changes with those upon reaction with NaOH indicates that TBPH is acting as an acid. Phenols are known to act as moderate acids to protonate superoxide ion in aprotic solvents and induce its efficient disproportionation ($K = 10^{18}$) (*53*). The protonation of superoxide present in an equilibrium solution causes its decay (*50,53*), shifting the equilibrium from $[Fe^{III}(Por)(O_2^{2-})]^-$ to $[Fe^{II}(Por)]$ (Scheme 16, Fig. 13). Therefore, this reaction, in which acid is a trap for superoxide anions, can be used for direct determination of the rate constant of the back reaction, k_{off}, in a stopped-flow experiment in which a 1×10^{-5} M

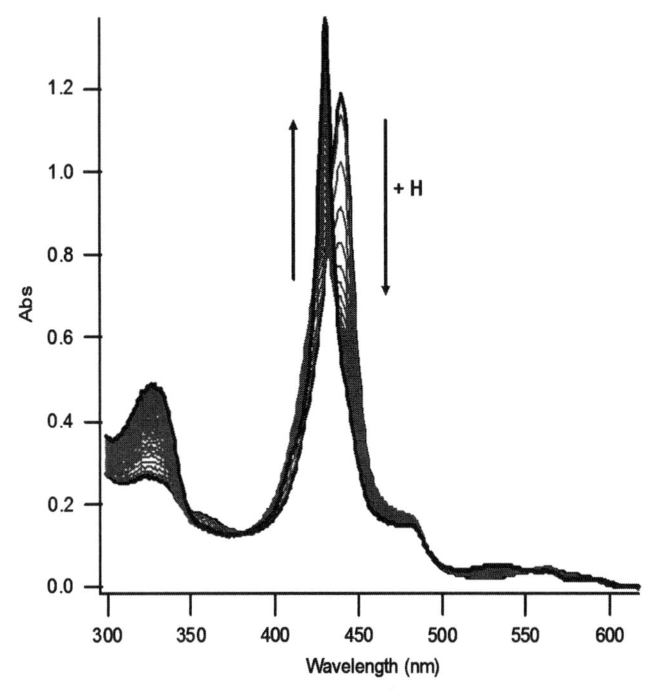

FIG. 13. Time-resolved spectra for the reaction of $K[Fe^{III}(Porph)(O_2^{2-})]$ and TBPH (2,4,6-tri(*t*-butyl)phenol) at 25°C in DMSO.

solution of Fe(III)-peroxo porphyrin was mixed with the acid solution in DMSO at 25°C.

It should be mentioned that the high stability of the Fe(III)-peroxo complex enables such an experiment. For the quantitative determination of k_{off}, HOTf was chosen since it is known that strong acids react with superoxide in aprotic solvents extremely rapidly ($k' > 1 \times 10^7\ M^{-1}\ s^{-1}$) (*53*), which makes them very efficient O_2^- scavengers even at lower concentrations. The observed first-order rate constant for the reaction of $[Fe^{III}(Porph)(O_2^{2-})]^-$ and HOTf does not depend on HOTf concentration in the range of 5×10^{-5}–5×10^{-4} M and was found to be $k_{obs} = k_{off} = 0.21 \pm 0.01\ s^{-1}$ (*61*). From the obtained second-order rate constant k_{on} and the first-order rate constant k_{off}, the kinetically determined equilibrium constant for the binding of superoxide is $KO_2^- = (1.7 \pm 0.2) \times 10^5\ M^{-1}$ at 25°C (*61*). The reversible binding of O_2^- to $[Fe^{II}(Porph)]$ and the relative stability of the $[Fe^{III}(Porph)(O_2^{2-})]^-$ complex allowed further investigation of the binding thermodynamics. The equilibrium constant for this reaction was determined by titrating an electrochemically prepared solution of $[Fe^{II}(Porph)]$ with O_2^- under a nitrogen atmosphere and both thermodynamically and kinetically determined equilibrium constants for superoxide binding are in a good agreement (*61*).

B.3. Fe(III)-peroxo or Fe(II)-superoxo?

An interesting question is whether the peroxo ligand in our $[Fe^{III}(Porph)(O_2^{2-})]^-$ complex is coordinated in a side-on or end-on fashion. Taking into consideration the DMSO coordination and the electrophilic potassium cation in the crown ether lying above the peroxo ligand, $[Fe^{III}(Porph)(O_2^{2-})]^-$ may in a way represent a model for the proposed (*59*) nucleophilic attack of the end-on peroxo form, with an axially coordinated solvent molecule, to an electron-deficient substrate (Scheme 14).

To understand better the role of axially coordinated DMSO in the structural and electronic properties of our peroxo complex we have performed the DFT calculations (UBP86/LACVP* *Jaguar 6.0*). The obtained preliminary results are quite intriguing. Namely, they show that when DMSO is coordinated to the iron center, the energy difference between side-on and end-on peroxo structures is only $3.0\ kcal\ mol^{-1}$, strongly suggesting that the equilibrium between these two structures can exist (Fig. 14). But even more interesting is an observation that the end-on structure has iron(II)-superoxo character different from the side-on structure with iron(III)-peroxo nature. This now shed a new light on the Mössbauer spectrum of our iron(III)-peroxo complex in DMSO, where we have observed both high-spin Fe(III) ($\delta = 0.41\ mm\ s^{-1}$, $\Delta Eq = 0.51\ mm\ s^{-1}$) and low-spin Fe(II) ($\delta = 0.36\ mm\ s^{-1}$, $\Delta Eq = 1.38\ mm\ s^{-1}$) (*61*). Whether the observed Fe(II) species is a rest of the starting low-spin $[Fe^{II}Porph(DMSO)_2]$

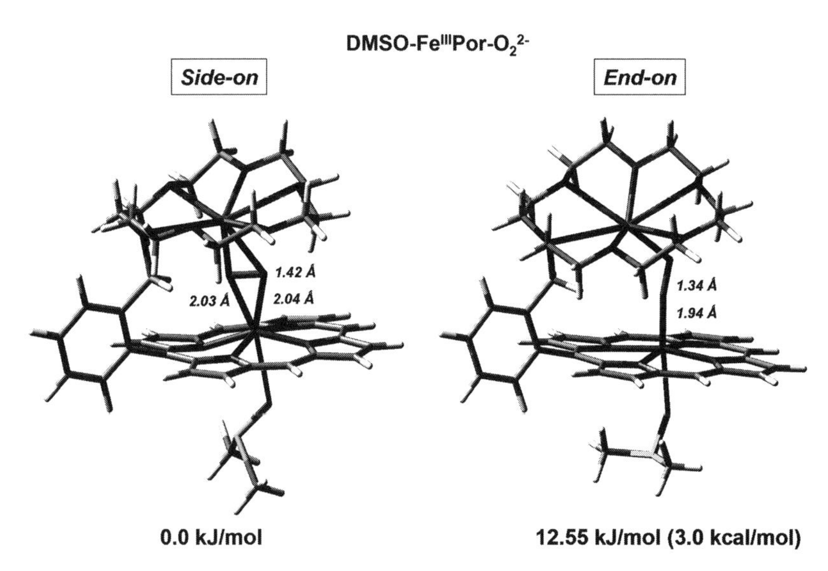

FIG. 14. With DFT methods (B3LYP/LANL2DZp) calculated structures for the side-on and end-on form of K[Fe^{III}(Porph)(O_2^{2-})(DMSO)].

complex or it is the end-on iron(II)-superoxide adduct postulated by DFT calculations, remains to be revealed in future experiments. This again shows that the nature of the Fe(II)-superoxide adduct cannot be simply defined as an Fe(III)-peroxo species, and that it most probably depend on the nature of the porphyrin ligand, solvents etc.

This also opens a question about the nature of the active species in, for example, oxidative nucleophilic reactions: is it peroxo or superoxo form which nucleophilically attacks an organic substrate? These are challenging questions which motivate future investigations.

B.4. Conclusion

Our studies demonstrate some new aspects of the reactions between superoxide and metalloporphyrins.

Firstly, superoxide can be reversibly bind to the metal center forming quite stable M(III)-peroxo species, which consequently can serve as a source of superoxide, releasing it to form an M(II) species by fine tuning of the proton concentration. This type of superoxide reactivity is a novelty of a general chemical, as well as biological importance, since it shows that upon proton addition the M(III)-peroxo species (known as the intermediates in various enzymatic processes) does not necessarily dissociate to hydrogen peroxide and an Fe(III) species (as in the case of the SOD active enzymatic and mimetic systems) or undergo O–O bond cleavage to form a high-valent oxo-iron species (as in the case of cytochrome P450) (Scheme 2). Whether this type of reactivity is a result of the unique structural feature of our Fe(III)-peroxo porphyrin (viz. the presence of the nearby K^+-crown ether moiety), or it is a more general feature that could not be observed before because of the much lower stability of the previously studied Fe(III)-peroxo porphyrin complexes, remains to be seen.

Secondly, there is an indication that metal(III)-peroxo side-on complexes are in general in equilibrium with corresponding metal(II)-superoxo end-on species. The position of such equilibrium could depend on various factors as structural and electronic properties of the porphyrin ligand, coordination of an axial ligand *trans* to peroxide/superoxide, solvent medium, temperature and involvement of coordinated peroxide/superoxide in possible hydrogen bonding or electrostatic interactions. These are interesting questions which should be addressed in future studies.

IV. Summary

Here we presented two general aspects of the interactions between superoxide and metal centers. One is the catalytic decomposition of superoxide by non-heme metal centers (Scheme 9) and the role of the ligand structure in it, and another is the reversible binding of superoxide to the heme metal center and the nature of the product metal(III)-peroxo species (Scheme 17). In both cases through the same redox reaction steps a metal(III)-peroxo species is formed as the intermediate (Scheme 9), in the catalytic cycle, or the product of stoichiometric reaction (Scheme 17). The crucial difference is in the protonation step. If the protonation of peroxo species is followed by efficient release of hydrogen peroxide (and not O–O bond cleavage,

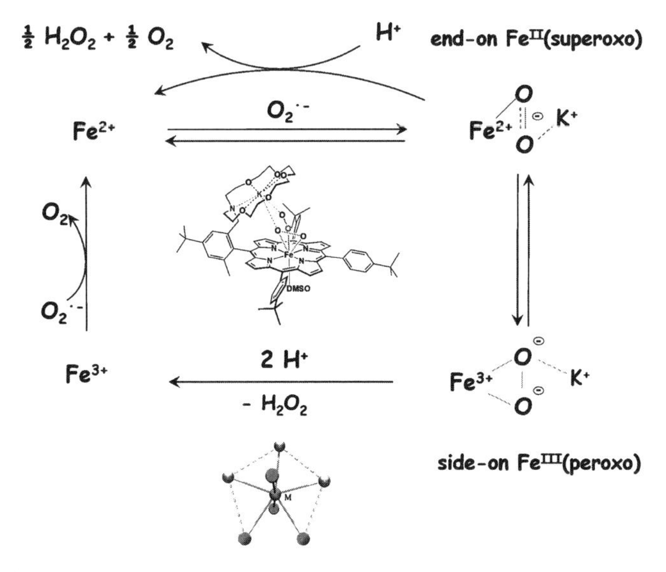

SCHEME 17.

Scheme 2), catalytic cycle is closed. However, if the protonation is not efficient, instead of catalytic disproportionation of superoxide a metal(III)-peroxo species is generated as the reaction product. As a novelty, we have shown that protonation of the Fe(III)-peroxo species does not necessarily lead to dissociation of hydrogen peroxide and formation of the Fe(III) complex (as in the case of the SOD active enzymatic and mimetic systems) or O–O bond cleavage and formation of a high-valent oxo-iron species (as in the case of cytochrome P450). Namely, we have demonstrated that superoxide can react with a metal center in a reversible manner forming quite stable M(III)-peroxo species, which can release superoxide to form an M(II) species by fine tuning of the proton concentration. This type of reactivity of the Fe(III)-peroxo species and reversible binding of superoxide, could be of significant biological importance since it could operate in a specific biological environment as well. Additionally, the existence of Fe(II)-superoxo species and its equilibrium with Fe(III)-peroxo form, which we postulate, can certainly explain the reversible nature of superoxide binding and may also be an explanation for some reaction mechanisms in which M(III)-peroxo species are involved. Further investigations should shed more light on these phenomena.

Acknowledgment

The author gratefully acknowledges financial support from the Deutsche Forschungsgemeinschaft within SFB 583 "Redox-active Metal Complexes".

References

1. (a) Fridovich, I. *J. Biol. Chem.* **1989**, *264*, 7761–7764. (b) Fridovich, I. *Ann. NY Acad. Sci.* **1999**, *893*, 13–18.
2. Stanbury, D. M. *Adv. Inorg. Chem.* **1989**, *33*, 70–138.
3. (a) Lah, M. S.; Dixon, M. M.; Pattridge, K. A.; Stallings, W. C.; Fee, J. A.; Ludwig, M. L. *Biochemistry* **1995**, *34*, 1646–1660. (b) Emerson, J. P.; Coulter, E. D.; Cabelli, D. E.; Phillips, R. S.; Kurtz, D. M., Jr. *Biochemistry* **2002**, *41*, 4348–4357.
4. Sisemore, M. F.; Selke, M.; Burstyn, J. N.; Valentine, J. S. *Inorg. Chem.* **1997**, *36*, 979–984.
5. (a) Mathe′, C.; Mattioli, T. A.; Horner, O.; Lombard, M.; Latour, J.-M.; Fontecave, M.; Nivie‘ re, V. *J. Am. Chem. Soc.* **2002**, *124*, 4966–4967. (b) Hsu, J.-L.; Hsieh, Y.; Tu, C.; O'Connor, D.; Nick, H. S.; Silverman, D. N. *J. Biol. Chem.* **1996**, *271*, 17687–17691. (c) Hearn, A. S.; Tu, C.; Nick, H. S.; Silverman, D. N. *J. Biol. Chem.* **1999**, *274*, 24457–24460.
6. (a) Momenteau, M.; Reed, C. A. *Chem. Rev.* **1994**, *94*, 659–698. (b) Schlichting, I.; Berendzen, J.; Chu, K.; Stock, A. M.; Maves, S. A.; Benson, D. E.; Sweet, R. M.; Ringe, D.; Petsko, G. A.; Sligar, S. G. *Science* **2000**, *287*, 1615–1622. (c) Jensen, K. P.; Ryde, U. *J. Biol. Chem.* **2004**, *279*, 14561–14569. (d) Wertz, D. L.; Valentine, J. S. *Struct. Bonding (Berlin)* **2000**, *97*, 37–60. (e) Stuehr, D. J.; Santolini, J.; Wang, Z.-Q.; Wei, C.-C.; Adak, S. *J. Biol. Chem.* **2004**, *279*, 36167–36170. (f) Kim, E.; Chufan, E. E.; Kamaraj, K.; Karlin, K. D. *Chem. Rev.* **2004**, *104*, 1077–1133.

7. Kühnel, K.; Derat, E.; Terner, J.; Shaik, S.; Schlichting, I. *Proc. Natl. Acad. Sci. USA* **2007**, *104*, 99–104.
8. VanAtta, R. B.; Strouse, C. E.; Hanson, L. K.; Valentine, J. S. *J. Am. Chem. Soc.* **1987**, *109*, 1425–1434.
9. Bull, C.; Niederhoffer, E. C.; Yoshida, T.; Fee, J. A. *J. Am. Chem. Soc.* **1991**, *113*, 4069–4076.
10. Kaim, W.; Schwederski, B. *"Bioinorganic Chemistry: Inorganic Elements in the Chemistry of Life"*; Wiley: Chichester, England, **1994**.
11. Fridovich, I. *J. Biol. Chem.* **1995**, *264*, 7761–7764.
12. (a) Dix, T. A.; Hess, K. M.; Medina, M. A.; Sullivan, R. W.; Tilly, S. L.; Webb, T. L. L. *Biochemistry* **1996**, *35*, 4578–4583. (b) Macarthur, H.; Westfall, T. C.; Riley, D. P.; Misko, T. P.; Salvemini, D. *Proc. Natl. Acad. Sci. USA* **2000**, *97*, 9753–9758. (c) Aikens, J.; Dix, T. A. *Arch. Biochem. Biophys.* **1993**, *305*, 516–525. (d) Riley, D. P. *Chem. Rev.* **1999**, *99*, 2573–2587. (e) Gryglewski, R. J.; Palmer, R. M. J.; Moncada, S. *Nature* **1986**, *320*, 454–456. (f) Salvemini, D.; Muscoli, C.; Riley, D. P.; Cuzzocrea, S. *Pulm. Pharmacol. Ther.* **2002**, *15*, 439–447.
13. (a) Cass, A. E. G. Superoxide dismutases. In: *"Metalloproteins, Part I: Metal Proteins with Redox Roles"*; Ed. Harrison, P; Verlag Chemie: Weinheim, **1985**, pp. 121–156. (b) Barondeau, D. P.; Kassmann, C. J.; Bruns, C. K.; Tainer, J. A.; Getzoff, E. D. *Biochemistry* **2004**, *43*, 8038–8047.
14. (a) Edeas, M. A.; Emerit, I.; Khalfoun, Y.; Lazizi, Y.; Cernjavski, L.; Levy, A.; Lindenbaum, A. *Free Radic. Biol. Med.* **1997**, *23*, 571–578. (b) Mollace, V.; Nottet, H. S. L. M.; Clayette, P.; Turco, M. C.; Muscoli, C.; Salvemini, D.; Perno, C. F. *Trends Neurosci.* **2001**, *24*, 411–416. (c) Church, S. L.; Grant, J. W.; Ridnour, L. A.; Oberley, L. W.; Swanson, P. E.; Meltzer, P. S.; Trent, J. M. *Proc. Natl. Acad. Sci. USA* **1993**, *90*, 3113–3117. (d) Safford, S. E.; Oberley, T. D.; Urano, M.; St. Clair, K. K. *Cancer Res.* **1994**, *54*, 4261–4265. (e) Yoshizaki, N.; Mogi, Y.; Muramatsu, H.; Koike, K.; Kogawa, K.; Niitsu, Y. *Int. J. Cancer* **1994**, *57*, 287–292. (f) Borgstahl, G. E. O.; Parge, H. E.; Hickey, M. J.; Johnson, M. J.; Boissinot, M.; Hallewell, R. A.; Lepock, J. R.; Cabelli, D. E.; Tainer, J. A. *Biochemistry* **1996**, *35*, 4287–4297.
15. Lefaix, J.-L.; Delanian, S.; Leplat, J.-J.; Tricaud, Y.; Martin, M.; Nimrod, A.; Pr. Baillet, F.; Daburon, F. *Int. J. Radiat. Oncol. Biol. Phys.* **1996**, *35*, 305–312.
16. (a) Muscoli, C.; Cuzzocrea, S.; Riley, D. P.; Zweier, J. L.; Thiemermann, C.; Wang, Z.-Q.; Salvemini, D. *Brit. J. Pharmacol.* **2003**, *140*, 445–460. (b) Ohtsu, H.; Shimazaki, Y.; Odani, A.; Yamauchi, O.; Mori, W.; Itoh, S.; Fukuzumi, S. *J. Am. Chem. Soc.* **2000**, *122*, 5733–5741. (c) Li, D.; Li, S.; Yang, D.; Yu, J.; Huang, J.; Li, Y.; Tang, W. *Inorg. Chem.* **2003**, *42*, 6071–6080. (d) Durackova, Z.; Labuda, J. *J. Inorg. Biochem.* **1995**, *58*, 297–303. (e) Liao, Z.; Xiang, D.; Li, D.; Mei, F.; Yun, F. *Synth. React. Inorg. Met.-Org. Chem.* **1998**, *28*, 1327–1341.
17. Batinic-Haberle, I.; Spasojevic, I.; Hambright, P.; Benov, L.; Crumbliss, A. L.; Fridovich, I. *Inorg. Chem.* **1999**, *38*, 4011–4022.
18. Zhang, D.; Busch, D. H.; Lennon, P. L.; Weiss, R. H.; Neumann, W. L.; Riley, D. P. *Inorg. Chem.* **1998**, *37*, 956–963.
19. Yamaguchi, S.; Kumagai, A.; Funahashi, Y.; Jitsukawa, K.; Masuda, H. *Inorg. Chem.* **2003**, *42*, 7698–7700.
20. Shearer, J.; Long, L. M. *Inorg. Chem.* **2006**, *45*, 2358–2360.
21. Durot, S.; Lambert, F.; Renault, J.-P.; Policar, C. *Eur. J. Inorg. Chem.* **2005**, 2789–2793.
22. Riley, D. P.; Weiss, R. H. *J. Am. Chem. Soc.* **1994**, *116*, 387–388.
23. Salvemini, D.; Wang, Z.-Q.; Zweier, J. L.; Samouilov, A.; Macarthur, H.; Misko, T. P.; Currie, M. G.; Cuzzocrea, S.; Sikorski, J. A.; Riley, D. P. *Science* **1999**, *286*, 304–306.
24. Riley, D. P.; Schall, O. F. *Adv. Inorg. Chem.* **2007**, *59*, 233–263.
25. Aston, K.; Rath, N.; Naik, A.; Slomczynska, U.; Schall, O. F.; Riley, D. P. *Inorg. Chem.* **2001**, *40*, 1779–1789.

26. Cuzzocrea, S.; Riley, D. P.; Caputi, A. P.; Salvemini, D. *Pharmacol. Rev.* **2001**, *53*, 135–159.
27. Salvemini, D. *PCT Int. Appl. WO 98/58636.*
28. Salvemini, D.; Mazzon, E.; Dugo, L.; Riley, D. P.; Serraino, I.; Caputi, A. P.; Cuzzocrea, S. *Br. J. Pharmacol.* **2001**, *132*, 815–827.
29. (a) Riley, D. P.; Rivers, W. J.; Weiss, R. H. *Anal. Biochem.* **1991**, *196*, 344–349. (b) Weiss, R. H.; Flickingerl, A. G.; Rivers, W. J.; Hardyq, M. M.; Aston, K. W.; Ryanll, U. S.; Riley, D. P. *J. Biol. Chem.* **1993**, *268*(31), p. 23049.
30. (a) Stallings, W. C.; Pattridge, K. A.; Strong, R. K.; Ludwig, M. L. *J. Biol. Chem.* **1985**, *260*, 16424–16432. (b) Tierney, D. L.; Fee, J. A.; Ludwig, M. L.; Penner-Hahn, J. E. *Biochemistry* **1995**, *34*, 1661–1668.
31. Riley, D. P.; Lennon, P. J.; Neumann, W. L.; Weiss, R. H. *J. Am. Chem. Soc.* **1997**, *119*, 6522–6528.
32. Riley, D. P.; Henke, S. L.; Lennon, P. J.; Aston, K. *Inorg. Chem.* **1999**, *38*, 1908–1917.
33. Ducommun, Y.; Newman, K. E.; Merbach, A. E. *Inorg. Chem.* **1980**, *19*, 3696–3703.
34. Dees, A.; Zahl, A.; Puchta, R.; van Eikema Hommes, N. J. R.; Heinemann, F. W.; Ivanovic-Burmazovic, I. *Inorg. Chem.* **2007**, *46*, 2459–2470.
35. Riley, D. P.; Henke, S. L.; Lennon, P. J.; Weiss, R. H.; Neumann, W. L.; Rivers, W. J., Jr.; Aston, K. W.; Sample, K. R.; Rahman, H.; Ling, C.-S.; Shieh, J.-J.; Busch, D. H.; Szulbinski, W. *Inorg. Chem.* **1996**, *35*, 5213–5231.
36. Jiménez-Sandoval, O.; Ramírez-Rosales, D.; Rosales-Hoz, M.; Sosa-Torres, M. E.; Zamorano-Ulloa, R. *J. Chem. Soc. Dalton Trans.* **1998**, 1551–1556.
37. (a) Nelson, S. M. *Pure Appl. Chem.* **1980**, *52*, 2461–2476. (b) Cairns, C.; McFall, S. G.; Nelson, S. M.; Drew, M. G. B. *J. Chem. Soc. Dalton Trans.* **1979**, 446–453. (c) Cook, D. H.; Fenton, D. E. *Inorg. Chim. Acta* **1977**, *25*, L95–L96. (d) Pedrido, R.; Romero, M. J.; Bermejo, M. R.; Gonzalez-Noya, A. M.; Maneiro, M.; Rodriguez, M. J.; Zaragoza, G. *Dalton Trans.* **2006**, 5304–5314.
38. Zetter, M. S.; Grant, M. O.; Wood, E. J.; Dodgen, H. W.; Hunt, J. P. *Inorg. Chem.* **1972**, *11*, 2701–2706.
39. Zetter, M. S.; Dodgen, H. W.; Hunt, J. P. *Biochemistry* **1973**, *12*, 778–782.
40. Grant, M.; Dodgen, H. W.; Hunt, J. P. *Inorg. Chem.* **1971**, *10*, 71–73.
41. Balogh, E.; He, Z.; Hsieh, W.; Liu, S.; Toth, E. *Inorg. Chem.* **2007**, *46*, 238–250.
42. Liu, G.; Dodgen, H. W.; Hunt, J. P. *Inorg. Chem.* **1977**, *16*, 2652–2653.
43. Troughton, J. S.; Greenfield, M. T.; Greenwood, J. M.; Dumas, S.; Wiethoff, A. J.; Wang, J.; Spiller, M.; McMurry, T. J.; Caravan, P. *Inorg. Chem.* **2004**, *43*, 6313–6323.
44. (a) Ivanovic-Burmazovic, I.; Hamza, M. S. A.; van Eldik, R. *Inorg. Chem.* **2002**, *41*, 5150–5161. (b) Ivanovic-Burmazovic, I.; Hamza, M. S. A.; van Eldik, R. *Inorg. Chem.* **2006**, *45*, 1575–1584.
45. (a) Andjelkovic, K.; Bacchi, A.; Pelizzi, G.; Jeremic, D.; Ivanovic-Burmazovic, I. *J. Coord. Chem.* **2002**, *55*, 1385–1392. (b) Ivanovic-Burmazovic, I.; Hamza, M. S. A.; van Eldik, R. *Inorg. Chem.* **2002**, *41*, 5150–5161. (c) Ivanovic-Burmazovic, I.; Hamza, M. S. A.; van Eldik, R. *Inorg. Chem.* **2006**, *45*, 1575–1584.
46. Sarauli, D.; Meier, R.; Liu, G.-F.; Ivanovic-Burmazovic, I.; van Eldik, R. *Inorg. Chem.* **2005**, *44*, 7624–7633.
47. Sarauli, D.; Popova, V.; Zahl, A.; Puchta, R.; Ivanović-Burmazović, I. *Inorg. Chem.* **2007**, *46*, 7848–7860.
48. Miller, A.-F.; Padmakumar, K.; Sorkin, D. L.; Karapetian, A.; Vance, C. K. *J. Inorg. Biochem.* **2003**, *93*, 71–83.
49. Liu, G.-F.; Filipović, M.; Heinemann, F. W.; Ivanović-Burmazović, I. *Inorg. Chem.* **2007**, *46*, 8825–8835.
50. For the O_2 detection see: Karlin, K. D.; Cruse, R. W.; Gultneh, Y.; Farooq, A.; Hayes, J. C.; Zubieta, J. *J. Am. Chem. Soc.* **1987**, *109*, 2668–2679. For the H_2O_2

detection a peroxide indicator paper suitable for the organic solvents (QUANTOFIX-Peroxide 100) was used.

51. Ivanovic-Burmazovic, I.; Andjelkovic, K. *Adv. Inorg. Chem.* **2004**, *55*, 315–360.

52. (a) Mccords, J. M.; Fridovich, I. *J. Biol. Chem.* **1969**, *244*, 6049–6055. (b) Policar, C.; Durot, S.; Lambert, F.; Cesario, M.; Ramiandrasoa, F.; Morgenstern-Badarau, I. *Eur. J. Inorg. Chem.* **2001**, 1807–1818. (c) Fu, H.; Zhou, Y.-H.; Chen, W.-L.; Deqing, Z.-G.; Tong, M.-L.; Ji, L.-N.; Mao, Z.-W. *J. Am. Chem. Soc.* **2006**, *128*, 4924–4925.

53. Chin, D.-H.; Chiericato, G., Jr.; Nanni, E. J., Jr.; Sawyer, D. T. *J. Am. Chem. Soc.* **1982**, *104*, 1296–1299.

54. (a) McCandlish, E.; Miksztal, A. R.; Nappa, M.; Sprenger, A. Q.; Valentine, J. S.; Stong, J. D.; Spiro, T. G. *J. Am. Chem. Soc.* **1980**, *102*, 4268–4271. (b) Burstyn, J. N.; Roe, J. A.; Miksztal, A. R.; Shaevitz, B. A.; Lang, G.; Valentine, J. S. *J. Am. Chem. Soc.* **1988**, *110*, 1382–1388.

55. Shirazi, A.; Goff, H. M. *J. Am. Chem. Soc.* **1982**, *104*, 6318–6322.

56. (a) Afanas'ev, I. B.; Prigoda, S. V.; Khenkin, A. M.; Shteinman, A. S. *Dokl. Akad. Nauk SSSR* **1977**, *236*, 641–644. (b) Kol'tover, V. K.; Koifman, O. I.; Khenkin, A. M.; Shteinman, A. S. *Izu. Akad. Nauk SSSR Ser. Khim.* **1978**, 1690–1691.

57. (a) Reed, C. A. In: "*Electrochemical and Spectrochemical Studies of Biological Redox Components*"; Ed. Kadish, K.M.; American Chemical Society: Washington, DC, **1982**; *Adu. Chem. Ser. No. 201*, 333–356. (b) Welborn, C. H.; Dolphin, D.; James, B. R. *J. Am. Chem. Soc.* **1981**, *103*, 2869–2871.

58. Selke, M.; Sisemore, M. F.; Valentine, J. S. *J. Am. Chem. Soc.* **1996**, *118*, 2008–2012.

59. Selke, M.; Valentine, J. S. *J. Am. Chem. Soc.* **1998**, *120*, 2652–2653.

60. (a) Park, M. J.; Lee, J.; Suh, Y.; Kim, J.; Nam, W. *J. Am. Chem. Soc.* **2006**, *128*, 2630–2634. (b) Seo, M. S.; Kim, J. Y.; Annaraj, J.; Kim, Y.; Lee, Y.-M.; Kim, S.-J.; Kim, J.; Nam, W. *Angew. Chem. Int. Ed.* **2007**, *46*, 377–380.

61. Duerr, K.; Macpherson, B. P.; Warratz, R.; Hampel, F.; Tuczek, F.; Helmreich, M.; Jux, N.; Ivanovic-Burmazovic, I. *J. Am. Chem. Soc.* **2007**, *129*, 4217–4228.

62. Helmreich, M. Ph. D. thesis, University of Erlangen-Nürnberg, 2005.

63. Jones, S. E.; Srivatsa, G. S.; Sawyer, D. T. *Inorg. Chem.* **1983**, *22*, 3903–3910.

64. (a) Traylor, T. G.; Lee, W. A.; Stynes, D. V. *J. Am. Chem. Soc.* **1984**, *106*, 755–764. (b) Traylor, T. G.; Lee, W. A.; Stynes, D. V. *Tetrahedron* **1984**, *40*, 553–568.

TRIPODAL *N,N,O*-LIGANDS FOR METALLOENZYME MODELS AND ORGANOMETALLICS

NICOLAI BURZLAFF

Department of Chemistry and Pharmacy and Interdisciplinary Center for Molecular Materials (ICMM), University of Erlangen-Nürnberg, Egerlandstraße 1, D-91058 Erlangen, Germany

I. The '2-His-1-Carboxylate Facial Triad' in Non-Heme Iron Oxygenases

In the past decade there has been an enormous progress in the field of non-heme iron oxygenases. Almost annually, new reviews are published showing clearly an increasing interest in this field (*1–18*).

ADVANCES IN INORGANIC CHEMISTRY
VOLUME 60 ISSN 0898-8838 / DOI: 10.1016/S0898-8838(08)00004-4

The number of known or presumed mononuclear, non-heme iron oxygenases and related enzymes continues to grow. This is due to intensive biochemical research and especially based on sequence data derived from genome research projects (*14*). For several of these enzymes structural data are available by now from protein crystallography (*12–14*). In many of the iron oxygenases the iron is facially bound by two histidines and one carboxylate donor, either glutamic acid or aspartic acid. Thus, the term '2-His-1-carboxylate facial triad' has been introduced by L. Que Jr. for this motif (*19*).

A. Isopenicillin N Synthase

Iron enzymes with such a 2-His-1-carboxylate facial triad are thought to be as important as the group of heme proteins or enzymes with iron–sulfur clusters. Site-directed mutagenesis proved that this triad is essential for the reactivity of these enzymes (*20*). Furthermore, the geometry of the metal-binding residues seems to be constant through the catalytic cycle, as shown by pseudokinetic protein crystallography (*21–23*). These studies were performed with the enzyme isopenicillin N synthase (IPNS), a ferrous dependent oxygenase with 2-His-1-carboxylate facial triad and the key enzyme in the biosynthesis of penicillins (*20–30*). IPNS transforms the tripeptide δ-(L-α-aminoadipoyl)-L-cysteinyl-D-valine (ACV) in an oxidative cyclization to isopenicillin N (IPN). Most of the penem and cephem antibiotics are derived from this. For example, epimerization by other enzymes transforms IPN to penicillin N, derivatization to penicillin G and ring expansion to cephalosporins. IPNS requires Fe(II) and consumes one equivalent of molecular oxygen as cosubstrate during the reaction cycle. Both oxygen atoms are reduced to two molecules of water during the reaction (Scheme 1).

Due to the importance of penicillins and cephalosporins, which certainly saved the lives of hundreds of millions of people, especially the mechanism of the penicillin biosynthesis was intensively investigated (*24*). Although total syntheses of penicillins are known now for more than 50 years, this reaction in which both rings of the penicillin structure are formed in one step is still impossible to reproduce for

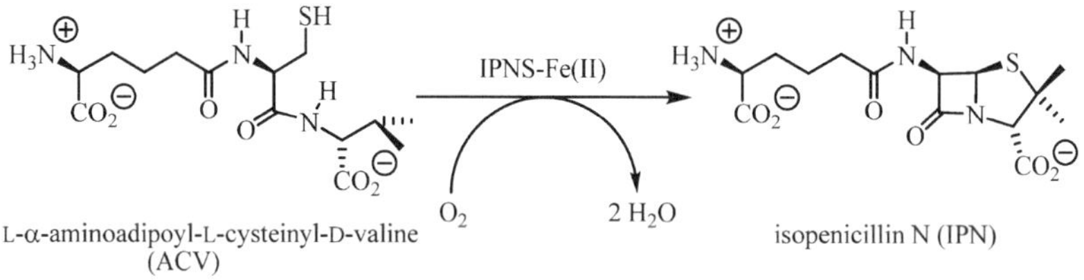

Scheme 1. Cyclisation of L-α-aminoadipoyl-L-cysteinyl-D-valine (ACV) to isopenicillin N (IPN) by the enzyme isopenicillin N synthase (IPNS).

synthetic chemists and is a special synthetic challenge in organic synthesis. By now there is no penicillin synthesis known that is as efficient as this unique biocatalytic reaction. Based on incubation experiments with hundreds of derivatives of the natural substrates ACV (*24*) including isotopic labeled ACV substrates (*25*) and with the help of several protein X-ray structure determinations (*21–23,26–29*), Sir J. E. Baldwin proposed the following mechanism for the penicillin biosynthesis (Scheme 2) (*23,28*).

When the reaction starts, Gln330 of the C-terminus leaves its octahedral position at the Fe(II) center, allowing the ACV substrate to coordinate via the cysteine of the tripeptide as thiolate. The hydrophobic D-valine of the tripeptide induces the release of one of the two coordinated water molecules from Fe(II) and another coordination site is now free for the cosubstrate O_2. When O_2 is bound, a superoxide complex Fe(III)-$O_2^{-\bullet}$ is formed. H-abstraction from the SCH_2 group and a single electron transfer (SET) then results in a hydroperoxide complex Fe(II)–OOH with a coordinated thioaldehyde ligand. Due to the α-effect and the proximity of the hydroperoxide ligand to the amide proton of the ACV substrate a deprotonation and formation of an amide anion should be possible. This N-nucleophile might then attack the thiocarbonyl group of the thioaldehyde ligand and thus form the β-lactam ring in a 4-exo-trig reaction. In this step a molecule of water is released and a Fe(IV)=O species is formed, while the β-lactam ring

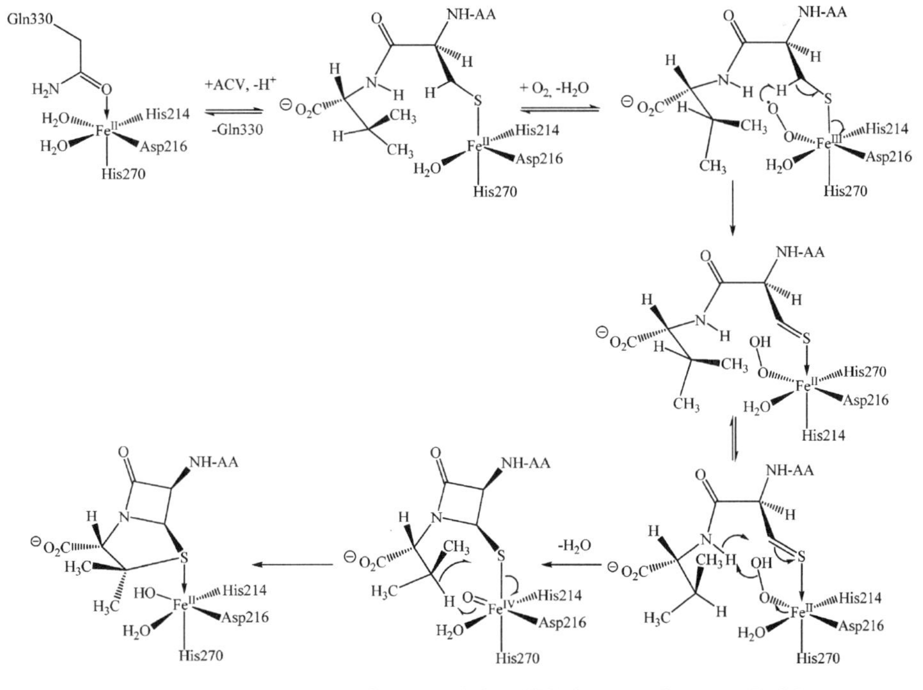

SCHEME 2. Bio-catalytic pathway of the IPNS according to Baldwin (*28*).

is still coordinated as a thiolate. The Fe(IV)=O species accomplishes now in a radical reaction the second ring closure reaction to the penem product. A ferrous center is formed to which the IPN product coordinates as a thioether ligand. Finally, the IPN product is released and the catalytic cycle is closed.

Several steps in this mechanism are supported by protein structures of IPNS. In 1997 P. L. Roach from the Baldwin group obtained a molecular structure of IPNS with Fe(II) and a bound ACV substrate (Fig. 3) by crystallization of IPNS under anaerobic conditions in a glove box (*29*). The coordination site for O_2 was proven by exposing these protein crystals to NO as an O_2 analog (*29*). In pseudokinetic protein crystallography experiments with the enzyme IPNS (*21*) crystals of the IPNS:Fe(II):substrate complex grown under anaerobic conditions were pressurized with pure oxygen up to 40 bar to run the IPN biosynthesis in a crystal. After a certain time the pressure was released and the reaction was trapped by freezing the crystals on liquid nitrogen. Pressurizing at 40 atmospheres for 320 min resulted in a product structure IPNS:Fe(II):IPN (Fig. 1) (*21*).

All the pseudokinetic protein crystallographic experiments performed so far with IPNS clearly show severe changes in the electron density of the ACV substrate throughout the catalytic cycle but no change of electron density at all was observed for the residues of the 2-His-1-carboxylate facial triad (*21–23*). This implies, but does not unequivocally prove, a rather fixed geometry of the three iron-binding residues during this catalytic cycle. This makes the 2-His-1-carboxylate facial triad a rather promising candidate to be mimicked by bioinorganic model complexes. This applies even more, since several non-heme iron oxygenases exhibit the same motif.

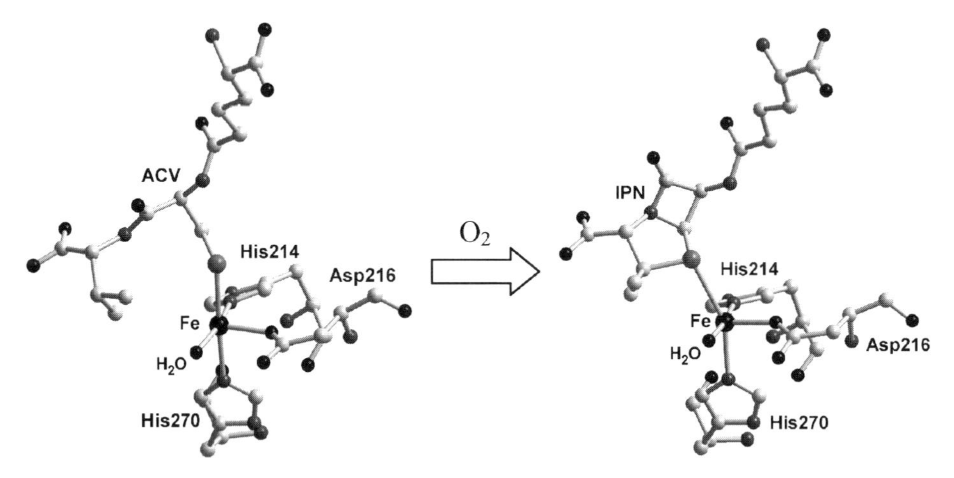

FIG. 1. Active site of isopenicillin N synthase (IPNS) derived from *Aspergillus nidulans*: turn over of Fe(II) bound ACV substrate (PDB-Code: 1BKO) (*29*) in the crystal to Fe(II) bound IPN product (PDB-Code: 1QJE) (*21*).

B. 2-Oxoglutarate Dependent Fe(II) Enzymes

Especially, the family of the 2-oxoglutarate (2-OG) (α-ketoglutarate) dependent iron(II) enzymes forms the most prominent group in these mononuclear non-heme iron oxygenases (*18*). In many of these enzymes the 2-His-1-carboxylate facial triad can be detected by simple sequence alignment (*26*). Several protein structures of 2-OG dependent iron(II) enzymes were solved in recent years, some of them with very high resolution (*18*). Thus, structural data are available for deacetoxycephalosporin C synthase (DAOCS), clavaminic acid synthase (CAS), carbapenem synthase (CarC), proline 3-hydroxylase (P-3-H), human HIF inhibiting factor (FIH), taurine dioxygenase (TauD) (Fig. 2) and the anthocyanidin synthase (ANS).

Most of the 2-OG dependent iron enzymes are hydroxylases. The hydroxylation of a C–H bond is coupled to an oxidative decarboxylation of the 2-OG to succinate and CO_2 (*8*). One oxygen atom of O_2 forms the hydroxyl group of the product (Scheme 3). The second oxygen atom ends up in the succinate (*8,18*).

In 2-OG dependent enzymes ferrous iron is bound in the active site by the 2-His-1-carboxylate facial triad. The carboxylate is either an aspartic acid or a glutamic acid. In the beginning the slightly distorted

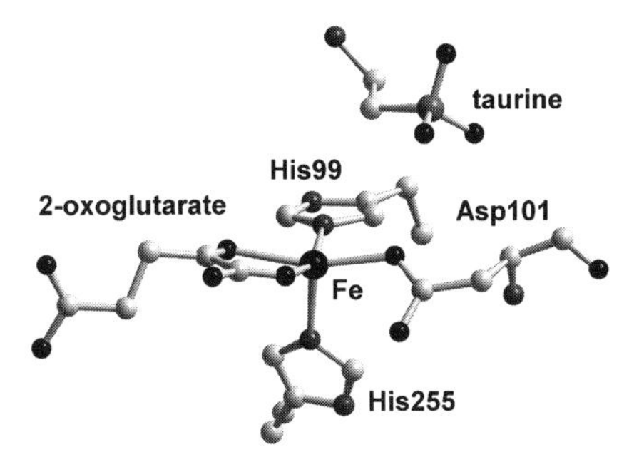

Fig. 2. Active site of taurine dioxygenase (TauD) with 2-oxoglutarate bound to Fe(II) and a taurine substrate (PDB-Code: 1GY9) (*31*).

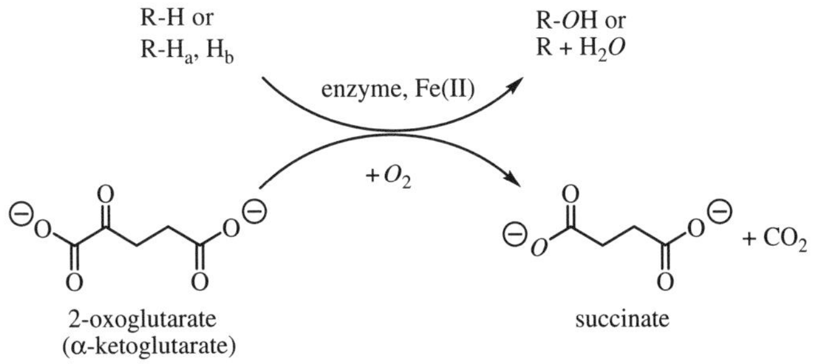

R-H or
$R\text{-}H_a, H_b$

R-*O*H or
$R + H_2O$

enzyme, Fe(II)

$+O_2$

2-oxoglutarate
(α-ketoglutarate)

succinate

$+ CO_2$

Scheme 3. Reaction of 2-oxoglutarate dependent iron enzymes (*8*).

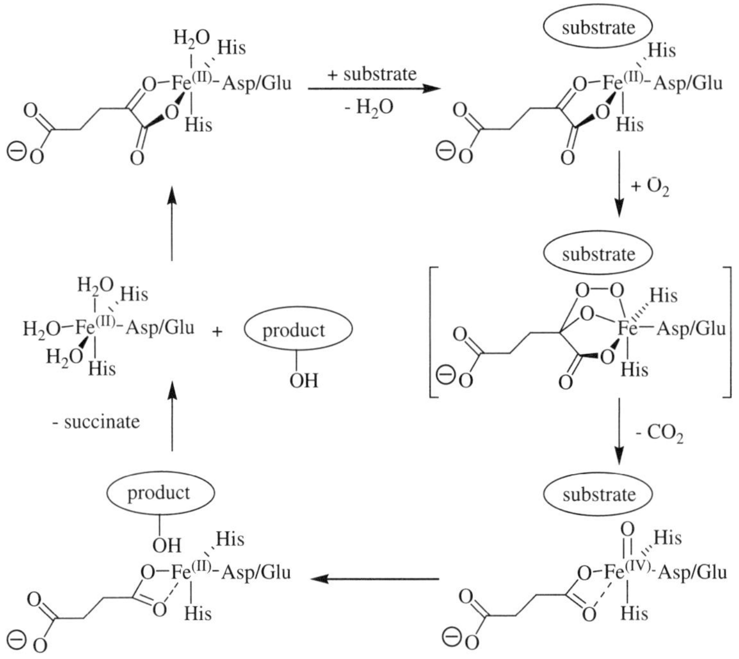

SCHEME 4. Mechanism of the 2-oxoglutarate dependent iron oxygenases (*8*).

octahedral coordination is completed by three water molecules (Fig. 2) (*8,18*).

Then the cosubstrate 2-OG is bound by κ^2 as a bidentate ligand with the carboxylate donor and the keto group (*8,18*). The resulting 6C octahedral geometry is still rather unreactive because of the remaining water molecule (Scheme 4) (*8*). To start a reaction with O_2, a substrate (e.g., taurine) is bound in the enzyme pocket but not coordinated to Fe(II). The remaining water molecule leaves the octahedral coordination site and a 5C center with almost square pyramidal geometry is formed (*8*). O_2 now binds at the free site and a superoxide anion is obtained by electron transfer from Fe(II) to oxygen (*8,18*). Due to a partially anionic character, this might attack the keto group in a nucleophilic reaction (*8*) which has been activated by the Lewis-acidic metal center. Then the peroxo-intermediate decays by decarboxylation and heterolytic cleavage of one O–O-bond to the Fe(IV)=O intermediate. Formation of CO_2 is most likely the first irreversible reaction step that drives this reaction forward. The Fe(IV)=O-intermediate then abstracts a H-atom from the substrate and hydroxylates the substrate in the case of the hydroxylases in a *radical rebound* mechanism (*8*).

In most protein structures that have been published so far, the 2-OG cosubstrate is bound as a chelate ligand to Fe(II) with the 2-oxo group *trans* to the aspartate (*32*). Until now in none of these has a geometry

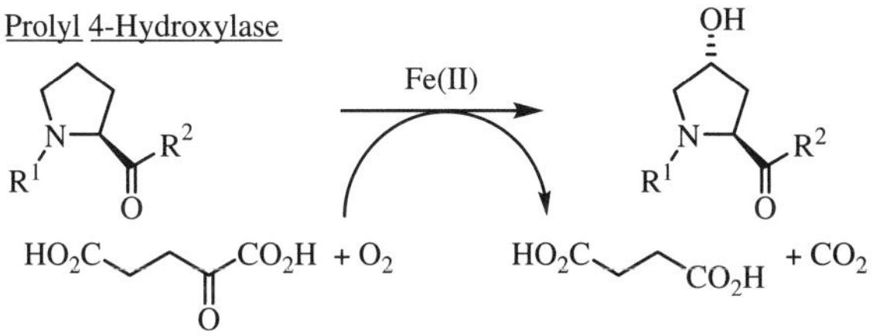

SCHEME 5. Reaction catalyzed by prolyl 4-hydroxylase.

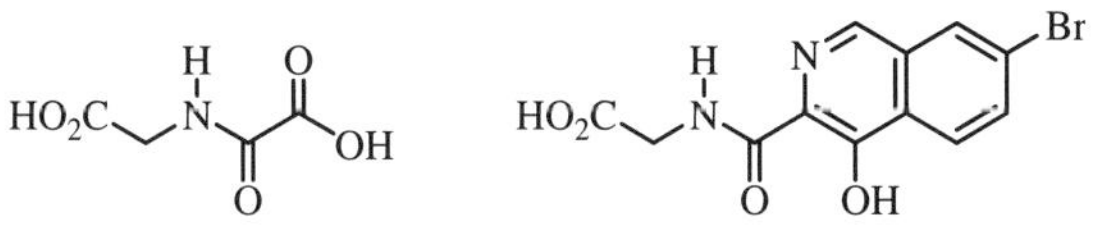

FIG. 3. *N*-Oxalylglycine lead structure and a related inhibitor for prolyl-4 hydroxylase (*36*).

with the 2-oxo group *trans* to a histidine been observed. The activity of DAOCS is limited to 2-OG and 2-oxoadipic acid as cosubstrate. But the DAOCS mutant R258Q also works with other 2-oxocarboxylic acids such as pyruvate or 2-oxo-3-methylbutyric acid (*33*). This shows that the 2-oxo group of the cosubstrate is essential for the catalytic cycle.

Another important 2-OG dependent oxygenase in mammals is prolyl-4 hydroxylase, which catalyzes the hydroxylation of the proline residue in collagen (Scheme 5). This reaction is essential for the structure of the collagen triple helices (*9,34–36*). An overproduction of collagen is related to fibrotic diseases such as rheumatic arthritis. Thus collagen prolyl-4 hydroxylase is a target for therapeutics (*34,36*).

Inhibitors for 2-OG dependent iron oxygenases thus have a huge pharmaceutical potential as future drugs. In particular inhibitors for prolyl 4-hydroxylase could help to control collagen production (*34,36*). Such an inhibitor should mimic the $\kappa^2 O^1,O^2$ chelate coordination of 2-OGs but should not react with O_2. *N*-Oxalylglycine (Fig. 3) is such an inhibitor for prolyl 4-hydroxylase ($IC_{50} = 3\ \mu M$) (*34,36*).

II. *N,N,O*-Ligands as Mimics for the '2-His-1-Carboxylate Facial Triad'

A. BIS(PYRAZOL-1-YL)ACETIC ACIDS

In our quest to find suitable *N,N,O*-ligands as models for the active sites of the facial 2-His-1-carboxylate triad in iron and zinc containing enzymes, heteroscorpionate ligands like bis(3,5-dimethylpyrazol-1-yl)acetic acid Hbdmpza (**3b**) seem to be suitable as a mimic for the 2-His-1-carboxylate facial triad. In these ligands the two pyrazol-1-yl donors mimic the two histidines and the carboxylate donor stands for aspartic acid or glutamic acid.

The first synthesis for such a bis(pyrazol-1-yl)acetate ligand was described in 1999 by A. Otero and coworkers, who reported on several group four and group five transition metal complexes bearing the bis(3,5-dimethylpyrazol-1-yl)acetato (bdmpza) ligand. The multistep synthesis was achieved via bis(3,5-dimethylpyrazol-1-yl)methane (**2b**), which is available from 3,5-dimethylpyrazole and CH_2Cl_2. Deprotonation at the bridging CH_2 group and subsequent reaction with CO_2 resulted in a lithium bdmpza salt $\frac{1}{4}$[{Li(H_2O)(bdmpza)}] (*37*). Since the introduction of bis(pyrazol-1-yl)acetato ligand class to coordination chemistry, a broad spectrum of transition and main group metal complexes bearing these ligands has been investigated (*37–69*). Recent reviews already cover some of these investigations (*70,71*).

In addition to Otero's multistep ligand synthesis we found a simple one-step synthesis for Hbdmpza (**3b**) starting from commercially available reagents (Scheme 6) (*41*).

Treatment of dibromo or dichloroacetic acid with two equivalents of 3,5-dimethylpyrazole and excess of potassium hydroxide, potassium carbonate and a small portion of TEBA phase-transfer catalyst gives Hbdmpza (**3b**) in reasonable yield after acidification and extraction with diethylether. This synthesis can also be applied for the unsubstituted pyrazole but not for sterically hindered pyrazoles (*40*).

It has to be emphasized that the resulting bis(pyrazol-1-yl)acetic acid Hbpza (**3a**) (Fig. 4) would not be available via the Otero route, since his synthesis is restricted to pyrazoles with substituents in 5-position of the pyrazoles. Other bis(pyrazol-1-yl)methanes would be deprotonated at the CH_2 bridge but also in position five of the pyrazole, due to *ortho* metallation. For the sterically more hindered bis(3,5-di-*tert*-butylpyrazole)acetic acid (**3c**) we followed Otero's synthetic pathway. In analogy to the synthesis of bis(3,5-dimethylpyrazol-1-yl)methane (**2b**) (*72*) 3,5-*tert*-butylpyrazole reacted with dichloromethane, base and phase-transfer catalyst to bis(3,5-*tert*-butylpyrazol-1-yl)methane (bd*t*bpzm) (**2c**) (*41*).

As in the synthesis of bdmpza by Otero *et al.*, the scorpionate ligand bis(3,5-di-*tert*-butylpyrazol-1-yl)acetate Hbd*t*bpza (**3c**) can be synthesized from bis(3,5-*tert*-butylpyrazol-1-yl)methane (bd*t*bpzm) (**2c**) by

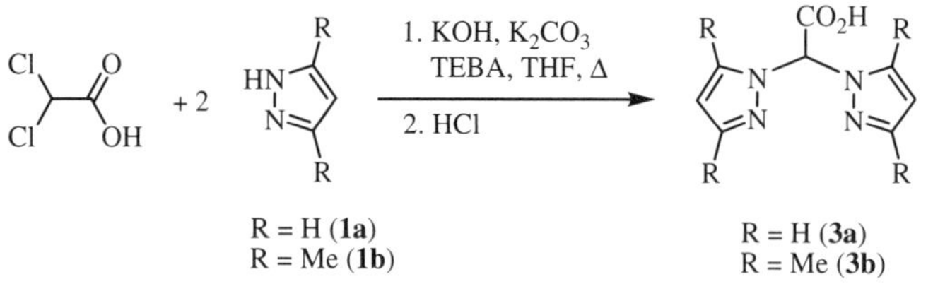

Cl, Cl, O, OH + 2 HN, N, R, R — 1. KOH, K_2CO_3 TEBA, THF, Δ; 2. HCl → CO_2H, R, N, N, R

R = H (**1a**)
R = Me (**1b**)

R = H (**3a**)
R = Me (**3b**)

SCHEME 6. Synthesis of bis(pyrazol-1-yl)acetic acid Hbpza (3a) and bis(3,5-dimethylpyrazol-1-yl)acetic acid Hbdmpza (3b) from dichloroacetic acid.

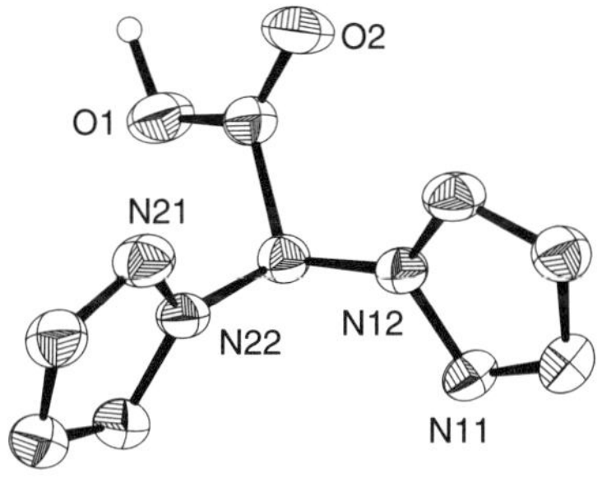

FIG. 4. Molecular structure of bis(pyrazol-1-yl)acetic acid Hbpza (**3a**) (*40*).

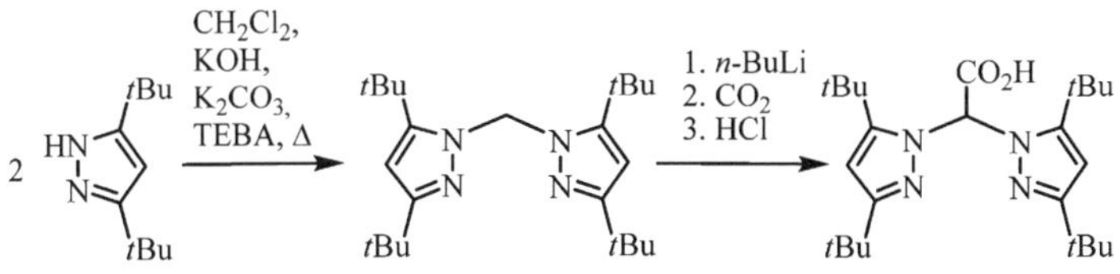

SCHEME 7. Synthesis of bis(3,5-di-*tert*-butylpyrazol-1-yl)acetic acid Hbd*t*bpza (**3c**) (*41*).

deprotonation with *n*-butyllithium and subsequent addition of carbon dioxide (Scheme 7) (*41*).

B. FERROUS COMPLEXES BEARING BIS(PYRAZOL-1-YL)ACETATO LIGANDS

To test the coordination properties of the bis(pyrazol-1-yl)acetate ligands with respect to the bioinorganic model concept, we first tried reactions of the bdmpza anion to form ferrous model complexes. In a first attempt deprotonation of Hbdmpza and reaction with $FeCl_2$ afforded the 2:1 bisligand complex [Fe(bdmpza)$_2$] (**4b**) (*41,49*). The double coordination of bdmpza to the metals demonstrates clearly that the steric hindrance of Hbdmpza (**3b**) is too small. Even equimolar amounts of Hbdmpza (**3b**), base and anhydrous $FeCl_2$ yielded only complexes with two heteroscorpionate ligands bound to the metal. The target compound [Fe(bdmpza)Cl] could not be detected. The formation of a small amount of this 1:1 complex should be detectable by MS by the typical isotopic fingerprint of the chlorine atom. The infrared spectra of [Fe(bdmpza)$_2$] (**4b**) shows an intense $\nu_{as}(CO_2^-)$ absorption at 1659 cm^{-1} and medium intensity bands at 1558 cm^{-1} due to $\nu(C{\equiv}N)$. The ferrous high-spin center in **4b** causes a paramagnetic ^{1}H-NMR spectrum with four single signals at $\delta(^1H) = 1.2$, 7.0, 14.4 and 55.7. These resonances have been assigned to H^4 at the pyrazolyl groups, the two methyl groups and the proton at the bridging carbon atom.

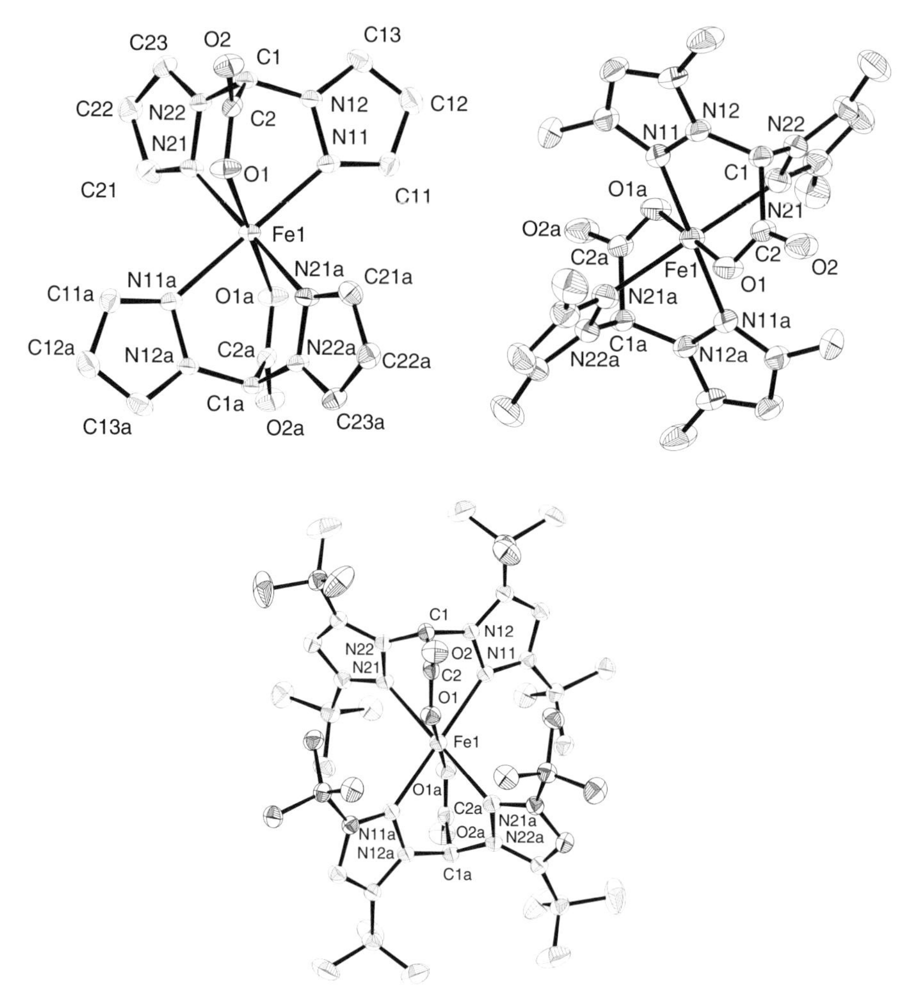

FIG. 5. Molecular structures of (a) [Fe(bpza)$_2$] (**4a**), (b) [Fe(bdmpza)$_2$] (**4b**) and (c) [Fe(bd*t*bpza)$_2$] (**4c**); thermal ellipsoids are depicted at the 50% probability level (*41,49*).

Attempts to exchange one of the bdmpza ligands by a thiolate ligand have not been successful. An X-ray structure determination of the complex [Fe(bdmpza)$_2$] (**4b**) finally confirmed the deduced molecular structure (Fig. 5).

The coordination sphere around the ferrous iron center is almost octahedral with the heteroscorpionate 'clamp' causing a deviation of 4–6° from the ideal 90° angle. A comparison with very high-resolution protein structures such as IPNS or CAS clearly shows a good correspondence of structure **4b** in distances and angles with the active sites of the non-heme iron enzymes (Table I).

TABLE I

COMPARISON OF THE FE–N AND FE–O DISTANCES IN THE ACTIVE SITES OF IPNS, DAOCS AND CAS WITH $[Fe(bdmpza)_2]$ (**4B**)

	IPNS (*28,29*)	DAOCS (*32*)	CAS (*73*)	$[Fe(bdmpza)_2]$ (*41*)
d(Fe–N), Å	2.19, 2.09	2.2	2.15, 2.14, 2.25	2.169(3)
d(Fe–N), Å	2.20, 2.24	2.2	2.12, 2.12, 2.28	2.212(3)
d(Fe–O), Å	2.06, 2.09	2.2	2.06, 2.07, 2.16	2.080(3)

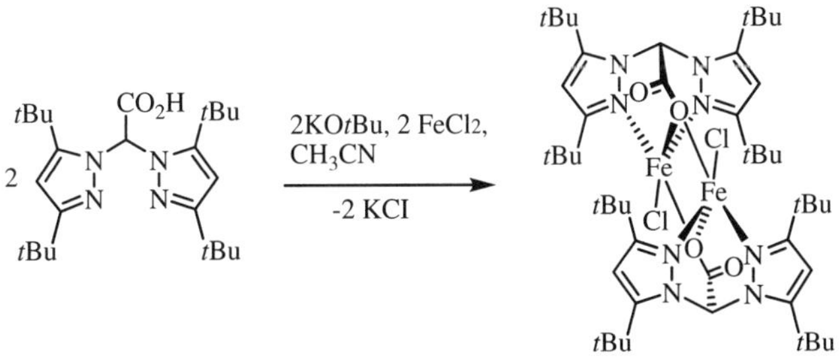

2KO*t*Bu, 2 $FeCl_2$, CH_3CN

-2 KCl

3c **5**

SCHEME 8. Synthesis of the ferrous $[Fe(bdtbpza)Cl]_2$ (**5**) dimer (*41,49*).

This emphasizes that iron(II) complexes bearing more bulky bis(pyrazol-1-yl)acetate ligands should be good structural models to mimic mononuclear non-heme iron dependent enzymes.

Former studies of Moro-oka *et al.* with model complexes of the trispyrazolylborate ligands $Tp^{t\text{-}Bu,iPr}$ and Tp^{iPr_2} have already shown, that sterically more hindered pyrazoles can prevent the Tp ligands from coordinating twice to the metal (*74,75*). Therefore, in our further studies we tried to prevent double coordination by introducing substituents R larger than the methyl group at the pyrazole rings.

As in the previous reaction with Hbdmpza (**3b**) the ligand Hbd*t*bpza (**3c**) was treated with base and anhydrous $FeCl_2$. Indeed, the bdtbpza anion coordinates only once to iron(II), so that a species '[Fe(bdtbpza)Cl]' could be isolated, characterized and tested for its chemical reactivity (*41,49*). The complex displays a $\nu_{as}(CO_2^-)$ absorption around 1680 cm^{-1}. '[Fe(bd*t*bpza)Cl]' in contrast to $[Fe(bdmpza)_2]$ (**4b**) is not stable under air and against moisture. It has to be handled under an atmosphere of argon or nitrogen (*49*). Finally, an X-ray structure determination reveals this species to be a dimer $[Fe(bdtbpza)Cl]_2$ (**5**) (Scheme 8) (Fig. 6) (*49*).

The geometry of the ferrous iron is trigonal bipyramidal. Thus, this structure confirms the tendency of iron complexes to coordination numbers higher than four. The two N-donors and one of the carboxylates of **5** occupy the equatorial positions. In good agreement

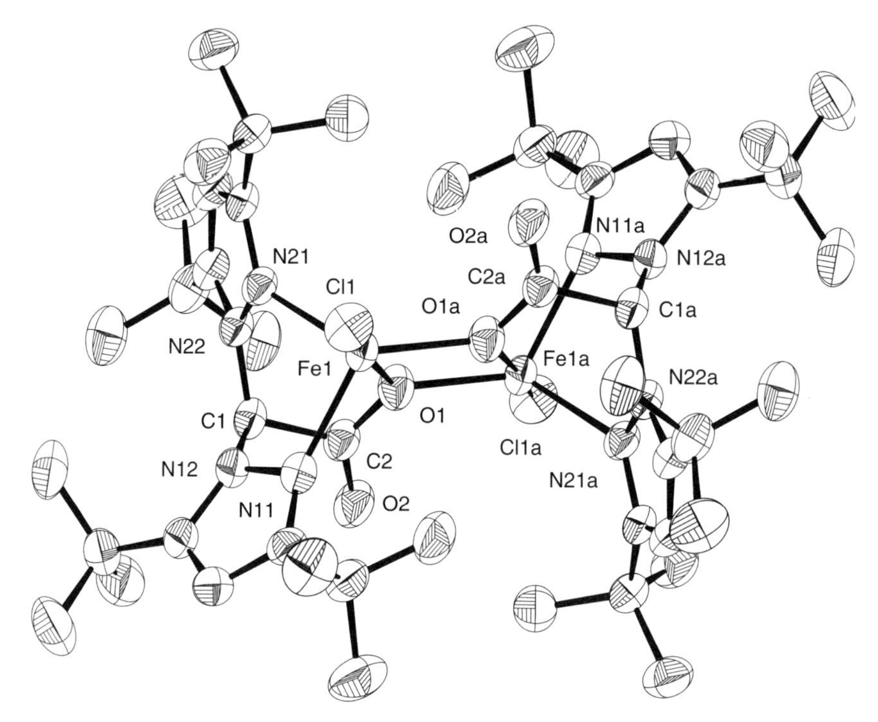

FIG. 6. Molecular structure of [Fe(bd*t*bpza)Cl]$_2$ (**5**); thermal ellipsoids are drawn at the 50% probability level. Selected distances (Å) and angles (°): Fe–N(11) 2.172(4), Fe–N(21) 2.196(4), Fe–O(1) 2.393(4), Fe–O(1a) 2.046(3), Fe–Cl(1) 2.320(4), Fe–Cl(2) Fe–Cl(3), Fe–Fe 3.670(4), O(2)–C(2) 1.227(5), O(1)–C(2) 1.307(5), C(1)–C(2) 1.559(5); N(11)–Fe–N(21) 90.94(16), O(1)–Fe–N(11) 78.44(12), O(1)–Fe–N(21) 80.40(17), O(1)–Fe–Cl(1) 170.83(7), O(1a)–Fe–Cl(1) 103.94(16), N(21)–Fe–Cl(1) 108.72(16), Fe–O(1)–Fe 111.28(17) (*49*).

with the Kepert model (*76,77*) this equatorial Fe–O distance [2.046 (3) Å] is much shorter than the axial Fe–O distance [2.393 (4) Å]. The equatorial Fe–O bond is formed by the *syn* lone pair of the oxygen atom which is the most frequent binding lone pair in metal coordinating carboxylate donors (*78*). This bridging μ-acetato-κ-*O*-binding modus might explain the asymmetric carboxylate IR absorption $\nu_{as} = 1681\,cm^{-1}$, which is $22\,cm^{-1}$ higher than that of [Fe(bdmpza)$_2$] ($\nu_{as} = 1659\ cm^{-1}$) (*41*). The most striking feature of [Fe(bd*t*bpza)Cl]$_2$ (**5**) is the unusually long Fe–Fe distance of 3.670 (4) Å, which is almost unprecedented, apart from two other examples with similar long Fe–Fe distances [3.724 (1) (*79*) and 3.645 (4) Å (*80*)]. Thus, dimer **5** might be a relevant structural model for the active site of dinuclear iron enzymes such as methane monooxygenase (MMO) or ribonucleotide reductase (R2).

The molecular structure of the species '[Fe(bd*t*bpza)Cl]' unambiguously reveals it to be a ferrous dimer $[Fe(bd\textit{t}bpza)Cl]_2$ (**5**) (Fig. 6) with bridging oxygen atoms and a trigonal bipyramidal geometry at the iron atoms. Therefore, this dimer represents two unfavored tetracoordinated '[Fe(bd*t*bpza)Cl]' species, which are now stabilized by this bridging oxygen and the formation of more favored penta-coordinated iron centers. A penta- or hexa-coordinated monomeric complex $[Fe(bd\textit{t}bpza)X(solv)_n]$ ($n = 1,2$) would mimic the 2-His-1-carboxylate facial triad of non-heme iron dependent very closely. We therefore focused on attempts to split this ferrous dimer in order to obtain good structural models. Especially the coordination of thiolato or 2-oxocarboxylato ligands would be a challenge on the way to structural models for the non-heme iron dependent oxygenases. Moro-oka *et al.* have already shown, that the access of small molecules to the reactive metal center can be affected a lot by the sterical properties e.g., of a $Tp^{t\text{-}Bu,iPr}$ ligand (*81*). Thus, a model complex $[Fe(Tp^{t\text{-}Bu,iPr})(O_2CMe)]$ showed almost no affinity to oxygen in contrast to the sterically less hindered and more reactive complex $[Fe^{II}(Tp^{iPr_2})(O_2CMe)]$. Since the dimer $[Fe(bd\textit{t}bpza)Cl]_2$ (**5**) is rather air sensitive a good affinity to O_2, an essential property for a possible functional ferrous model complex, can be assumed. An advantage of the bis(pyrazol-1-yl)acetato ligands might be the fact that the sterical hindrance of the metal ion by two pyrazolyl donors and a carboxylate donor should be much smaller than that by three pyrazolyl donors in a Tp ligand. Therefore, we tested a reaction of the dimeric complex **5** with a 2-oxocarboxylate to obtain a model for the active sites of 2-OG dependent iron oxygenases. Such a complex would mimic the 5C octahedral reaction step of the biocatalytic pathway, with κ^2 coordinated 2-OG cosubstrate but having already released the water molecule and might thus be well suited to activate oxygen as the enzyme does (Fig. 7).

Actually the reaction of the dimer $[Fe(bd\textit{t}bpza)Cl]_2$ (**5**) with thallium benzoylformate results in the purple benzoylformato complex $[Fe(bd\textit{t}bpza)(O_2CC(O)Ph)]$ (**6**). The purple color is caused by a visible absorption at 544 nm (Fig. 8).

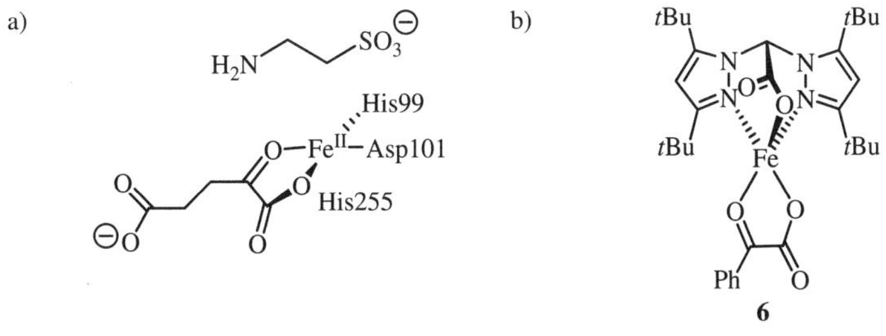

FIG. 7. (a) Active site of Taurine dioxygenase TauD after activation by the taurine substrate and (b) a benzoylformato model $[Fe(bd\textit{t}bpza)(O_2CC(O)Ph)]$ (**6**).

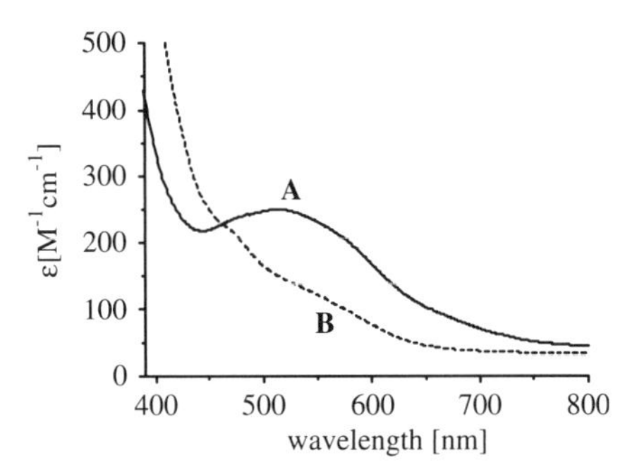

FIG. 8. UV/Vis spectra of (A) [Fe(bd*t*bpza)(O_2CC(O)Ph)] (**6**) (—) in N_2 atmosphere and (B) exposed to air for 6 min (- - -).

This absorption is characteristic for ferrous model complexes bearing $\kappa^2 O^1,O^2$ bound 2-oxocarboxylato ligands and arises from a metal to ligand charge transfer (MLCT) in the coordinated 2-oxo group (*81–88*). Similar MLCT bands have been reported before for related model complexes by Moro-oka *et al.* and Que Jr. *et al.* such as the benzoylformato complexes [Fe($Tp^{t\text{-Bu},i\text{Pr}}$)($O_2$CC(O)Ph)] (*81,75*), [Fe(Tp^{Ph2})(O_2CC(O)Ph)] (*86,88*) and [Fe(O_2CC(O)Ph)(TPA)] [TPA: tris(2-pyridylmethyl)amine] (*82–84*). Such a MLCT at 530 nm or 520 nm in the presence of taurine has also been observed for 2-OG dependent TauD, if ferrous iron and 2-OG are added to the enzyme under anaerobic conditions (*87*).

The IR spectrum of compound **6** shows the bd*t*bpza $\nu_{asym}(CO_2^-)$ band at 1674 cm^{-1}. A shoulder at 1688 cm^{-1} and a band at 1632 cm^{-1} have been assigned to the ν(C=O) and to the $\nu_{asym}(CO_2^-)$ vibration of the benzoylformato ligand. The phenyl ν(C=C) absorption is observed at 1595 cm^{-1}. In acetonitrile colorless plate shaped crystals are formed within some days at –30°C which unfortunately were not suitable for an X-ray structure determination due to crystal solvent and the plate form. The lack of the purple color in the crystals indicates a loss of the $\kappa^2 O^1,O^2$ coordination of the benzoylformato ligands. This might be the result of a change to a $\kappa^2 O^1,O^{1'}$ coordination, as it has been observed before by Y. Moro-oka *et al.* for [Fe($Tp^{t\text{-Bu},i\text{Pr}}$)($O_2$CC(O)Ph)] (*81,75*). Another explanation might be the formation of a dimeric species [Fe(bd*t*bpza)(O_2CC(O)Ph)]$_2$ similar to the dimer [Fe(bd*t*bpza)Cl]$_2$ (**5**) with $\kappa^1 O^1$-coordination of the benzoylformato ligand. This could explain a mass peak at 1091 (FAB-MS) which fits to [Fe_2(bd*t*bpza)$_2$ (O_2CC(O)Ph)]$^+$ and thus backs a dimer [Fe(bd*t*bpza)(O_2CC(O)Ph)]$_2$. The compound **6** is extremely air sensitive. Once exposed to air a solution in CH_2Cl_2 loses its color within a few minutes (Fig. 8). This reactivity is much higher compared to other functional models, which usually react with O_2 within one hour (*82–84,86,88*). So far only the benzoylformato complex [Fe(Tp^{Me2})(O_2CC(O)Ph)(CH_3CN)] by J. S. Valentine *et al.* is of comparable reactivity (*85*).

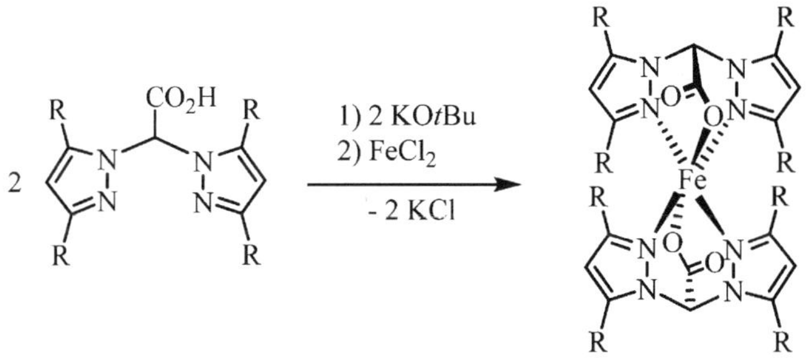

R = H (**3a**), R = Me (**3b**)

R = H (**4a**), R = Me (**4b**)

SCHEME 9. Synthesis of ferrous bisligand complexes [Fe(bpza)$_2$] (**4a**) and [Fe(bdmpza)$_2$] (**4b**) (*49*).

To detect any functional activity of complex **6** with O_2, the air-exposed solution was acidified and analyzed by GC. Unfortunately, we have not been able to detect any benzoic acid until now, which should have been formed from benzoylformic acid and O_2 in analogy to the catalytic cycle of the 2-OG dependent iron oxygenases.

As mentioned above, the sterically less hindered bis(3,5-dimethylpyrazol-1-yl)acetic acid **3b** forms with $FeCl_2$ in a similar reaction a 2:1 complex [Fe(bdmpza)$_2$] (**4b**) (*41*). We found a similar behavior for the even less hindered Hbpza (**3a**). The potassium salt of **3a** reacts with $FeCl_2$ to a bisligand complex [Fe(bpza)$_2$] (**4a**) (Scheme 9, Fig. 5b).

The bpza ligand in **4a** coordinates with the *anti* lone pair of the carboxylate donor to the ferrous center. Thus, the asymmetric carboxylate IR absorption ($\nu_{as} = 1653\,cm^{-1}$) appears at even lower wavenumbers compared to that of **4b**.

In order to assign more IR signals of **4a**, *ab initio* calculations on Hbdmpza (**3b**) and **4a** were performed. It is well known for the chosen HF/6-31G* basis set that calculated harmonical vibrational frequencies are typically overestimated compared to experimental data. These errors arise from the neglecting anharmonicity effects, incomplete incorporation of electron correlation and the use of finite basis sets in the theoretical treatment (*89*). In order to achieve a correlation with observed spectra a scaling factor (approximately 0.84–0.90) has to be applied (*90*). The calculations were calibrated on the asymmetric carboxylate ν_{asym} at $1653\,cm^{-1}$. We were especially interested in the symmetric carboxylate vibration ν_{sym}. Pursuant to literature the difference $\Delta(\nu_{asym} - \nu_{sym})$ for a unidentate carboxylate group should be $\geq 200\,cm^{-1}$ (*91*). Therefore, for several related transition metal bis(pyrazol-1-yl)acetato and bis(3,5-dimethylpyrazol-1-yl)acetato complexes absorptions around $1460\,cm^{-1}$ were assigned to ν_{sym} (*37–39*). Surprisingly, according to our calculations, this absorption at $1453\,cm^{-1}$ belongs to a $C \leftrightarrow N$ vibration as we could confirm by an IR of 3,5-dimethylpyrazol-1-ylmethane ($1451\,cm^{-1}$). Instead, the symmetric carboxylate vibration ν_{sym} shows up at 1350 cm^{-1}, according to the calculations. This is supported by the complex [Cu(bdmpza)$_2$], for which Reedijk and coworkers assigned ν_{sym} unambiguously to this region of the spectrum (*52*). These considerations imply a $\Delta(\nu_{asym} - \nu_{sym})$ around 300 cm^{-1} for the group of unidentate bis(pyrazol-1-yl)acetato ligands.

The **4a** related purple complex {Fe[HB(pz)$_3$]$_2$} has a low-spin t_{2g}^6 configuration at ambient temperature but exhibits a spin crossover to a $t_{2g}^4e_g^2$ high-spin state and a color change from purple to white at 393 K (*92*). On the other hand, {Fe[HB(3,5-Me$_2$pz)$_3$]$_2$} is at ambient temperature a colorless high-spin complex (*92*). Upon cooling down to below 200 K it changes gradually to a low-spin state (*92*). The long Fe–N distances in the molecular structure of **4a** [2.2054 (17) Å and 2.1542 (18) Å] and the lack of color indicate that **4a** is a high-spin ferrous complex at ambient temperature. The SQUID data indicate that **4a** as well as **4b** are high-spin complexes in the temperature range from 5 to 350 K. Both samples show Curie law behavior ($\mu_{eff} = 5.12\,\mu_B$). Therefore, a spin crossover upon cooling can be excluded so far. Although bd*t*bpza is a bulky ligand, a 2:1 bisligand complex [Fe(bd*t*bpza)$_2$] (**4c**) was also obtained by adding Fe[BF$_4$]$_2$ × 6 H$_2$O to K[bd*t*bpza]. Due to the bulky ligand the Fe–N distances in the molecular structure of the colorless complex **4c** [2.3400 (17) Å and 2.353 (2) Å] are much longer than those of **4a** and **4b** [**4b**: 2.169 (3) Å and 2.212 (3) Å] (Fig. 5c) (*41*). This causes a weak ligand field and makes of course a crossover to a low-spin state very unlikely. All three [FeL$_2$] complexes are stable under aerobic conditions even at $T > 200°C$.

C. Ferric Complexes Bearing Bis(pyrazol-1-yl)acetato Ligands

Subsequently, we focused on iron(III) complexes. Reaction of the ferric precursor [NEt$_4$]$_2$[Cl$_3$FeOFeCl$_3$] (*93*) with Hbpza (**3a**) or Hbdmpza (**3b**) yielded two equivalents of [NEt$_4$][Fe(bpza)Cl$_3$] (**3a**) and [NEt$_4$][Fe(bdmpza)Cl$_3$] (**3b**) by the loss of one equivalent of water (Scheme 10). The structure of **4b** is depicted in Fig. 9.

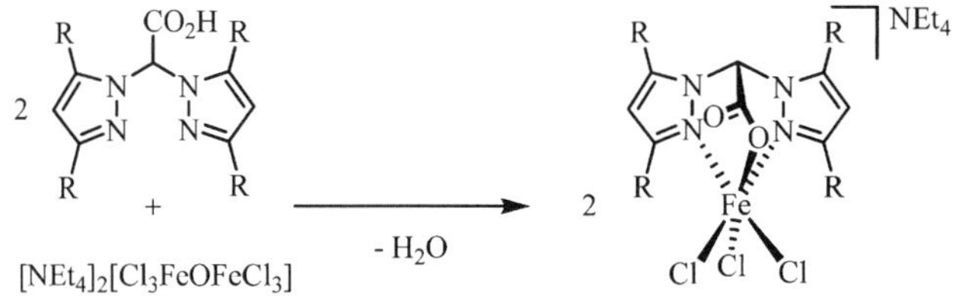

2 + [NEt$_4$]$_2$[Cl$_3$FeOFeCl$_3$] → (- H$_2$O) 2 [NEt$_4$][Fe(L)Cl$_3$]

3a (R = H), **3b** (R = Me) **7a** (R = H), **7b** (R = Me)

Scheme 10. Synthesis of the two complex salts [NEt$_4$][Fe(bpza)Cl$_3$] (**7a**) and [NEt$_4$][Fe(bdmpza)Cl$_3$] (**7b**) starting from [NEt$_4$]$_2$[Cl$_3$FeOFeCl$_3$] (*49*).

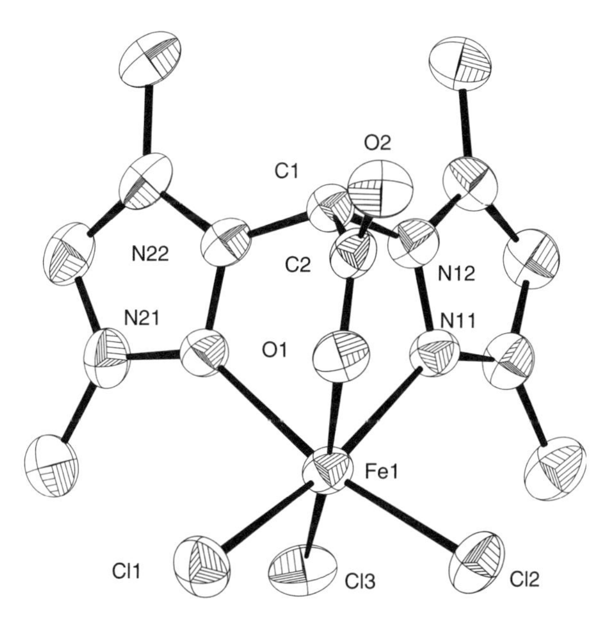

FIG. 9. Molecular structure of the anion $[Fe(bdmpza)Cl_3]^-$ of **7b**; NEt_4^+ cation omitted for better view; thermal ellipsoids are drawn at the 50% probability level (*49*).

The synthesis of similar complexes $[NEt_4][FeLCl_3]$ [L = $HB(pz)_3$ and $HB(3,5\text{-}Me_2pz)_3$] from $[NEt_4]_2[Cl_3FeOFeCl_3]$ and $K[HB(pz)_3]$ or K[HB $(3,5\text{-}Me_2pz)_3]$ was reported earlier (*94*). The Fe–Cl distances [2.326 (4), 2.334 (3) and 2.338 (3) Å] correspond well with those reported for $[NEt_4][Fe(HB(pz)_3)Cl_3]$ [2.335 (2), 2.325 (2) and 2.291 (2) Å] (*94*). The IR signals of **7b** were assigned by performing *ab initio* calculations on **7b** as described earlier. Again the symmetric carboxylate absorption $\nu_{sym}(CO_2^-)$ (1394 cm^{-1}) is found at rather low energy giving a difference $\Delta(\nu_{asym} - \nu_{sym})$ of ~260 cm^{-1}.

7a as well as **7b** show Curie law behavior with μ_{eff} of 5.85 μ_B (**7a**) and μ_{eff} of 5.68 μ_B (**7b**). These values are in good agreement with $t_{2g}^3e_g^2$ high-spin iron(III). No side reactions to ferric 2:1 complexes have been observed so far, as was reported by Kim *et al.* for $[NEt_4]$ $[Fe(HB(pz)_3)Cl_3]$ (*95*). The formation of $[NEt_4][Fe(bd\textit{t}bpza)Cl_3]$ in a reaction of the sterically hindered Hbd*t*bpzaH (**3c**) with $[NEt_4]_2$ $[Cl_3FeOFeCl_3]$ was not successful. In future experiments a reduction of these ferric complexes **7a** and **7b** to ferrous complexes should be possible, since the related $[NEt_4][Fe(Tp)Cl_3]$ were reduced by various methods (*94*). Thus the ferric complexes **7a** and **7b** might be useful precursors on the way to model complexes with natural substrates. In analogy to the synthesis of the complexes **7a** and **7b** the homochiral ligand $Hbpa^{4cam}$ (**8**) was reacted with $[NEt_4]_2[Cl_3FeOFeCl_3]$.

The synthesis of the enantiopure ligand $Hbpa^{4cam}$ (**8**) will be described later in this review (*48*). The molecular structure, derived

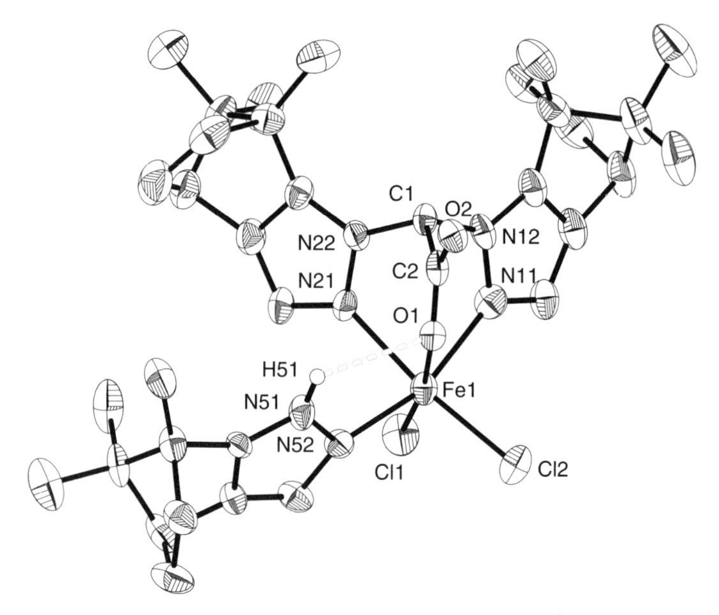

FIG. 10. Molecular structure of the complex [Fe(bpa^{4cam})(Cl)$_2$(pz^{cam}H)] (**9**); thermal ellipsoids are drawn at the 50% probability.

from crystals that have been obtained from an acetonitrile solution, does not show the expected product [NEt$_4$][Fe(bpa^{4cam})Cl$_3$], but a neutral complex [Fe(bpa^{4cam})(Cl)$_2$(pz^{cam}H)] (**9**) (Fig. 10, CCDC-232645). Obviously, one chlorido ligand was exchanged by a camphorpyrazole pz^{cam}H (**10**) with loss of [NEt$_4$]Cl.

Almost all distances between iron and the ligand are slightly shorter compared to those in the above described complex [NEt$_4$][Fe(bdmpza) Cl$_3$] (**7b**), due to the missing methyl substituents at position three of the pyrazolyl donors. Only the distance Fe–O is with 2.088 (7) Å bridge between the proton at N52 of the additional camphorpyrazolyl ligand and the carboxylate donor (d(O1-N51): 2.851 (9) Å). The additional camphorpyrazol pz^{cam}H (**10**) probably derives from traces in the ligand Hbpa^{4cam} (**8**) or from a slight degradation of the ligand Hbpa^{4cam} (**8**). Nevertheless, this result clearly shows, that substitution reactions are possible with complexes of the type [NEt$_4$][Fe(L)Cl$_3$] (L = bpza, bdmpza, bpa^{4cam}) and that this is a suitable pathway to new structural model complexes.

III. Structural Zinc Models

A. THE '2-HIS-1-CARBOXYLATE MOTIF' IN ZINCINS AND CARBOXYPEPTIDASES

In many zinc enzymes, the metal center is bound by two histidines and one aspartic or glutamic acid, which is rather similar to many mononuclear iron oxygenases. The 2-His-1-carboxylate facial triad is thus also a recurring motif for the zinc enzymes due to the broad spectrum of coordination numbers and geometries of the active sites of zinc enzymes. Some zinc peptidases are almost identical in their geometry to the iron oxygenases discussed earlier. As an example, cutouts of the active sites from Phenylalanine hydroxylase (PheOH) and from Thermolysin, a zinc peptidase out of the gluzincin class, are depicted (Fig. 11). It has been suggested that this similarity might be caused by a common purpose of their ancestor enzymes in the dawn of life (*96*). Before stromatolites during the Proterozoic Era evolved photosynthesis, that set free tremendous amounts of oxygen, iron(II) in enzymes could have been important as Lewis acid for the polarization of water, as zinc(II) is now in zinc peptidases. There is about 1000 times more iron (6.2%) in the earth's crust than zinc (76 ppm). Thus, it is quite plausible that billions of years ago iron(II) might have had the same function in metallopeptidases as zinc(II) does today.

Then later in an aerobic world with mainly iron(III) both, zinc peptidases and iron oxygenases, evolved from these ancestors. Today nature mainly uses zinc centers in metalloenzymes for the hydrolyses of peptide bonds (*99,100*). In the commonly accepted mechanism for zinc peptidases, zinc(II) has two tasks: It polarizes the carbonyl functionality of the peptide bond that is going to be cleaved and it supports the deprotonation of the coordinated water nucleophile.

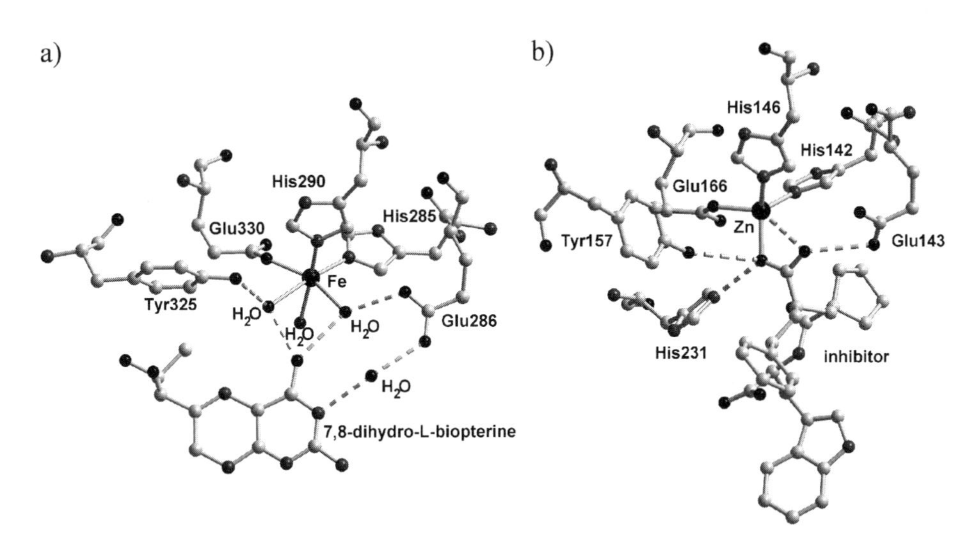

FIG. 11. (a) Active site of phenylalanine hydroxylase PheOH (PDB-Code: 1DMW) (*97*) and (b) active site Zentrum of thermolysin (THL) (PDB-Code: 1THL) (*98*).

B. Classification

One straightforward classification of zinc peptidases is based on the metal binding or catalytic relevant amino acids. Thus, many zinc peptidases exhibit a conserved sequence segment HExxH, with two zinc-binding histidines and a glutamic acid, which plays an essential role for the catalytic activity (*99,100*). This glutamic acid accepts one proton from a zinc(II)-bound water molecule and transfers it to the NH group of the amide functionality that is going to be cleaved. Zinc peptidases with such a HExxH sequence feature are called zincins. The zincins are distinguished further into the three subdivisions: metzincins, gluzincins or aspzincins (Fig. 12) depending on the third zinc-binding amino acid. Depending on the subdivision this third residue is a histidine (metzincins), a glutamic acid (gluzincins) or an aspartic acid (aspzincins). Of these the metzincins and gluzincins are the major ones. Beside the different groups of zincins there are also zinc peptidases without a HExxH motif. The most important group consists of the carboxypeptidases. They contain a HxxE motif, with a histidine and a glutamic acid binding zinc(II).

C. Gluzincins

We will now discuss in more detail the gluzincins according to the structures of the important examplary members such as the angiotensin converting enzyme (ACE) and the lethal factor of anthrax (LF). Many enzymes relevant for therapeutic or pathogen reasons are gluzincins, for instance, the ACE. ACE inhibitors have been used for almost 30 years as drugs for the treatment of hypertension. ACE cleaves the C-terminal dipeptide His-Leu from the decapeptide angiotensin I. The resulting octapeptide angiotensin II causes an increase in the blood pressure. Moreover, ACE catalyzes the

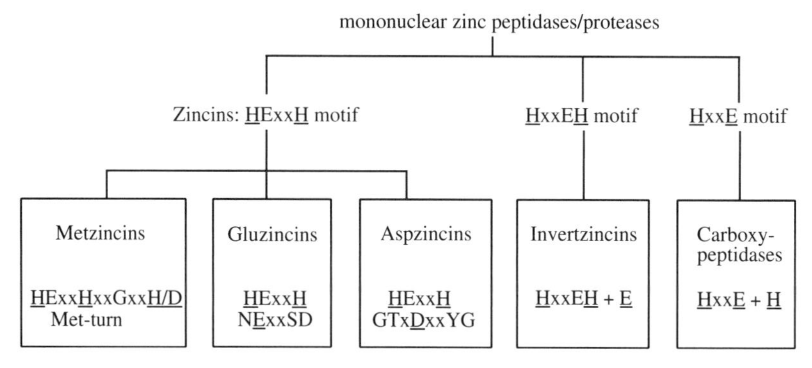

Fig. 12. Classification of mononuclear zinc peptidases (*100*).

degradation of the nonapeptide bradykinin, a vasodilator substance, thus affecting a further rise of the blood pressure.

Captopril (Capoten™), Enalapril and Lisinopril (Fig. 13) are examples of ACE inhibitors that are rather successful as pharmaceutical drugs against hypertension. In most cases these inhibitors are bound via zinc binding groups (ZBGs) such as thiolates, carboxylates or hydroxamates to the zinc(II) center of the active site.

Very recently the protein structures of ACE with the bound inhibitors Lisinopril (Fig. 4) and Captopril were published (*101,102*). Also the protein structure of the LF from *Bacillus anthracis* (PDB-Code: 1J7N) caused a sensation, which is now available to the public (Fig. 14b) (*103*). LF is part of the toxic exotoxin complex composed of three distinct proteins (protective antigen PA, the lethal factor LF and the edema factor EF), and is thought to be the most toxic

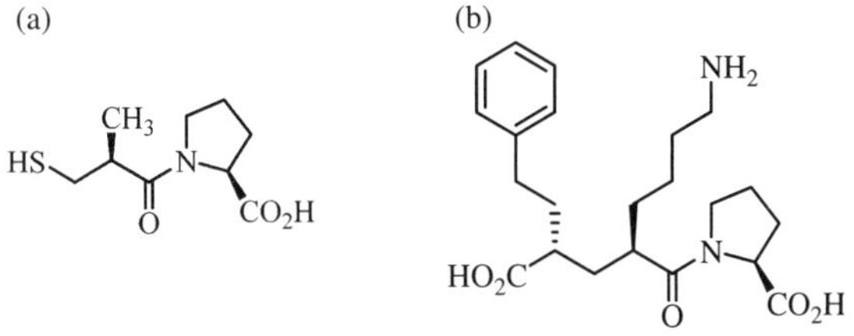

FIG. 13. The ACE inhibitors (a) Captopril and (b) Lisinopril.

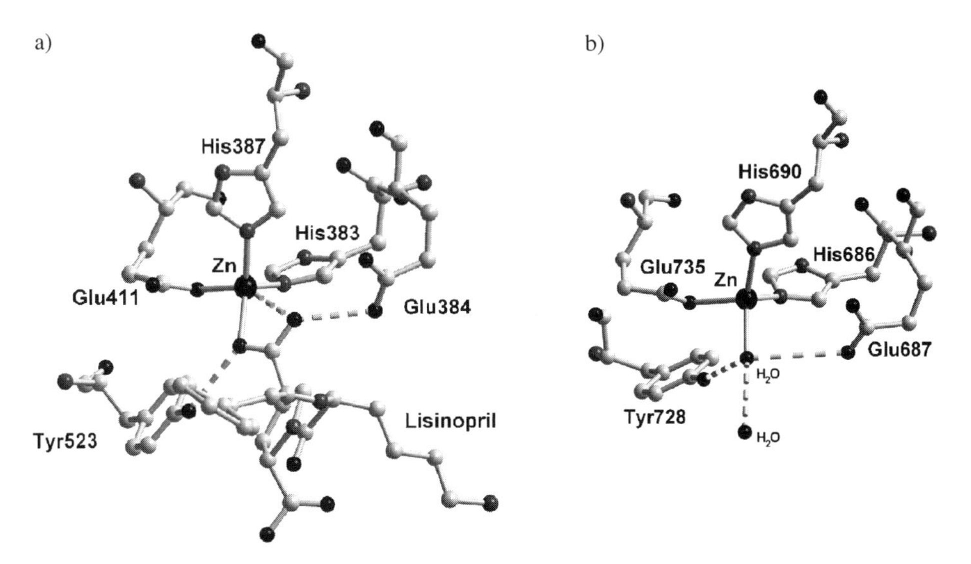

FIG. 14. Active site (a) of the angiotensin converting enzyme (ACE) with Lisinopril inhibitor bound to zinc(II) (PDB-Code: 1O86) (*101*), (b) of the lethal factor of *Bacillus anthracis* (PDB-Code: 1J7N) (*103*).

component of these three proteins. LF is a highly specific metalloprotease that cleaves members of the mitogen-activated protein kinase kinase (MAPKK) family especially in macrophage cell lines. A comparison of the ACE structure with that of LF (Fig. 14) shows a rather striking similarity between both metalloproteases. During the 'anthrax letter' assaults in 2001, G. Weissmann suggested the use of ACE inhibitors or LTA4-hydrolase inhibitors such as captopril or bestatin to assist in the ciprofloxacin (Cipro) treatment of anthrax victims (*104*).

D. Zinc Model Complexes Bearing Bis(pyrazol-1-yl)acetato Ligands

Model complexes for zincins have been investigated for many years by various groups. Especially the extensive studies by H. Vahrenkamp, G. Parkin, C. J. Carrano and recently S. Cohen have to be mentioned in this respect. A recent and very comprehensive review by Parkin covers almost all publications in this field (*105*). Most of these studies focused either on metzincin models or did not model the 2-His-1-carboxylate motif with a proper carboxylate based *N,N,O*-ligand. Therefore, there are only few examples of *N,N,O*-tripod ligands that mimic this motif. Carrano and coworkers used the scorpionate ligand (3-*tert*-butyl-2-hydroxy-5-methylphenyl)-bis(3,5-dimethylpyrazol-1-yl)methane to generate model complexes of zinc enzymes (*106*). Instead of a carboxylate *O*-donor, this ligand contained a phenolate *O*-donor. By insertion of formaldehyde or carbon dioxide into a B–H bond of zinc bis(pyrazol-1-yl)hydroborate complexes Parkin *et al.* accessed also *N,N,O* model complexes such as $[Zn\{\kappa^3\text{-}(HCO_2)Bp^{t\text{-}Bu,iPr}\}Cl]$ (*107*).

Thus, we decided to investigate the coordination properties of bis(pyrazol-1-yl)acetic acids towards the biologically relevant zinc(II). As expected from the formation of $[Fe(bdmpza)_2]$ (**4b**) we described earlier, a reaction of the sterically less hindered Hbdmpza ligand (**3b**) with base and $ZnCl_2$ afforded a zinc complex $[Zn(bdmpza)_2]$ (**11**) with two heteroscorpionate ligands bound to zinc(II). Again the target compound [(bdmpza)ZnCl] could not be detected via MS or NMR. $[Zn(bdmpza)_2]$ (**11**) is isostructural to $[Fe(bdmpza)_2]$ (**4b**) (Fig. 5a) and thus the IR spectra are almost identical.

For the four spectroscopically equivalent pyrazolyl donors only one set of signals is detected in the ^1H and ^{13}C NMR spectra of $[Zn(bdmpza)_2]$ (**11**).

By using the new ligand Hbd*t*bpza (**3c**) we tried to prevent bisligand 2:1 complex formation. Deprotonation of the sterically more hindered Hbd*t*bpza (**3c**) and subsequent reaction with $ZnCl_2$ resulted in a 1:1 complex [Zn(bdtbpza)Cl] (**12**) which displayed a $\nu_{as}(CO_2^-)$ IR absorption at 1680 cm^{-1} (Scheme 11).

The structure of the complex [Zn(bdtbpza)Cl] (**12**) is confirmed by X-ray crystallography (Fig. 15). Compared to the structures of the

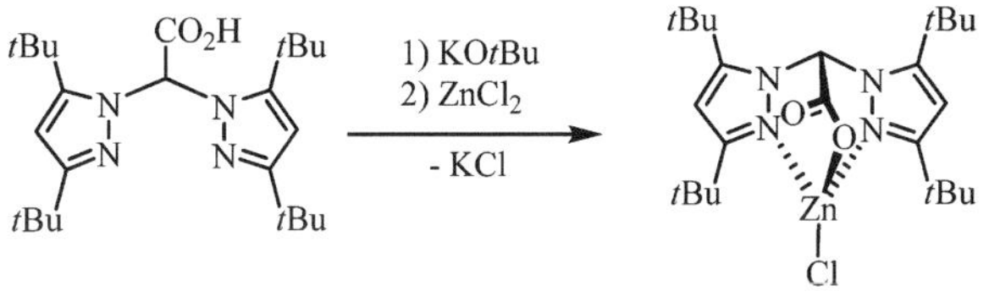

1) KO*t*Bu
2) $ZnCl_2$
- KCl

3c **12**

SCHEME 11. Synthesis of the zinc chlorido complex [Zn(bd*t*bpza)Cl] (**12**) (*41*).

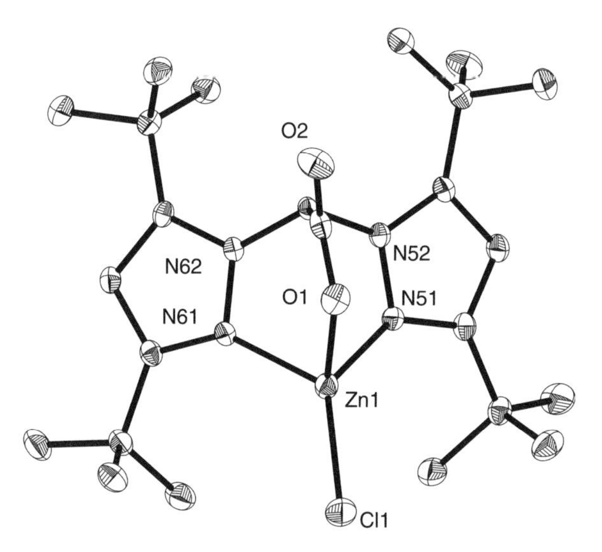

FIG. 15. Molecular structure of [Zn(bd*t*bpza)Cl] (**12**); thermal ellipsoids are drawn at the 50% probability level (*41*).

bisligand complex **11** the angle O–Zn–N in **12** is widened by 8–9°. The distance Zn–Cl and the angles O–Zn–N and N–Zn–N are in good agreement with those of the structure of [Zn{κ^3-(HCO_2)$Bp^{t\text{-Bu},i\text{Pr}}$}Cl] previously published by Parkin (*107*).

The tetrahedral complex geometry is distorted. The zinc atom sticks out of the N(51)–N(61)–Cl(1) plane by only 0.45 Å. The Zn–O distance [1.990 (2) Å] is significantly shorter compared to [Zn{κ^3-(HCO_2)$Bp^{t\text{-Bu},i\text{Pr}}$}Cl] [2.065 (4) Å], due to a real carboxylato donor instead of a –O–CH=O donor.

The goal of the experiments we report was to create new structural model complexes for gluzincins or carboxypeptidases. With [Zn(bdtbpza)Cl] (**12**) for the first time a tetrahedral zinc complex with a monoanionic *N,N,O*-tridentate using a carboxylate *O*-donor was synthesized (*41*). A comparison of the molecular structure of **12** with the coordination environment of the enzymes indicates its significance as a structural model for these active sites. To date [Zn(bdtbpza)Cl] (**12**) might be among the best structural analogues for the gluzincin enzymes such as ACE, LF or thermolysin.

In future biomimetic studies the quest for new possible ZBGs suitable for peptidase inhibitors is a challenging project. Binding studies with zinc model complexes should be rather helpful in finding such new ZBGs, since the coordination of the test compounds is not affected by a binding into the enzyme pocket. Thus, the pure coordination properties of the ZBGs might be deduced from these studies. Recent work by S. Cohen and coworkers focuses on the exploration of new ZBGs for matrix metalloproteinases (MMPs) by using zinc Tp complexes (*108–110*). Similar studies should now be possible for the gluzincins by using zinc model complexes bearing the bd*t*bpza ligand. As mentioned earlier thiolato, carboxylato and hydroxamato functionalities are typical ZBGs for ACE inhibitors. Thus we tested the reactivity of [Zn(bd*t*bpza)Cl] (**12**) towards exchange of the chlorido ligand by a thiolato ligand. [Zn(bd*t*bpza)Cl] (**12**) was treated with benzyl mercaptan and base to yield the benzylthiolato complex [Zn(bd*t*bpza)SCH_2Ph] (**13**) (Scheme 12).

The reaction is quantitative but the workup proved to be difficult due to an excess of thiol. The spectroscopic properties of the thiolato complex **13** are almost identical to the chlorido complex **12** apart from additional ^{1}H- and ^{13}C-NMR signals of the benzylthiolato ligand. Particularly characteristic for this thiolato ligand is a single resonance at $\delta(^1H) = 3.93$ which is assigned to the benzyl CH_2 group. The $\nu_{as}(CO_2^-)$ absorption appears at 1677 cm^{-1}. Over the years monomethyl zinc complexes, such as Tp zinc methyl complexes, have become very common precursors in the synthesis of zinc model complexes. Usually they react with protic ligands (e.g., thiols, carboxylic acids etc.) to release CH_4 in a very pure reaction. Former work by Vahrenkamp *et al.* (*111*) and Parkin *et al.* (*112*) have already shown that four-coordinated hydrotris(pyrazol-1-yl)borato zinc alkyl complexes can be obtained by metathesis of ZnR_2 (R=CH_3, CH_2CH_3) with potassium or thallium salts of sterically demanding hydrotris(pyrazol-1-yl)borate ligands. Later Carrano and coworkers obtained monomethyl zinc

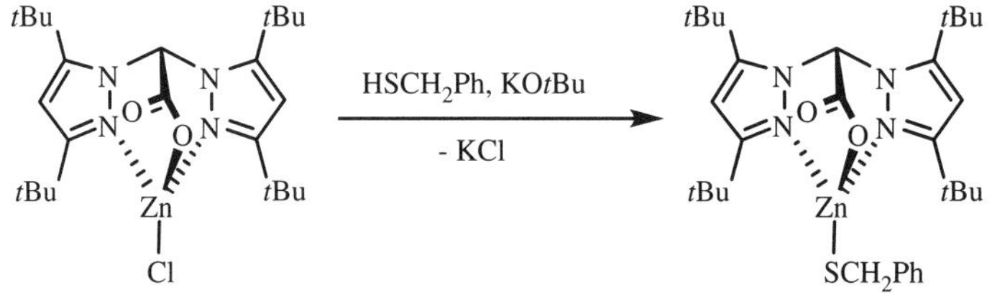

SCHEME 12. Synthesis of the thiolato complex [Zn(bd*t*bpza)(SCH_2Ph)] (**13**) (*41*).

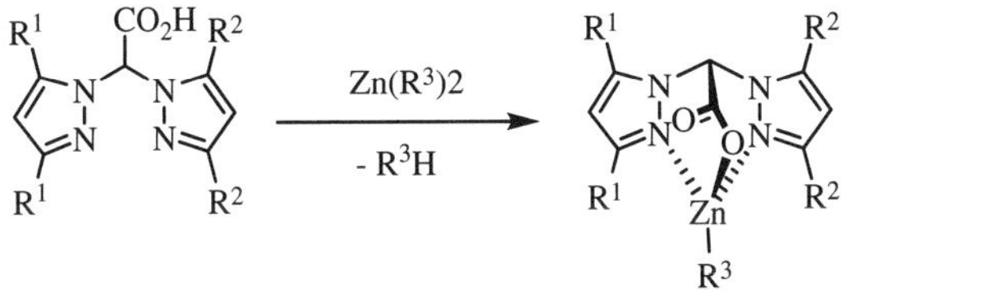

$Zn(R^3)2$

$- R^3H$

3c ($R^1 = R^2 = t$Bu)
3d ($R^1 =$ Me, $R^2 = t$Bu)

14a ($R^1 = R^2 = t$Bu, $R^3 =$ Me)
14b ($R^1 = R^2 = t$Bu, $R^3 =$ Et)
15 ($R^1 =$ Me, $R^2 = t$Bu, $R^3 =$ Me)

SCHEME 13. Synthesis of alkyl zinc complexes with bis(pyrazol-1-yl) acetato ligands (*47*).

complexes bearing the (3-*tert*-butyl-2-hydroxy-5-methylphenyl)-bis(3,5-dimethylpyrazol-1-yl)methane scorpionate ligand just by reacting it with $Zn(CH_3)_2$ (*106*). We found that with sterically demanding bis(pyrazol-1-yl)acetic acids monomeric four-coordinated bis(pyrazol-1-yl)acetato zinc alkyl complexes are obtained in a similar way. Reaction of bis(3,5-di-*tert*-butylpyrazol-1-yl)acetic acid (**3c**) with dialkyl zinc ZnR_2 (R$=$$CH_3$, CH_2CH_3) generates zinc alkyl complexes [Zn (bd*t*bpza)(CH_3)] (**14a**) and [Zn(bd*t*bpza)(CH_2CH_3)] (**14b**) (Scheme 13) by elimination of alkanes (*47*).

Since bis(pyrazol-1-yl)acetic acid and not an alkali carboxylate ligand salt was used for these syntheses, the workup is simple and zinc alkyl complexes are obtained in high yield and purity. [Zn(bd*t*bpza) (CH_3)] (**14a**) is characterized by resonances at $\delta = -0.35$ ppm (^{1}H NMR) and $\delta = -10.1$ ppm (^{13}C NMR), which are assigned to the Zn–Me group. For the Zn–Et group of [Zn(bd*t*bpza)(CH_2CH_3)] (**14b**) signals are observed in the ^{1}H NMR spectra at $\delta = 0.52$ ppm [q, ^3J(H,H) = 8.0 Hz, CH_2] and 1.29 ppm [t, ^3J(H,H) = 8.0 Hz, CH_3]. In the ^{13}C NMR spectrum the ethyl group was detected at $\delta = 3.3$ ppm (CH_2) and 11.9 ppm (CH_3). The remaining resonances have been assigned to the C_s symmetric bis(3,5-di-*tert*-butylpyrazol-1-yl)acetato zinc fragment.

One purpose of our work is to mimic the chiral environment of the enzymes. Therefore, we thought it a reasonable goal to supply chiral models for the active sites of metalloenzymes. This was achieved before by Alsfasser *et al.* (*113*) or Vahrenkamp *et al.* (*114*) via amino acids that have been incorporated into the ligand systems. Modification of Tp ligands by chiral pyrazoles derived from the chiral pool is another way to chiral *N,N,N* tripod ligands and has been achieved before by W. B. Tolman and coworkers (*115*).[1] Thus, first we focused on the synthesis of a racemic mixture of a chiral *N,N,O* scorpionate

[1]Due to otherwise extraordinary long abbreviations for compound **3d**, its precursor **2d**, and the ligands **8** and **25** the abbreviation system for Tp ligands introduced by S. Trofimenko was transferred to this problem. This resulted in the abbreviations $bpm^{t\text{-}Bu_2,Me_2}$ (**2d**), $Hbpa^{t\text{-}Bu_2,Me_2}$ (**3d**), $Hbpa^{4cam}$ (**8**) and $Hbpa^{4menth}$ (**25**).

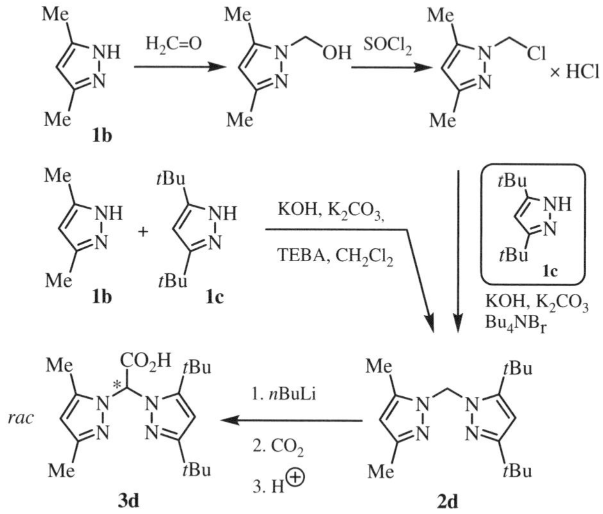

SCHEME 14. Synthesis of racemic (3,5-di-*tert*-butylpyrazol-1-yl)(3′,5′-dimethylpyrazol-1-yl)acetic acid (**3d**) (*47*).

ligand. The unsymmetrical 3,5-di-*tert*-butyl-1-[(3,5-dimethyl-1*H*-pyrazol-1-yl)methyl]-1*H*-pyrazole (bpm$^{t\text{-Bu}_2,\text{Me}_2}$) (**2d**) served as starting material (*116*). This precursor is either obtained by a phase-transfer catalyzed reaction of 1-chloromethyl-3,5-dimethylpyrazole hydrochloride with 3,5-di-*tert*-butylpyrazole (**1c**) or in a one-pot, phase-transfer catalyzed reaction of 3,5-dimethylpyrazole (**1b**) with 3,5-di-*tert*-butylpyrazole (**1c**), dichloromethane and base (Scheme 14).

Deprotonation of a methylene group in **2d** followed by reaction with carbon dioxide and acidic workup yielded a racemic mixture of (3,5-di-*tert*-butylpyrazol-1-yl)(3′,5′-dimethylpyrazol-1-yl)acetic acid (Hbpa$^{t\text{-Bu}_2,\text{Me}_2}$) (**3d**) (Scheme 14). Reaction of **3d** with base and anhydrous $ZnCl_2$ yielded [Zn(bpa$^{t\text{-Bu}_2,\text{Me}_2}$)Cl] that crystallized as a cross-linked dimer [(bpa$^{t\text{-Bu}_2,\text{Me}_2}$)ZnCl]$_2$ (**16**) (Scheme 15, Fig. 16).

The reaction of (3,5-di-*tert*-butylpyrazol-1-yl)(3′,5′-dimethylpyrazol-1-yl)acetic acid Hbpa$^{t\text{-Bu}_2,\text{Me}_2}$ (**3d**) with dimethyl zinc resulted in the chiral zinc methyl complex [Zn(bpa$^{t\text{-Bu}_2,\text{Me}_2}$)($CH_3$)] (**15**) (Fig. 17) as racemic mixture (*47*).

The structure of the methyl complex **15** represents one of the rare examples of monomeric four-coordinated zinc alkyl complexes and is the first one with a carboxylate *O*-donor ligand. The resonances assigned to the Zn–Me group are observed at $\delta = -0.42$ ppm (^{1}H NMR) and $\delta = -14.1$ ppm (^{13}C NMR). The Zn–CH_3 bond length in complex **15** [1.962 (8) Å] is almost identical to that in [ZnTp$^{\text{Me}_2}$(CH_3)] [1.981 (8) Å] (*112*). The molecular structure clearly shows that the methyl group is trying to avoid the bulky *tert*-butyl group. This observation is supported by two different angles C(3)–Zn–N(11) [119.8(3)°] and

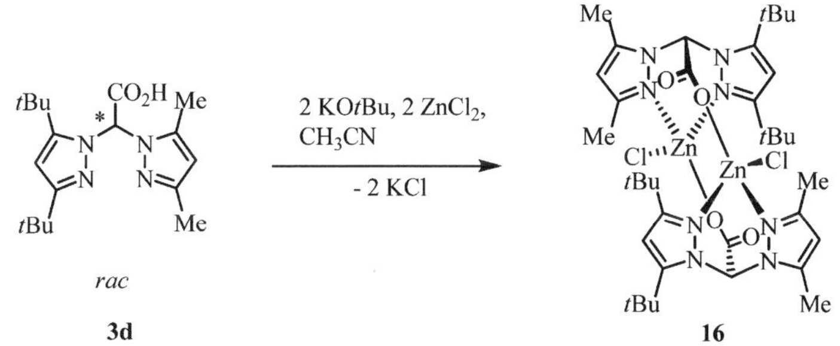

SCHEME 15. Synthesis of the chlorido complex $[Zn(bpa^{t\text{-}Bu_2,Me_2})Cl]_2$ (**16**) (*47*).

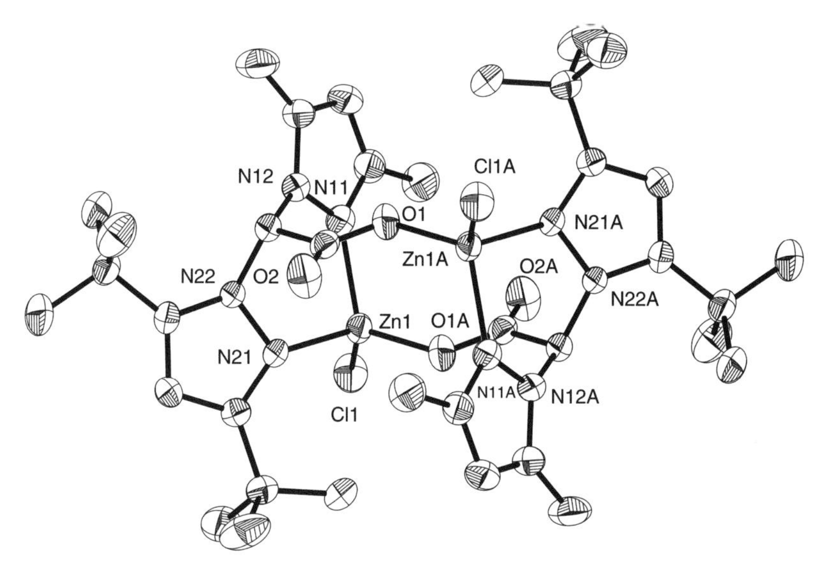

FIG. 16. Molecular structure of $[Zn(bpa^{t\text{-}Bu_2,Me_2})Cl]_2$ (**16**); thermal ellipsoids are drawn at the 50% probability level (*47*).

C(3)–Zn–N(21) [133.1(3)°] in the molecular structure of **15** and implies a possible stereo-selective induction for future alkyl transfer reactions.

The reactivity of the methyl complexes **14a** and **15** with carboxylic acid was examined to provide carboxylato complexes as useful precursors for enzyme models. For example, a slight excess of acetic acid does cleave the Zn–CH_3 bond by elimination of methane and yields [Zn(bd*t*bpza)O_2CCH_3] (**17**) (Scheme 16) and $[Zn(bpa^{t\text{-}Bu_2,Me_2})O_2CCH_3]$ (**18**).

Thus, the reactivity of the methyl complexes **14a** and **15** is rather similar to $[ZnTp^{Ph}(CH_3)]$ and $[ZnTp^{t\text{-}Bu}(CH_3)]$ (*111,112*). Most of the NMR signals of the acetato complex **17** are broad compared to those of [Zn(bd*t*bpza)Cl]. These broad resonances suggest a κ^1/κ^2 equilibrium of the carboxylato ligand. Two additional IR signals at 1633 cm^{-1} and 1602 cm^{-1} are assigned to $\nu_{as}(CO_2^-)$ of the η^1 and η^2 bound acetato

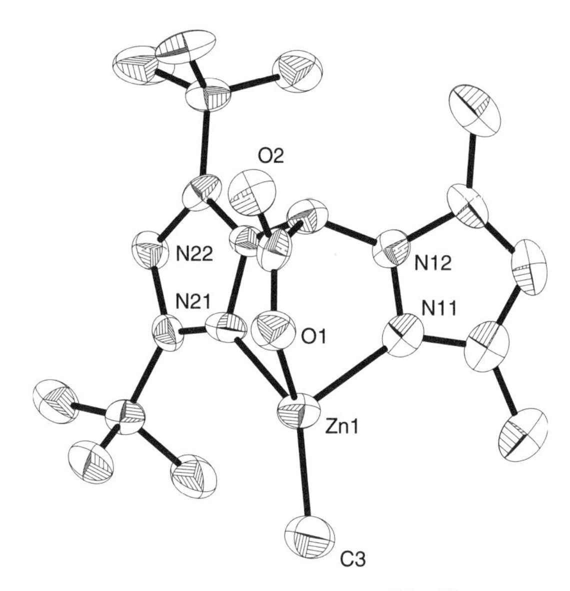

FIG. 17. Molecular structure of [Zn(bpa$^{t\text{-}Bu_2,Me_2}$)(CH_3)] (**15**); thermal ellipsoids are drawn at the 50% probability level (*47*).

ligands. The IR signals at 1681 cm^{-1} and 1467 cm^{-1} are assigned to $\nu_{as}(CO_2^-)$ and $\nu_{as}(CO_2^-)$ of the bd*t*bpza ligand and prove the monomeric structure of **17** since [Zn(bd*t*bpza)Cl] exhibits a similar $\nu_{as}(CO_2^-)$ at 1681 cm^{-1} (*41*). The difference $\Delta\nu_{as-s}$ indicates the binding mode of the carboxylate ligand. Values of $\Delta\nu_{as-s} \geq 200\ cm^{-1}$ are typical for a unidentate binding mode of acetates, which is in good agreement with the $\Delta\nu_{as-s}$ values of the bis(pyrazol-1-yl)acetato ligands in the complexes **17** and **18** (*91*).

The NMR spectra of carboxylato complexes [Zn(bpa$^{t\text{-}Bu_2,Me_2}$)O_2CCH_3] (**18**) are similar to those of **17** although two sets of signals for the two different pyrazolyl groups are visible. Most of the NMR signals of **18** are even broader than those of **17** again indicating a κ^1/κ^2 equilibrium of the carboxylato ligands. Various thiolato and hydroxamato complexes can also be prepared by reactions of [Zn(bpa$^{t\text{-}Bu_2,Me_2}$)(CH_3)] (**14a**) and [Zn(bpa$^{t\text{-}Bu_2,Me_2}$)(CH_3)] (**15**) with thiols or hydroxamic acids (Scheme 16). These hydroxamato and thiolato complexes **19**, **20a** and **20b** represent good structural models for the binding of typical zinc protease inhibitors to the zinc center, since the thiol, carboxylic acid and hydroxamic acid functionalities are the common ZBGs in such inhibitors. Often these complexes exhibit a κ^1/κ^2 equilibrium similar to those of the carboxylato complexes. Nevertheless, these complexes might be suitable to test the binding properties of new ZBGs relevant to future gluzincin inhibitors. Recently C. J. Carrano and coworkers reported on such zinc models bearing ZBGs based on the

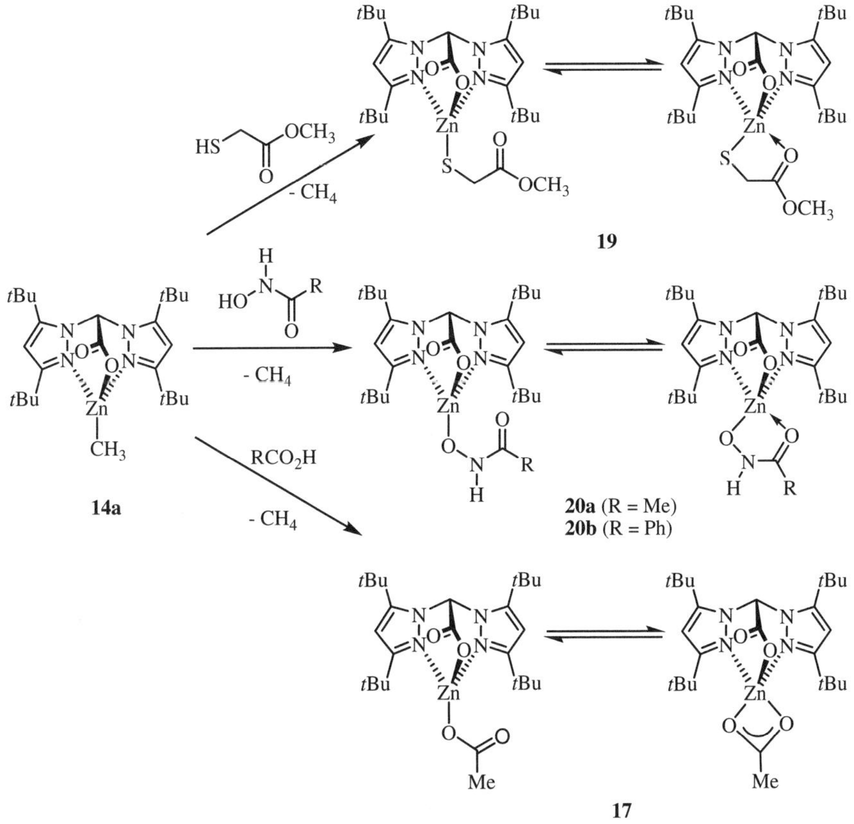

SCHEME 16. Synthesis of zinc carboxylato, thiolato and hydroxamato complexes.

bis(3-*tert*-butyl-5-methylpyrazol-1-yl)acetic acid and its zinc methyl complex (*50,51,59,61,69*).

IV. Homochiral Bis(pyrazol-1-yl)acetato Ligands

Modification of Tp ligands by chiral pyrazoles derived from the chiral pool is another way to chiral *N,N,N* tripod ligands and has been achieved before by Tolman and coworkers (*115*).

After investigating the chiral but racemic (3,5-di-*tert*-butylpyrazol-1-yl)(3′,5′-dimethylpyrazol-1-yl)acetic acid ligand (Hbpa$^{t\text{-}Bu_2,Me_2}$) (**3d**) we then focused on the design of enantiopure ligands. For asymmetric induction two problems have to be solved for such homochiral ligands – (a) enantiomeric purity and (b) stable configuration of the stereogenic centers. Separation of the two enantiomers from a racemic mixture often is an elaborate, low-yielding process and affords a validation of the enantiomeric purity. The problem is converted into a separation task of diastereomers, if one of the donor groups is chiral and enantiopure and/or derived from the chiral pool. This problem simplifies a lot, when C_3 symmetric ligands are constructed from three identical, chiral donor groups (*117*). Chiral pool derived ligands based on this concept such as $[OP(pz^{cam})_3]$ or $[BH(pz^{menth})_3]^-$ are well known examples (*118*). But due to the extreme steric hindrance by the three chiral donor groups the application of these ligands is limited especially in octahedral complexes. In our quest for new chiral ligands, we therefore developed a rather simple concept based on prochirality, that allows us to synthesize enantiopure facially binding tripod ligands from C_2 symmetric precursors without any additional separation of enantiomers or diastereomers (*48,55*).

C_2 symmetry is a common feature in chiral bidentate ligands (*119*). If two identical enantiopure donor groups Y* are connected by a tetrahedral bridging atom, any additional group Z at this atom will cause no additional stereo center but a prochiral center (Fig. 18a) (*120*). (*S*,*S*)- and (*R*,*R*)-Trihydroxyglutaric acid are textbook examples for such compounds (*120*). Inversion of configuration at C3 in (*S*,*S*)-trihydroxyglutaric acid (Fig. 18b) yields the identical compound.

A homochiral tripod ligand thus can be obtained from a C_2 symmetric precursor without additional separation of enantiomers or diastereomers. As an example for this concept we chose bis(camphorpyrazol-1-yl)methane as starting material to afford an enantiopure but C_2 symmetric bidentate ligand which is obtained in three steps from (+)-camphor (Scheme 17).

As reported by Steel *et al.* three structural isomers of bis(camphorpyrazol-1-yl)methane (**21a**, **21b** and **21c**) are formed by coupling of camphorpyrazole **10** [i.e., (4*S*,7*R*)-7,8,8-trimethyl-4,5,6,7-tetrahydro-4,7-methano-1(2)H-indazole] with CH_2Cl_2 (*121*). Isomer **21c** can be separated from the other two structural isomers by crystallization or column chomatography. Deprotonation at the bridging carbon atom, subsequent reaction with carbon dioxide and acidic workup yields the enantiopure bis(camphorpyrazol-1-yl)acetic acid $Hbpa^{4cam}$ (**8**) (Scheme 17, Fig. 19) (*116*). Due to missing substituents at the pyrazolyl carbon C5 and a hence likely *ortho* metallation, isomers **21a** and **21b** are not suited for his reaction (*72*).

As mentioned earlier, the ligand $Hbpa^{4cam}$ (**8**) reacts with $[NEt_4]_2[Cl_3FeOFeCl_3]$ to form a homochiral ferric complex $[Fe(bpa^{4cam})(Cl)_2$

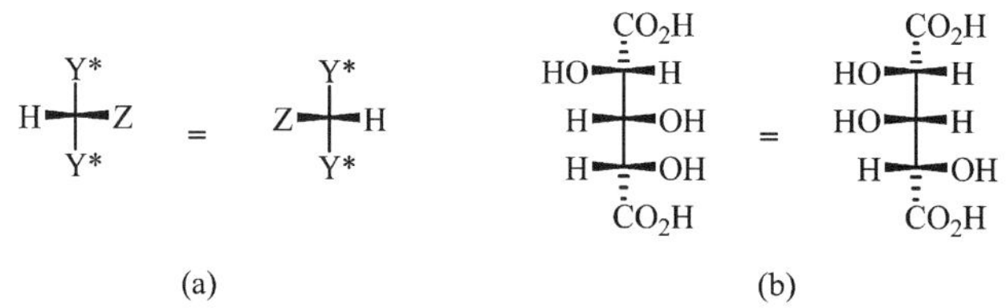

(a) (b)

FIG. 18. Prochiral center of (*S*,*S*)-trihydroxyglutaric acid (*120*).

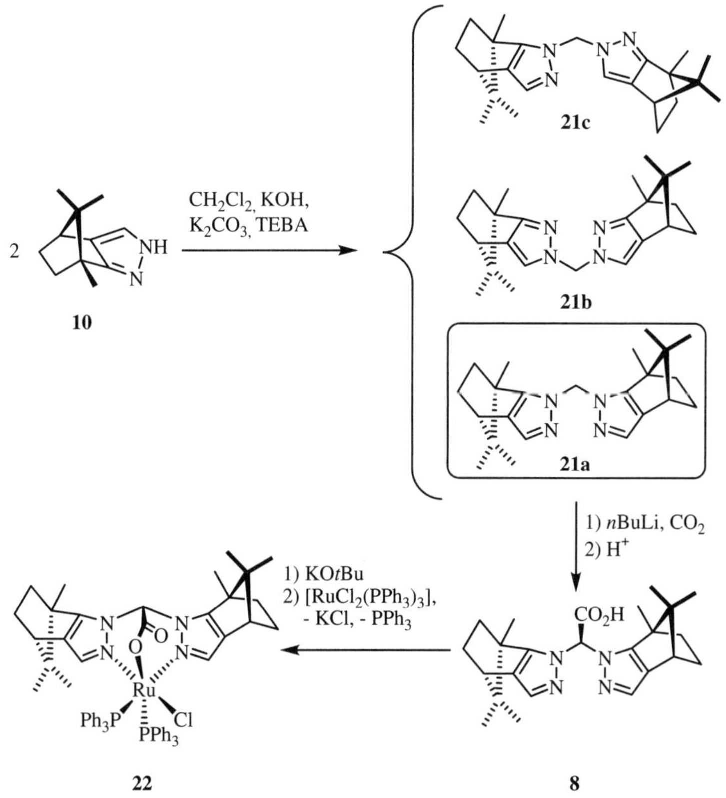

SCHEME 17. Synthesis and coordination of bis(camphorpyrazol-1-yl) acetic acid (*48*).

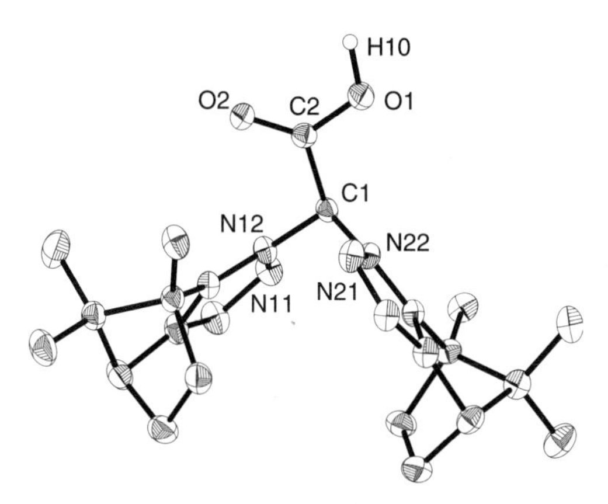

FIG. 19. Molecular structure of Hbpa^{4cam} (**8**); thermal ellipsoids at 50% probability level (*48*).

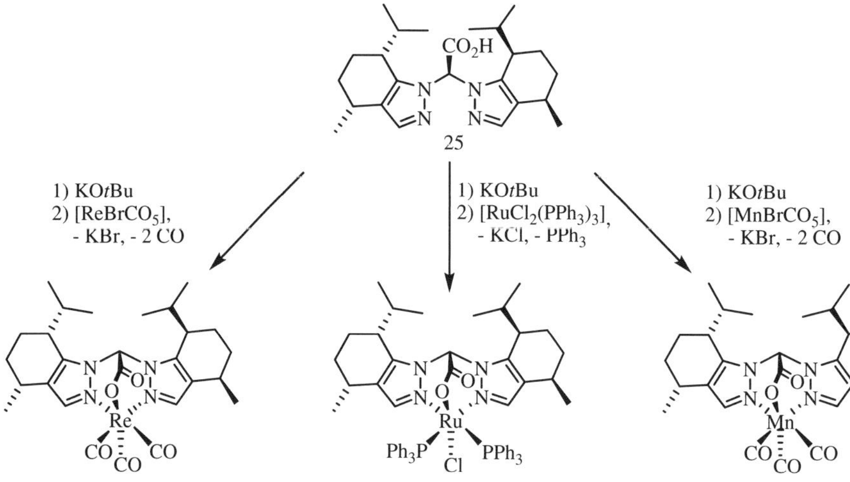

SCHEME 18. Synthesis of $[Ru(bpa^{4menth})Cl(PPh_3)_2]$ (**26**), $[Mn(bpa^{4menth})(CO)_3]$ (**27**) and $[Re(bpa^{4menth})(CO)_3]$ (**28**) (*55*).

$(pz^{cam}H)]$ (**9**) (Fig. 10). A reaction of the potassium carboxylate $K[bpa^{4cam}]$ with $[RuCl_2(PPh_3)_3]$ yields $[Ru(bpa^{4cam})Cl(PPh_3)_2]$ (**22**) (Scheme 17). The formation of the ruthenium complexes $[Ru(bpza)Cl(PPh_3)_2]$ (**23**) and $[Ru(bdmpza)Cl(PPh_3)_2]$ (**24**) takes place in a similar reaction, as reported before (*44*).

According to cross coupling observations in the COSY spectrum a geometry with one of the PPh_3 ligands *trans* to the carboxylate group and the other *trans* to the camphorpyrazolyl group is deduced for complex **22**. Although the camphorpyrazole donors are far from the ruthenium center, this chiral information is passed onto the PPh_3 ligands and hence amplified. This is not the case for an analogous bis(menthylpyrazol-1-yl)acetic acid ligand $Hbpa^{4menth}$ (**25**) (Scheme 18) (*55*). Here the reaction of $K[bpa^{4menth}]$ with $[RuCl_2(PPh_3)_3]$ results in a ruthenium(II) complex $[Ru(bpa^{4menth})Cl(PPh_3)_2]$ (**26**) (Scheme 10) in which both PPh_3 are placed *trans* to the pyrazole donors (*55*). A similar reaction with $[MnBr(CO)_5]$ and $[ReBr(CO)_5]$ gives the tricarbonyl complexes $[Mn(bpa^{4menth})(CO)_3]$ (**27**) and $[Re(bpa^{4menth})(CO)_3]$ (**28**) respectively (*55*).

V. Bis(pyrazol-1-yl)acetato Ligands in Coordination Chemistry and Organometallics

These examples show clearly that bis(pyrazol-1-yl)acetato ligands are well suited for various aspects of coordination chemistry and organometallics. This allows a comparison of these ligands with other common ligands. So far to our experience these ligands are as versatile to coordination chemistry, bioinorganic chemistry and organometallics as the well-established ligands Cp, Cp*, Tp and Tp^{Me_2} (Fig. 20). Some exemplary examples for this analogy will be presented below.

A. Carbonyl Complexes Bearing Bis(pyrazol-1-yl)acetato Ligands

The reaction with $[MnBr(CO)_5]$ and $[ReBr(CO)_5]$ is successful with all kinds of bis(pyrazol-1-yl)acetates provided that they are not too bulky (*40,55*). $[Mn(bdmpza)(CO)_3]$ (**29b**), $[Re(bpza)(CO)_3]$ (**30a**) and $[Re(bdmpza)(CO)_3]$ (**30b**) have been characterized by X-ray structure determination (Fig. 21) (*40*). The progress of the product formation is usually monitored by the three carbonyl IR signals (A′, A″ and A′) of the resulting piano stool type complexes. From a comparison of the IR-spectroscopic data and single-crystal X-ray analyses of **29b–30b** with the data of related cyclopentadienyl or Tp complexes one can deduce that for group VII metal carbonyls bpza and bdmpza are less electron donating than Tp, Tp^{Me_2}, Cp* and even Cp. In the case of $[Re(bdmpza)(CO)_3]$ (**30b**) a CO versus NO^+ exchange yielded

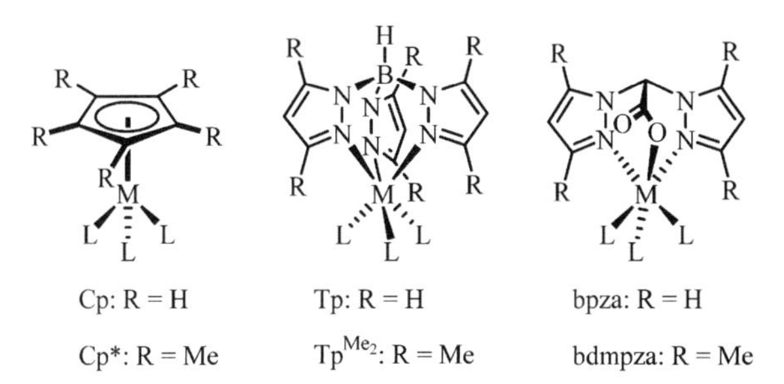

Fig. 20. Analogy between Cp, Cp*, Tp, Tp_2^{Me}, bpza and bdmpza.

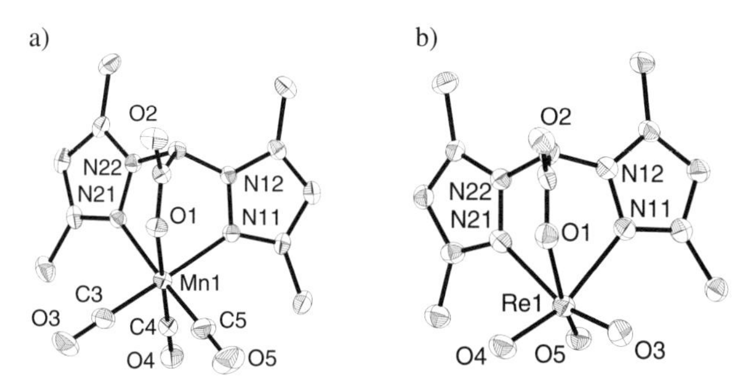

Fig. 21. Molecular structures (a) of $[Mn(bdmpza)(CO)_3]$ (**29b**) and (b) of $[Re(bdmpza)(O)_3]$ (**32b**); thermal ellipsoids are drawn at the 50% probability level (*40,43*).

$[Re(bdmpza)(CO)_2(NO)]BF_4$ (**31**), showing a reactivity similar to $[CpRe(CO)_3]$ (*40*).

B. Rhenium Trioxo Complexes Bearing Bis(pyrazol-1-yl)acetato Ligands

For many years rhenium tricarbonyl complexes $[LRe(CO)_3]$ have been used as starting materials in the synthesis of trioxorhenium complexes $[LReO_3]$ by oxidation of the carbonyl ligands (*122–124*). Other synthetic pathways start from Re_2O_7 e.g., for complexes bearing the Cp or Tp ligand (*125–131*). The pyrazolylborato complex $[TpReO_3]$ in particular stimulated our interest, raising the question whether the bis(pyrazol-1-yl)acetato ligand might also be a good tripod ligand for trioxorhenium complexes (*132,133*). The reaction of Re_2O_7 with anionic ligands often is low yielding due to the loss of one equivalent of ReO_4^-. This problem can be solved by using mixed anhydrides $[ReO_3(OC(O)R)]$ that are formed from Re_2O_7 and organic anhydrides $(RCO)_2O$ (*134*). Inspired by this, we investigated the reaction of bis(pyrazol-1-yl)acetic acid (**3a**) and bis(3,5-dimethylpyrazol-1-yl) acetic acid (**3b**) with perrhenic acid (Scheme 19).

We found that the formation of tripodal trioxorhenium bis(pyrazol-1-yl)acetato complexes $[Re(bpza)O_3]$ (**32a**) and $[Re(bdmpza)O_3]$ (**32b**) is favored enough to be generated by the reaction of perrhenic acid and bis(pyrazol-1-yl)acetic acids with loss of a water molecule (Fig. 21b) (*43*). Since perrhenic acid can be handled much more easily than Re_2O_7, this method provides a very simple access to trioxorhenium complexes of the form $[ReLO_3]$. The IR spectra (KBr pellet) exhibit several bands that have been assigned to ν(Re=O): 944 (w), 922 (s-br)cm^{-1} for **32a** and 945(m), 925 (s), 913(s) cm^{-1} for **32b** (*43*). These bands agree well with those in $[Re\{HB(pz)_3\}O_3]$: 944(m), 924(s), 911(s) and 894(m) cm^{-1} (*132*).

C. Ruthenium Cumulenylidene Complexes Bearing Bis(pyrazol-1-yl)acetato Ligands

Another example for the Tp versus bdmpza analogy is the synthesis of various bdmpza ruthenium cumulenylidene complexes. The

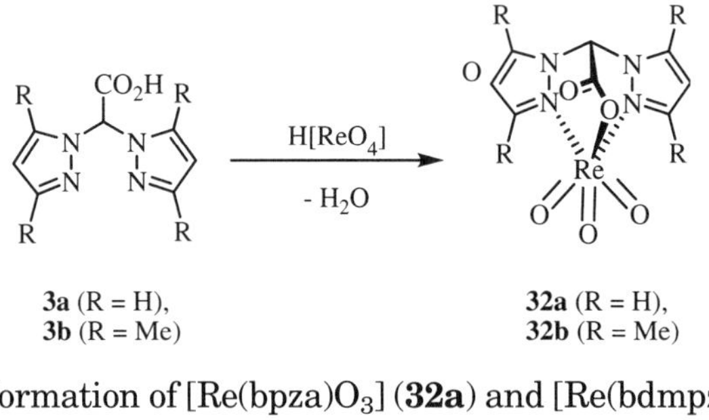

$H[ReO_4]$, $- H_2O$

3a (R = H), **3b** (R = Me) → **32a** (R = H), **32b** (R = Me)

Scheme 19. Formation of $[Re(bpza)O_3]$ (**32a**) and $[Re(bdmpza)O_3]$ (**32b**) (*43*).

hydrotris(pyrazol-1-yl)borate $HB(pz)_3^-$ (Tp) chemistry of ruthenium(II) has received considerable attention during the past decade (*135–139*). Several metallacumulenylidene ruthenium complexes (*140*) bearing tridentate Tp or Tp-related ligands have been reported (*141–147*). The focus of these studies has been their catalytic activity in Ring Closing Metathesis (RCM) or Ring Opening Metathesis Polymerization (ROMP) (*148–151*).

The ruthenium complex [Ru(bdmpza)Cl(PPh_3)$_2$] (**24**), which was mentioned earlier, easily releases one of the two phosphine ligands and allows the substitution not only of a chlorido but also of a triphenylphosphine ligand by κ^2-coordinating carboxylato or 2-oxocarboxylato ligands, as it will be discussed later on (*58*). In our effort to exploit the analogy between bis(pyrazol-1-yl)acetato and hydrotris(pyrazol-1-yl)borate (Tp) ligands, we thus focused on the synthesis of neutral ruthenium complexes [Ru(bdmpza)(Cl)(L)(PPh_3)] (L = carbene-, vinylidene- and allenylidene).

Three different structural isomers are conceivable for these complexes [Ru(bdmpza)(Cl)(L)(PPh_3)], which are (A) ligand L (L=C(OR')R, C=CHR, CO, C=C=CR_2) *trans* to pyrazole and phosphine *trans* to pyrazole, (B) ligand L *trans* to carboxylate and phosphine *trans* to pyrazole and (C) ligand L *trans* to pyrazole and phosphine *trans* to carboxylate (Fig. 22).

Due to the *trans* influence of the pyrazole acceptor ligands a coordination of the phosphine donor *trans* to the carboxylate (isomer **C**) is very unlikely. The remaining isomers **A** and **B** are both racemic mixtures of two enantiomers with two different sets of NMR signals for the two pyrazolyl groups.

C.1. Synthesis of vinylidene complexes [Ru(bdmpza)Cl(=C=CHR)(PPh$_3$)] (R = Ph, Tol, Pr, Bu) (33a–d)

The steric hindrance associated with the bis(3,5-dimethylpyrazol-1-yl)acetato ligand labilizes one of the PPh_3 ligands in **24**. Due to this labilization the reaction of [Ru(bdmpza)Cl(PPh_3)$_2$] (**33a**) with two

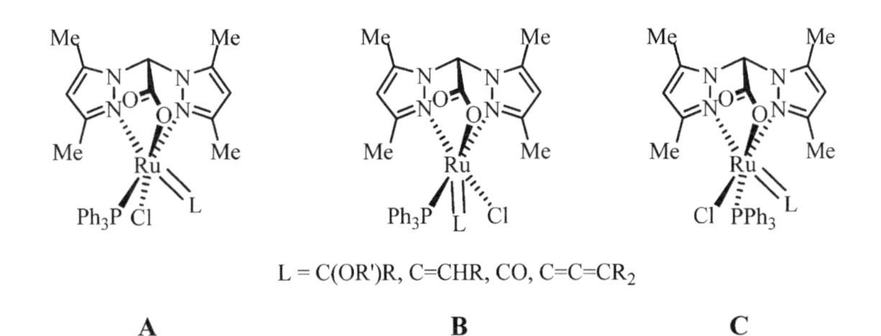

FIG. 22. Structural isomers of ruthenium complexes [Ru(bdmpza)(Cl)(L) (PPh_3)].

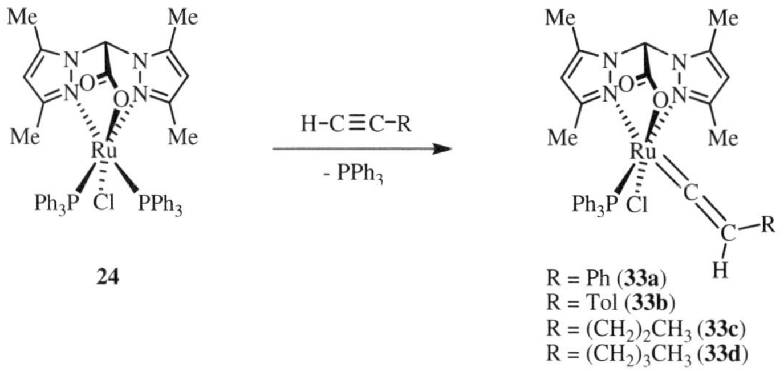

H–C≡C–R

- PPh_3

24

R = Ph (**33a**)
R = Tol (**33b**)
R = $(CH_2)_2CH_3$ (**33c**)
R = $(CH_2)_3CH_3$ (**33d**)

SCHEME 20. Syntheses of the ruthenium vinylidene complexes **33a–d** (*67*).

equivalents of phenylacetylene afforded the vinylidene complex [Ru(bdmpza)Cl(=C=CHPh)(PPh_3)] (**33a**) within a few hours (Scheme 20).

[Ru(bdmpza)Cl(=C=CHTol)(PPh_3)] (**33b**) can be prepared in the same way by reaction of **33a** with *para*-tolylacetylene (Scheme 20). The formation of the vinylidene complexes **33a** and **33b** proceeds much faster compared to the corresponding Tp and Cp* complexes [RuCl{κ^3-HB(pz)$_3$}(=C=CHPh)(PPh_3)] and [Ru(η^5-C_5Me_5)Cl(=C=CHPh)(PPh_3)] (*138,152*). The progress of the reaction can be monitored by IR spectroscopy of the THF reaction mixture, since the asymmetric carboxylate vibration of the tripodal bdmpza ligand shifts from 1672 cm^{-1} for **24** to 1678 cm^{-1} for the product **33a**. The vinylidene complexes show the typical β-H NMR resonance of the vinylidene ligand at 4.93 ppm (**33a**), splitted by a coupling $^4J_{HP}$ of 4.9 Hz. ROESY experiments indicate a coordination of the vinylidene ligand *trans* to a pyrazolyl ring of the bdmpza ligand and thus a type A isomer (see Fig. 22). The $^{13}C\{^1H\}$ NMR resonances of the β-carbons (**33a**: 113.1 ppm, **33b**: 115.1 ppm) are assigned by HMQC experiments and exhibit no J_{CP} coupling. However, the $^{13}C\{^1H\}$ NMR resonances of the α-C are found at 369.3 (**33a**) and 364.2 ppm (**33b**) with $^2J_{CP}$ couplings of 24.0 (**33a**) or 24.4 Hz (**33b**), respectively. These values are in good agreement with those of the vinylidene complex [RuCl{κ^3-HB(pz)$_3$}(=C=CHTol)(PPh_3)] [$^{13}C\{^1H\}$ NMR: 370 (d, $^2J_{CP} = 19.4$ Hz, C_α), 112 (C_β) ppm] as reported by A. F. Hill *et al.* (*138*).

Beside the spectroscopic evidence, the type A configuration is confirmed also by an X-ray structure determination of the vinylidene complex [Ru(bdmpza)Cl(=C=CHTol)(PPh_3)] (**33b**) (Fig. 23).

Beside the aromatic vinylidene complexes also alkyl vinylidene complexes [Ru(bdmpza)Cl(=C=CHPr)(PPh_3)] (**33c**) and [Ru(bdmpza)Cl(=C=CHBu)(PPh_3)] (**33d**) have been synthesized (Scheme 20), following the same procedure as earlier by using 1-pentyne or 1-hexyne. IR and NMR spectroscopic data of **33c** and **33d** are similar to those of **33a** and **33b**. Again only one major isomer was isolated in the case of [Ru(bdmpza)Cl(=C=CHPr)(PPh_3)] (**33c**), though the

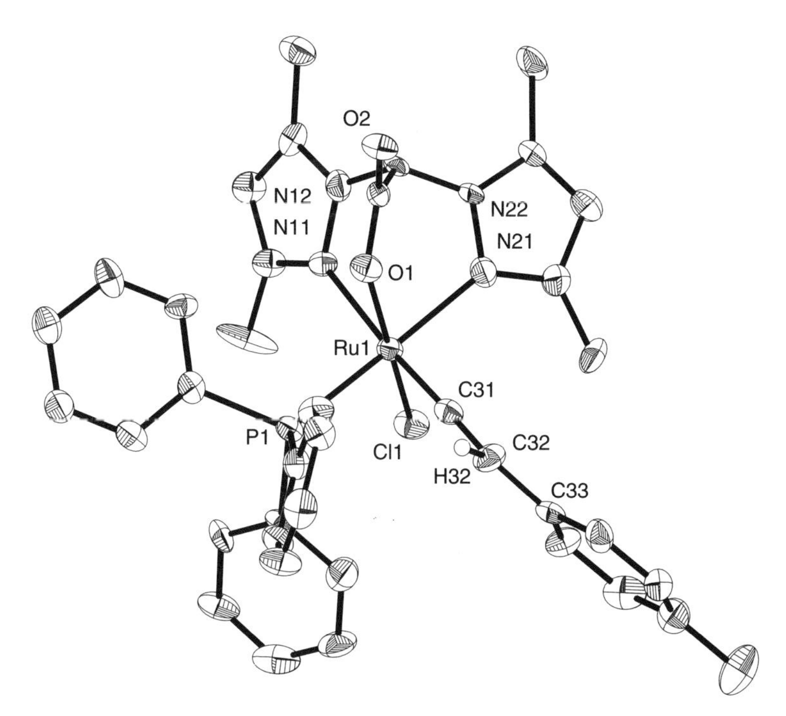

FIG. 23. Molecular structure of [Ru(bdmpza)Cl(=C=CHTol)(PPh_3)] (**33b**) with thermal ellipsoids drawn at the 50% probability level. Most hydrogen atoms and solvent molecules are omitted for clarity (*67*).

reaction mixture showed both isomers. For [Ru(bdmpza)Cl(=C=CHBu)(PPh_3)] (**33d**) a 1:1 ratio of isomers was found in the isolated product. An explanation might be the different solubility of complex **33d** compared to **33a–c**. In case of the complexes **33a–c** the reaction equilibrium seems to be shifted to the less soluble and thus precipitating isomer.

During several attempts to crystallize the vinylidene complexes **33a–d** the formation of a carbonyl complex [Ru(bdmpza)Cl(CO)(PPh_3)] (**34**) was observed. This carbonyl complex [Ru(bdmpza)Cl(CO)(PPh_3)] (**34**) can also be obtained in a direct synthesis by replacing a PPh_3 ligand of [Ru(bdmpza)Cl(PPh_3)$_2$] (**24**) for CO (Scheme 21, Fig. 24). An IR signal at 1969 cm^{-1} [CH_2Cl_2] and a doublet in the $^{13}C\{^1H\}$ NMR spectrum at 202.6 ppm ($^2J_{CP}$ = 19.8 Hz) were assigned to the carbonyl ligand. A degradation of ruthenium vinylidene complexes to form ruthenium carbonyl complexes have been reported by various authors before and was explained either by a reaction with dioxygen (*139,153,154*) or with H_2O (*155–157*). To distinguish between these two pathways, water was added to a sample of [Ru(bdmpza)Cl(=C=CHPh)(PPh_3)] (**33a**). The reaction mixture

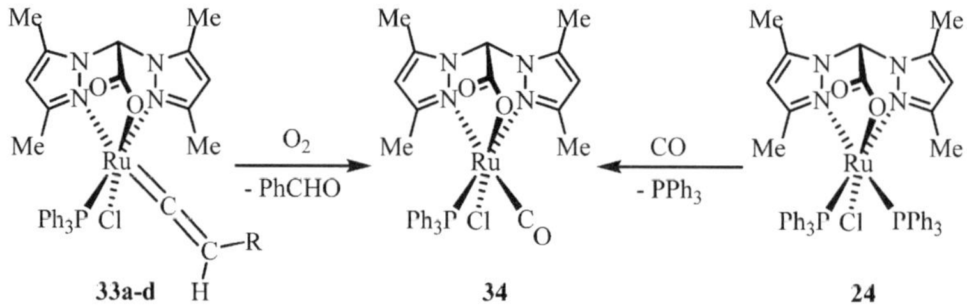

SCHEME 21. Formation of a ruthenium carbonyl complex **34** (*67*).

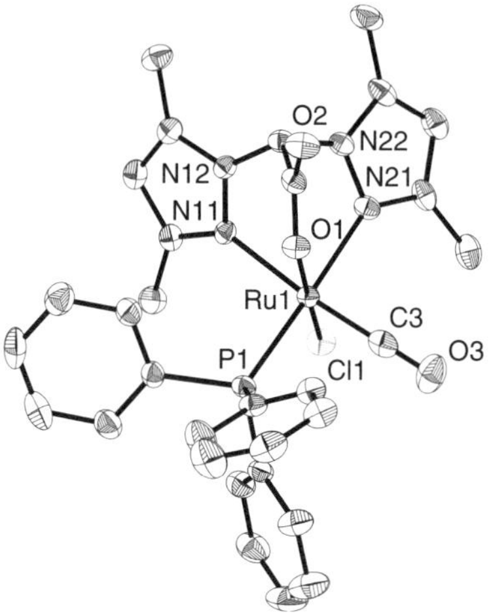

FIG. 24. Molecular structure of [Ru(bdmpza)Cl(CO)(PPh_3)] (**34**) with thermal ellipsoids drawn at the 50% probability level. Hydrogen atoms and solvent molecules are omitted for clarity (*67*).

showed no formation of **34** within 24 h under anaerobic conditions. Once exposed to air, formation of the carbonyl complex [Ru(bdmpza) Cl(CO)(PPh_3)] (**34**) was observed within three hours.

The oxidative cleavage of the C=C bond (Scheme 21) is also supported by the observation of benzaldehyde in the ^{1}H NMR of the reaction mixture.

C.2. Synthesis of cyclic Fischer carbene complexes [Ru(bdmpza)Cl(=C(CH$_2$)$_{n+2}$O)(PPh$_3$)] (n = 1, 2) (35a,b)

The formation of cyclic Fischer carbene complexes by ruthenium-mediated activation of 3-butyn-1-ol and 4-pentyn-1-ol has been reported before (*147,155,158–164*). Correspondingly, a reaction of [Ru(bdmpza)Cl($(PPh_3)_2$)] (**24**) with these terminal alkynols results

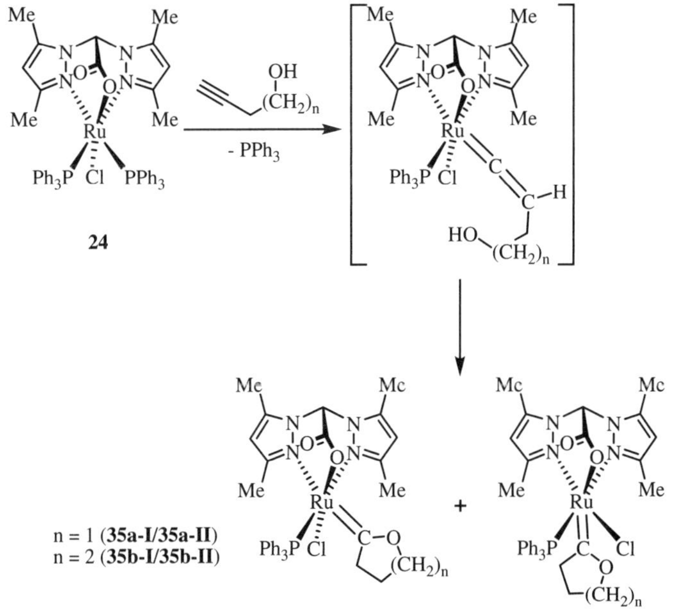

n = 1 (**35a-I/35a-II**)
n = 2 (**35b-I/35b-II**)

SCHEME 22. Syntheses of the ruthenium carbene complexes **35a** and **35b** *(67)*.

in the formation of the cyclic, heteroatom stabilized Fischer carbene complexes [Ru(bdmpza)Cl(=$\overline{C(CH_2)_{n+2}O}$)(PPh$_3$)] ($n = 1, 2$) (**35a, b**) (Scheme 22).

Obviously, the first intermediates in the syntheses with terminal alkynols are the vinylidene complexes [Ru(bdmpza)Cl(=C=CH(CH$_2$)$_{n+1}$OH)(PPh$_3$)] ($n = 1, 2$), which then react further via an intramolecular addition of the alcohol functionality to the α-carbon (Scheme 22), although in none of our experiments we were able to observe or isolate any intermediate vinylidene complexes. The subsequent intramolecular ring closure provides the cyclic carbene complexes with a five-membered ring in case of the reaction with but-3-yn-1-ol and with a six-membered ring in case of pent-4-yn-1-ol. For both products type A and type B isomers **35a-I/35a-II** and **35b-I/35b-II** are observed (Scheme 22, Fig. 22). The molecular structure shows a type A isomer **35b-I** with the carbene ligand and the triphenylphosphine ligand in the two *trans* positions to the pyrazoles and was obtained from an X-ray structure determination (Fig. 25).

The pentacyclic oxycarbene ligands of both isomers of **35a** are evidenced in the $^{13}C\{^1H\}$ NMR spectra by the resonances at 311.9 ($^2J_{CP} = 16.6$ Hz) (**35a-I**, isomer A) and 311.1 ($^2J_{CP} = 14.3$ Hz) ppm (**35a-II**, isomer B) for the carbene carbon atoms. Also the $^{13}C\{^1H\}$ NMR data of **35b** are consistent with a hexacyclic oxycarbene ligand with

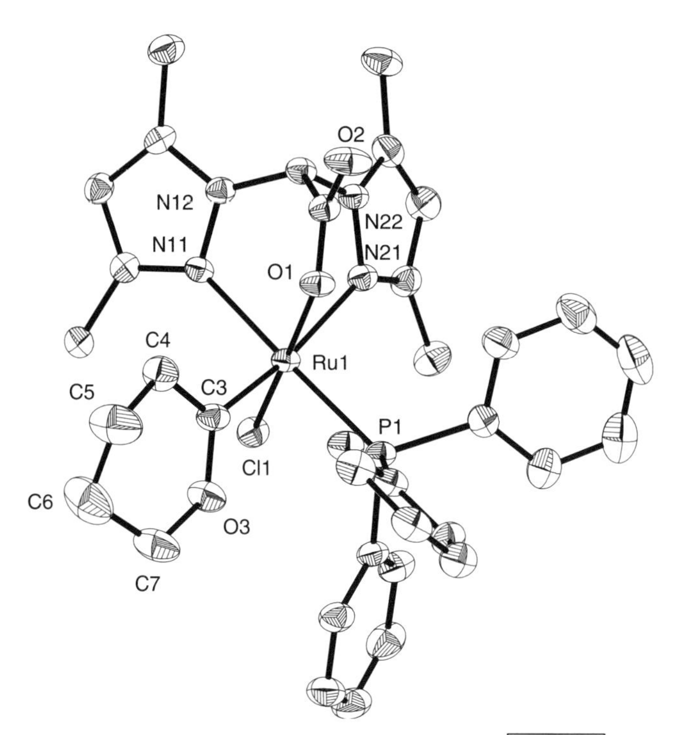

FIG. 25. Molecular structure of [Ru(bdmpza)Cl(=$\overline{C(CH_2)_4O}$)(PPh_3)] (**35b-I**) with thermal ellipsoids drawn at the 50% probability level. Hydrogen atoms are omitted for clarity (*67*).

resonances at 315.7 (**35b-I**) (d, $^2J_{CP} = 16.1$ Hz) and 314.5 (**35b-II**) ($^2J_{CP}$ not resolved) ppm.

C.3. Synthesis of allenylidene complexes [Ru(bdmpza)Cl(=C=C=CR$_2$)(PPh$_3$)] (R=Ph, Tol) (36a,b)

The preparation of allenylidene complexes was achieved by using an excess of propargylic alcohol as terminal alkyne according to the method of Selegue (*141*). A similar reaction of [RuCl{κ^3-HB(pz)$_3$}(PPh_3)$_2$] with HC≡C-CPh$_2$OH has been reported previously by Hill and coworkers (*138,143*). The formation of the allenylidene complexes [Ru(bdmpza)Cl(=C=C=CR$_2$)(PPh_3)] (R = Ph, Tol) (**36a**, **b**) via the intermediate vinylidene complexes [Ru(bdmpza)Cl(=C=CH–CR$_2$OH)(PPh_3)] is slow (Scheme 23) and usually takes more than 24 h at ambient temperature. Again no vinylidene intermediates were isolated, but a β-H NMR signal at 4.71 ppm ($^4J_{HP} = 4.9$ Hz) supports a vinylidene formation during the reaction in case of HC≡C–CPh$_2$OH.

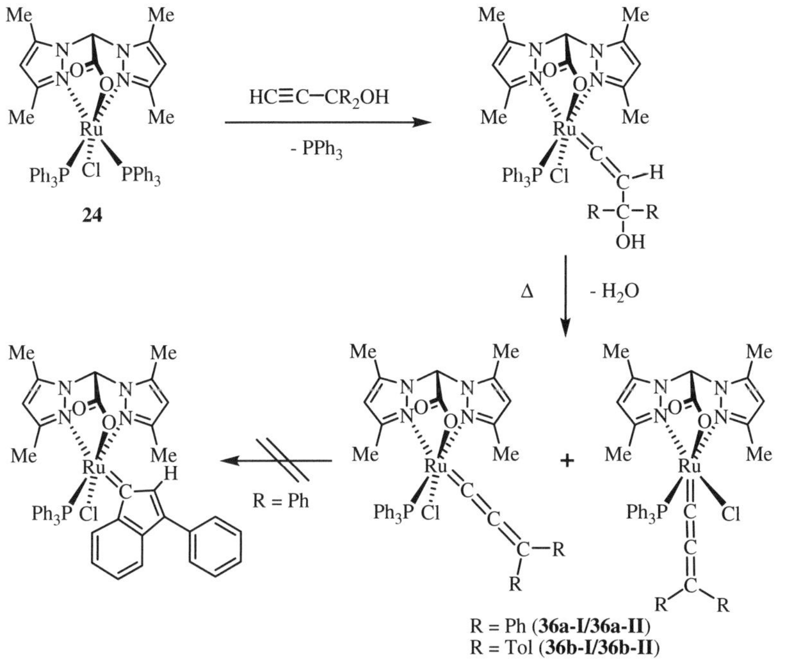

SCHEME 23. Formation of ruthenium allenylidene complexes [Ru(bdmpza)(Cl)(=C=C=CR$_2$)(PPh$_3$)] (**36a-I/36a-II**: R = Ph, **36b-I/36b-II**: R = Tol) (*67*).

The allenylidene complex formation is indicated by a color change from yellow to purple and can be monitored by the disappearance of the vinylidene β-H ^{1}H NMR resonance. The reaction is completed by heating under reflux for some hours. The neutral allenylidene complexes are rather stable towards oxygen and water. According to the ^{1}H, ^{13}C{^{1}H} and ^{31}P{^{1}H} NMR spectra, two isomers of the allenylidene complexes [Ru(bdmpza)(Cl)(=C=C=CR$_2$)(PPh$_3$)] (**36a-I/36a-II**: R=Ph, **36b-I/36b-II**: R = Tol) are formed, of which in each case one is violet and the other red. These quite intense colors [λ_{max} = 520 (**36a-I**), 495 (**36a-II**), 533 (**36b-I**) and 507 (**36b-II**) nm] have been assigned to MLCT absorptions. Column chromatography under aerobic conditions allows a separation of the isomers of which two-dimensional NMR studies indicate that the violet isomers **36a-I**, **36b-I** are of type A with the allenylidene ligand in *trans* position to a pyrazole donor and the red isomers **36a-II**, **36b-II** are of type B with a coordination of the allenylidene ligand *trans* to the carboxylate donor. The ratio of the isomers varies a lot depending on reaction time and temperature. The ^{13}C{^{1}H} NMR signals of the allenylidene ligands have been assigned to 142.1 (C$_\gamma$), 227.4 (C$_\beta$) and 305.5 (C$_\alpha$, $^2J_{CP}$ = 26.4 Hz) ppm for the violet isomer **36a-I** and to 142.5 (C$_\gamma$), 220.0 (C$_\beta$) and 299.0 (C$_\alpha$, $^2J_{CP}$ = 26.1 Hz) ppm for the violet isomer **36b-I.** The red type B isomers show $^{13}C\{^1H\}$ NMR allenylidene ligand signals at 149.2 (C_γ), 239.7 (C_β) and 314.7 (C_α, $^2J_{CP} = 19.1\,Hz$) ppm (**36a-II**) and at 150.0 (C_γ), 232.8 (C_β) and 311.0 (C_α, $^2J_{CP} = 18.8\,Hz$) ppm (**36b-II**), respectively. A rearrangement of the allenylidene complex to an indenylidene complex was excluded by two-dimensional NMR techniques (ROESY). Such a rearrangement has been reported recently in the attempt to obtain the related allenylidene complex $[Ru(Cl)_2(PPh_3)_2({=}C{=}C{=}CPh_2)]$ (*165,166*). The structural proposals for the allenylidene complexes [Ru(bdmpza)(Cl)$({=}C{=}C{=}CPh_2)(PPh_3)]$ (**36a-I/36a-II**) are confirmed by two X-ray crystal structural analyses (Figs. 26 and 27).

Finally, we investigated the catalytic activities of the vinylidene complexes **33a–d** and the allenylidene complexes **36a** and **36b** for RCM reactions. Unfortunately, none of the complexes showed any activity for RCM of diethyl diallylmalonate. Neither the addition of the phosphine scavenger CuCl nor heat induces any metathesis activity by these compounds. Nevertheless, our results indicate that a fruitful chemistry is accessible from the complex $[Ru(bdmpza)Cl(PPh_3)_2]$ (**24**) often with parallels to the chemistry of $[Ru(\eta^5\text{-}C_5H_5)Cl(PPh_3)_2]$ and especially $[RuCl\{\kappa^3\text{-}HB(pz)_3\}(PPh_3)_2]$. Three advantages should be noted in favor of the bdmpza ligand versus the $\{\kappa^3\text{-}HB(pz)_3\}$ or Cp ligands. First, the conditions under which **24** performs a ligand

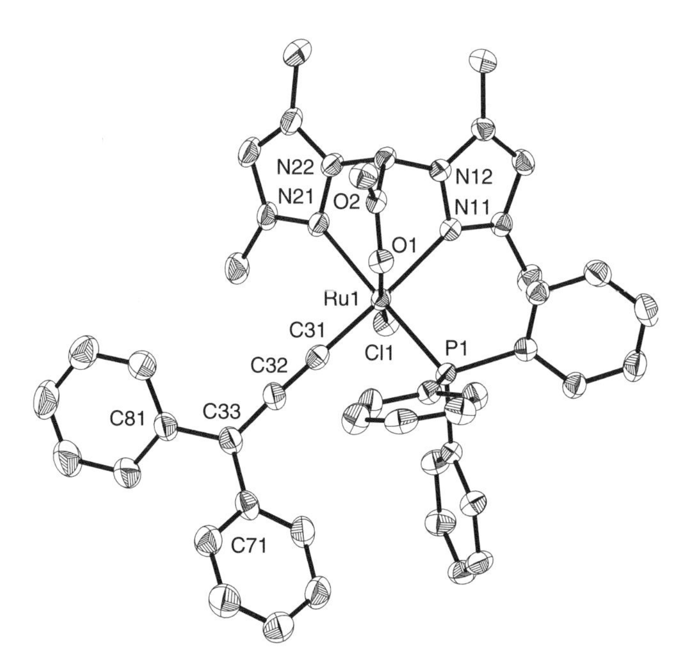

FIG. 26. Molecular structure of $[Ru(bdmpza)Cl({=}C{=}C{=}CPh_2)(PPh_3)]$ (**36a-I**) with thermal ellipsoids drawn at the 50% probability level. Hydrogen atoms and solvent molecules are omitted for clarity (*67*).

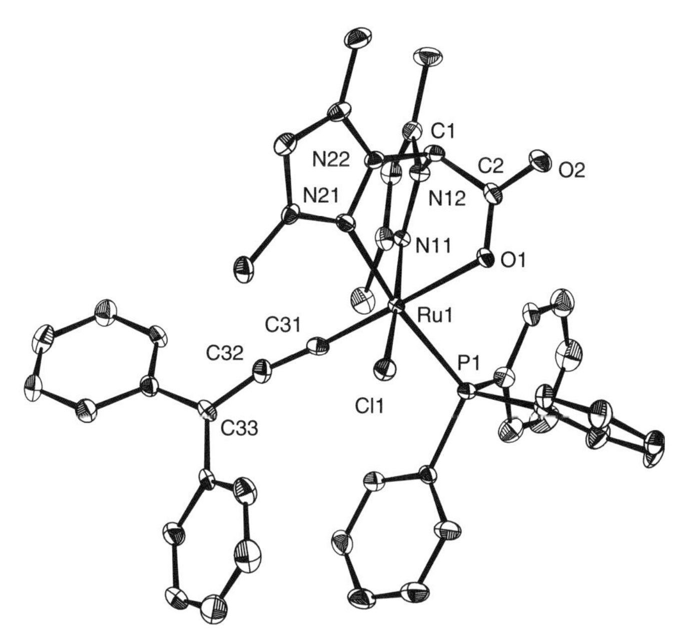

FIG. 27. Molecular structure of [Ru(bdmpza)Cl(=C=C=CPh_2)(PPh_3)] (**36a-II**) with thermal ellipsoids drawn at the 50% probability level. Hydrogen atoms and solvent molecules are omitted for clarity (*67*).

exchange are a lot milder than for [Ru(η^5-C_5H_5)Cl(PPh_3)$_2$] or [RuCl {κ^3-HB(pz)$_3$}(PPh_3)$_2$]. Second, the bdmpza ligand itself is not affected by hydrolytic cleavage, so that many of the cumulenylidene complexes are extraordinary robust and can be handled under aerobic and/or humid conditions. Third, the facial binding *N,N,O*-ligand causes a differentiation of the opposite coordination sites regarding their π-basic nature. This allows the formation and separation of structural isomers with different chemical and physical properties.

D. RUTHENIUM CARBOXYLATO COMPLEXES BEARING BIS(PYRAZOL-1-YL)ACETATO LIGANDS

As already mentioned earlier, the ruthenium complex [Ru(bdmpza) Cl(PPh_3)$_2$] (**24**) easily releases one of the two phosphine ligands and allows the substitution not only of a chlorido but also of a triphenylphosphine ligand for κ^2-coordinating carboxylato or 2-oxocarboxylato ligands (*58*). The purpose of these studies was to find structural ruthenium models for the active site of 2-OG dependent iron enzymes, since ruthenium(II) complexes are low spin and thus suitable for NMR characterization, whereas ferrous iron complexes with *N,N,O*-ligands are often difficult to investigate, due to their

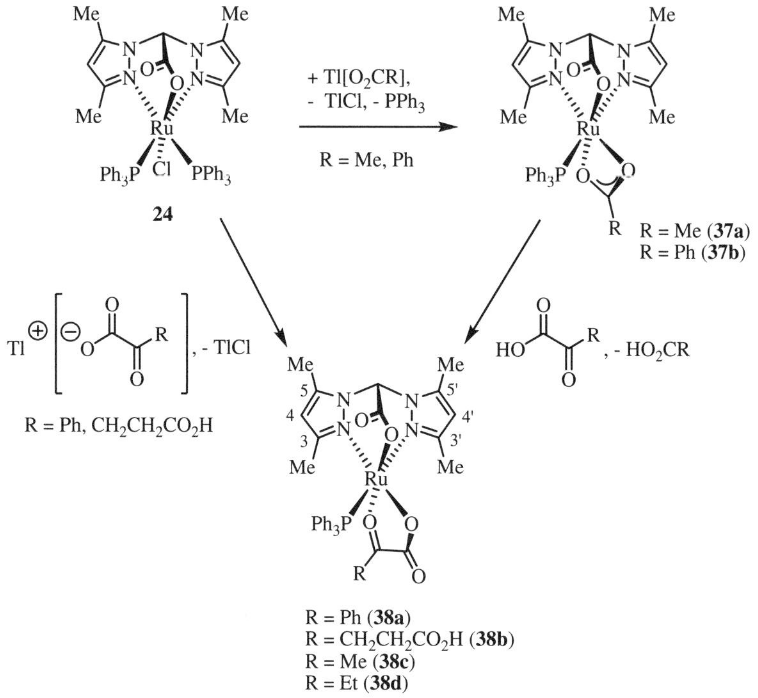

SCHEME 24. Formation of ruthenium(II) carboxylato and 2-oxocarboxylato complexes (*58*).

paramagnetic high-spin constitution. For example, does the complex [Ru(bdmpza)Cl(PPh_3)$_2$] (**24**) react with thallium carboxylates Tl[O_2CR] (R = Me, Ph) and forms $\kappa^2 O^1,O^{1'}$-carboxylato complexes [Ru(bdmpza)(O_2CR)(PPh_3)] (**37a**, **b**) (Scheme 24).

The thallium benzoate and thallium 2-oxocarboxylates are available via the reaction of thallium acetate with benzoic acid or the 2-oxocarboxylic acids, due to the lower pK_a values of these acids (benzoic acid: 4.22, benzoylformic acid: 1.2, 2-oxoglutaric acid: 2.31/5.14, pyruvic acid: 2.49) compared to acetic acid (4.76) (*167*). The free acetic acid is distilled off with water as an azeotrope and the thallium carboxylates are obtained in high purity.

The resulting carboxylato complexes [Ru(bdmpza)(O_2CR)(PPh_3)] (**37a**, **b**) exhibit a chiral geometry with a PPh_3 *trans* to one of the pyrazolyl groups. A single ^{31}P NMR signal for each of the two complexes at 62.0 ppm (**37a**) and 60.2 ppm (**37b**) and the carboxylato ^{13}C NMR signals at 189.0 (**37a**) and 183.8 (**37b**) ppm correspond well with those of [Ru(Tp)(η^2-O_2CCHPh_2)(PPh_3)] [^{31}P: 63.7 ppm, $^{13}C(CO_2^-)$: 186.97 ppm] (*168*). Adding three equivalents of water to a NMR sample of **37a** caused an additional set of signals beside those of **37a**, which, due to a high field shift of the acetate ^{13}C NMR signal

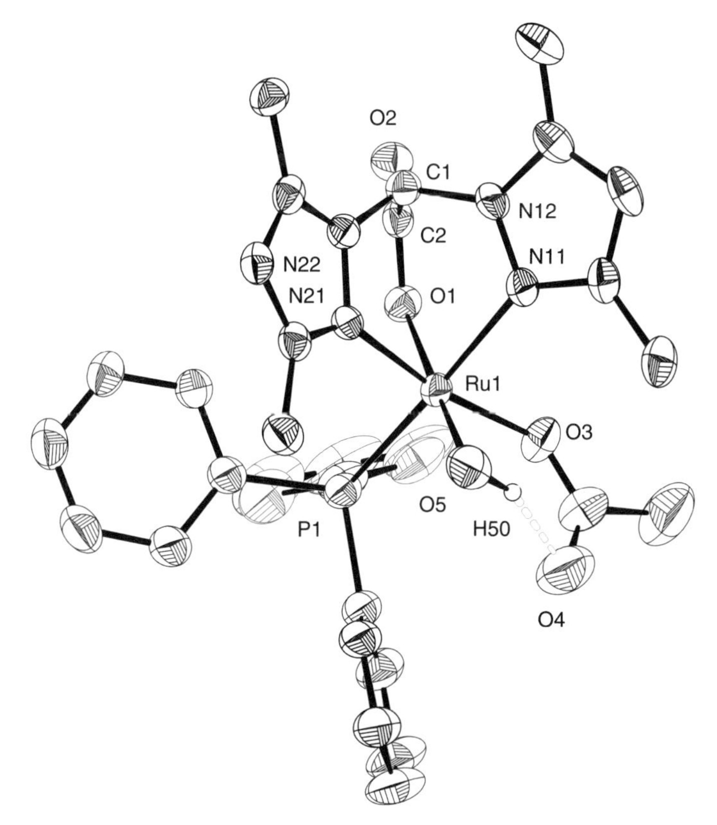

FIG. 28. Molecular structure of [Ru(bdmpza)(H_2O)(O_2CCH_3)(PPh_3)] (**37a** × H_2O); thermal ellipsoids are drawn at the 50% probability level (*58*).

(CO_2^-: 187.4 ppm), we explain by the formation of an adduct **37a** × H_2O. An X-ray structure determination of **37a** confirms this water adduct of **37a** × H_2O with H_2O coordinated to ruthenium(II) and a κ*O*-acetato ligand (Fig. 28).

The presence of water in the crystals can be explained by humid crystalization conditions and is proven by elemental analysis. The water adduct formation shows, that $\kappa^2 O^1,O^{1'}$-carboxylato ligands are hemilabile, chelating ligands, which can be easily displaced from one coordination site by a solvent. Thus, the carboxylato complexes **37a** and **37b** are equivalent to 16 VE fragments stabilized by the hemilabile carboxylato complexes.

E. RUTHENIUM 2-OXOCARBOXYLATO COMPLEXES BEARING BIS(PYRAZOL-1-YL)ACETATO LIGANDS

Reaction of [Ru(bdmpza)Cl(PPh_3)$_2$] (**24**) with thallium 2-oxocarboxylates Tl[O_2CC(O)R] (R = Ph, $CH_2CH_2CO_2H$) yields $\kappa^2 O^1,O^2$-2-oxo-carboxylato complexes (Scheme 24). Single ^{31}P NMR signals at

57.9 (**38a**) and 58.2 ppm (**38b**) prove the loss of one PPh_3 ligand. The chiral complexes exhibit one (**38a**) and two (**38b**) additional CO_2^- signals in the ^{13}C NMR spectra as well as signals of the 2-oxo groups (**38a**: 202.8 ppm, **38b**: 215.5 ppm). Due to the chirality in **38b**, an ABXY system in the ^1H NMR spectrum can be assigned to CH_2CH_2 of the 2-OG. The X-ray structure determination of **38a** establishes the chelate binding of the benzoylformate (Fig. 29).

The most striking feature in this molecular structure is the geometry with the 2-oxo group in *trans* position to the carboxylate donor of the bdmpza ligand. This structure corresponds well with the binding of the 2-OG in the active site of 2-OG dependent enzymes where the 2-oxo group is found in *trans* position to the iron-binding aspartate (Fig. 29). Complexes **38a** and **38b** can also be obtained in high yield and high purity by reacting benzoylformic acid and 2-oxoglutaric acid with the carboxylato complexes **37a** or **37b**. This exchange of a carboxylato ligand by a 2-oxocarboxylato ligand is alike the regeneration step of the biocatalytic pathway postulated for the 2-OG dependent enzymes. In the 2-OG dependent enzymes an iron(II)

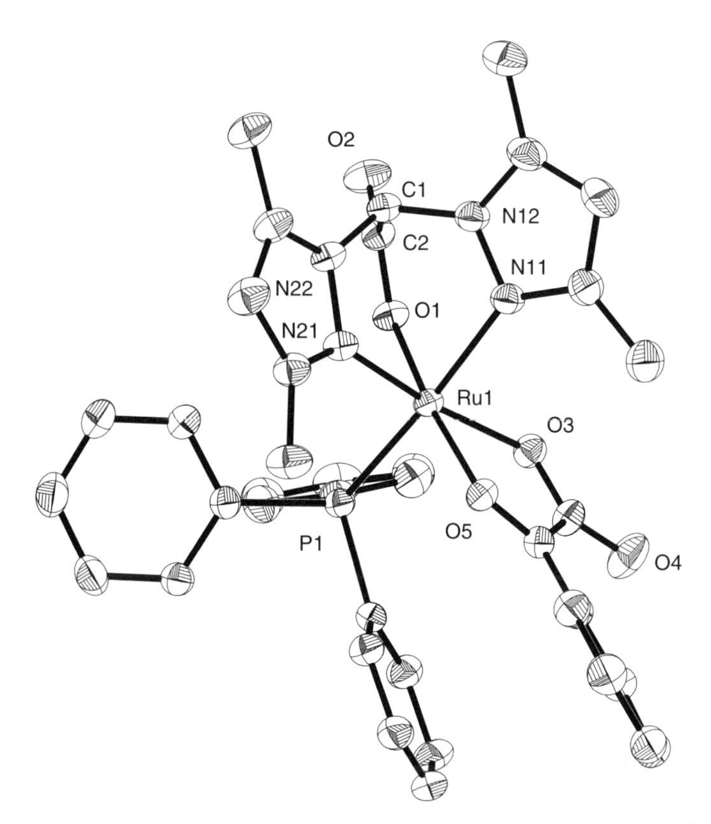

FIG. 29. Molecular structure of [Ru(bdmpza)($O_2CC(O)Ph$)(PPh_3)] (**38a**); thermal ellipsoids are drawn at the 50% probability level (*58*).

bound succinate is replaced by 2-OG to recharge the active site of the enzyme with the 2-OG cosubstrate (*31*).

In a combination of the two reactions, addition of thallium acetate and a 2-oxocarboxylic acid to the chlorido complex **24** affords the ruthenium(II) 2-oxocarboxylato complexes [Ru(bdmpza)($O_2CC(O)R$)(PPh_3)] (**38a**: R = Ph, **38b**: R = $CH_2CH_2CO_2H$, **38c**: R = Me, **38d**: R = Et) in a one-pot synthesis. The NMR spectra of the resulting pyruvato and 2-oxobutyrato complexes **38c** and **38d** agree well with the other 2-oxocarboxylato complexes **38a** and **38b** with ^{13}C NMR signals at 215.8 (**38c**) and 217.9 ppm (**38d**) assigned to the 2-oxo carbonyl groups. Again the complexes are chiral. Thus **38d** presents an ABX_3 system in the ^1H NMR spectrum that can be assigned to the CH_2CH_3 group. The binding of the 2-oxocarboxylato ligands is clearly indicated by a purple (**38a**) or brownish-red (**38b–38d**) color that has been attributed to a MLCT from the metal center to the coordinated 2-oxo group as we could verify by extended Hückel and DFT calculations. The HOMO is clearly localized at the ruthenium(II) metal center in **38a** as well as in **38c**. The LUMOs in **38a** and **38c** are localized at the 2-oxocarboxylato ligands. They are composed of p_z density on the 2-oxo group with some conjugation to the carboxylate p_z. This is in good agreement to the π^* LUMO of a coordinated pyruvate which was reported earlier by Solomon (*169*). In **38a** the LUMO is also delocalized into the aromatic substituent. The missing delocalization into an aromatic system can explain the hypsochromic shift of the longest wavelength absorption of **38c**. Although the ruthenium(II) complexes **38a** and **38c** have a low-spin t_{2g}^6 configuration they exhibit a HOMO–LUMO gap which is by chance almost identical in size to that of the $t_{2g}^4e_g^2$ iron centers of the enzymes.

An acetonitrile solution of the benzoylformate complex **38a** changes its color from purple to yellow upon heating, due to the formation of an adduct [Ru(bdmpza)($O_2CC(O)Ph$)($NCCH_3$)(PPh_3)] (**39**). A similar behavior was observed before for iron 2-oxocarboxylato complexes (*75,82–84*).

Although the benzoylformate complex **38a** seems to be rather stable an aerobic solution of **38a** in 1,2-dichloroethane decomposed within two weeks forming red crystals of the ruthenium(III) complex [Ru(bdmpza)Cl_2(PPh_3)] (**40**) (Fig. 30), which were also obtained in a direct synthesis with $RuCl_3 \times H_2O$, PPh_3 and Hbdmpza (*44*).

Our future work will focus on the coordination of typical inhibitors of 2-OG dependent iron enzymes. First ruthenium complexes bearing *N*-oxalylglycine or triketone ligands have already been synthesized.

VI. 3,3-Bis(1-methylimidazol-2-yl)propionic Acid as *N,N,O*-Ligand

Inspired by the 2-His-1-carboxylate facial triad our interest focused on the development of a new facially coordinating tripod ligand which

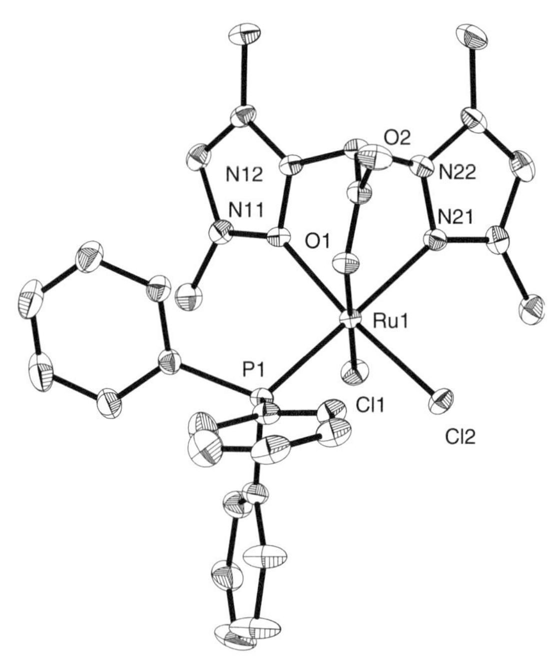

FIG. 30. Molecular structure of [Ru(bdmpza)(Cl)$_2$(PPh$_3$)] (**40**); thermal ellipsoids are drawn at the 50% probability level (*44*).

resembles the electronic properties of this binding motif more closely than bis(pyrazol-1-yl)acetato ligands. DFT calculations indicate that both 1-methylimidazol-2-yl and pyrazol-1-yl ligands possess, similar to histidine, σ-donor, π-donor and π-acceptor properties. 3,3-bis(1-methylimidazol-2-yl)propionato ligand would probably resemble the overall binding properties of histidine in the 2-His-1-carboxylate facial triad closer than a bis(pyrazol-1-yl)acetato ligand. So far research has focused on 3,3-bis(imidazol-2-yl)propionic acid (Hbip) and derivatives of it (*170–172*). For a direct comparison with bis(pyrazol-1-yl)acetic acids (Hbdmpza, Hbpza) it would have been desirable to synthesize 3,3-bis(1-methylimidazol-2-yl)acetic acid. However, it decarboxylates rather easily and is thus not suited for coordination chemistry. Therefore, we concentrated on the synthesis of the much more stable 3,3-bis(1-methylimidazol-2-yl)propionic acid ligand, which was synthesized in two steps starting from bis(1-methylimidazol-2-yl)methane (Scheme 25), following the preparation of 4,4-dimethyl-1,1-bis(1-methylimidazol-2-yl)pentan-3-one (*173*).

Deprotonation of 3,3-bis(1-methylimidazol-2-yl)methane (**41**) with *n*BuLi at the bridging carbon atom and subsequent reaction with methyl bromoacetate gives methyl 3,3-bis(1-methylimidazol-2-yl) propionate bmipme (**42**) in high yields. Saponification of bmipme with aqueous NaOH gives bis-3,3-(1-methylimidazol-2-yl)propionate (bmip)

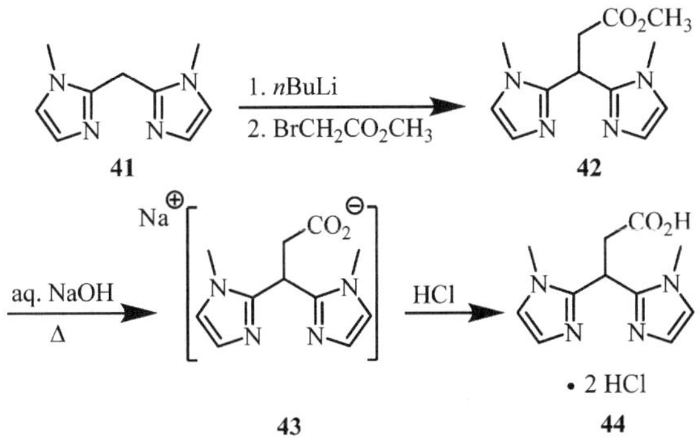

SCHEME 25. Synthesis of bmipme (**42**), Na[bmip] (**43**) and Hbmip × 2 HCl (**44**) (*174,175*).

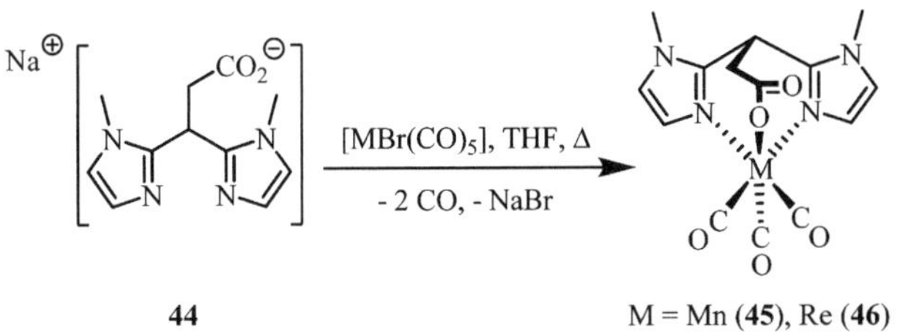

SCHEME 26. Synthesis of the complexes [M(bmip)$(CO)_3$] (M = Mn, Re) (**45**, **46**) (*174*).

in quantitative yield. The ligand was isolated either as its sodium salt Na[bmip] (**43**) or as a hydrochloride Hbmip · 2HCl (**44**). Since the binding properties of bmip and other known tripod ligands such as bpza, bdmpza, Tp and Tp_2^{Me} are best compared by IR spectroscopy and the molecular structures, we prepared manganese and rhenium tricarbonyl complexes of bmip. Complexes were synthesized and purified similar to [M(bpza)$(CO)_3$] and [M(bdmpza)$(CO)_3$], starting from **43** and the pentacarbonyl complexes [MnBr$(CO)_5$] and [ReBr$(CO)_5$] (Scheme 26) (*174*). The progress of the reactions was monitored by IR spectroscopy. The resulting tricarbonyl complexes [Mn(bmip)$(CO)_3$] (**45**) and [Re(bmip)$(CO)_3$] (**46**) are soluble in methanol, but not in THF and rather stable as solids under aerobic conditions. The ^{13}C NMR spectra of both complexes exhibit two signals for the carbonyl ligands at a ratio of 2:1 (**45**: $\delta = 197.8$ and 198.0 ppm, **46**: $\delta = 220.7$ and 225.3 ppm) and resonances at 171.3 ppm (**45**) and 172.9 ppm (**46**) respectively, assigned to the carboxylate donor.

The molecular structures of both complexes [Mn(bmip)$(CO)_3$] (**45**) and [Re(bmip)$(CO)_3$] (**46**) were revealed by single-crystal X-ray determination (Figs. 31a and 31b).

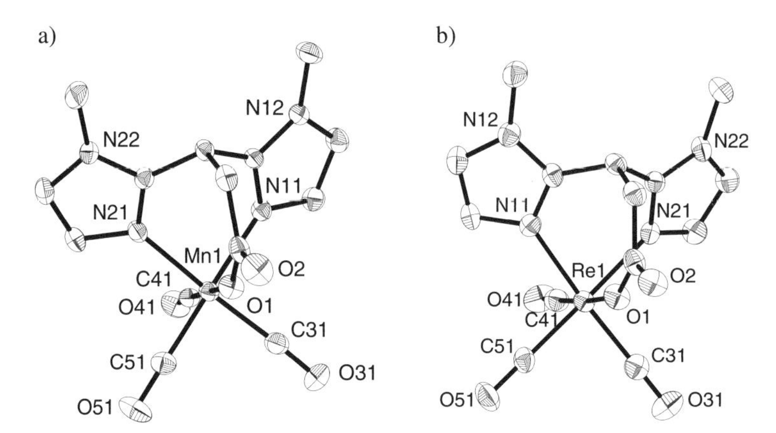

FIG. 31. Molecular structure of (a) [Mn(bmip)$(CO)_3$] (**45**) and (b) of [Re(bmip)$(CO)_3$] (**46**); thermal ellipsoids are drawn at a 50% probability level (*174*).

The metal–carbonyl distances are in good agreement with those of [Mn(bdmpza)$(CO)_3$], [Re(bdmpza)$(CO)_3$] and [Re(bpza)$(CO)_3$]. In both complexes **45** and **46** the metal–carbonyl distances M–C(31) and M–C(51) are longer than the metal–carbonyl distance M–C(41) *trans* to the carboxylate group, due to the *trans* influence of the two imidazolyl π acceptor groups onto the *trans* metal–carbonyl bonds. The IR spectra of **45** and **46** are those of 'piano stool' complexes with three strong A′, A″ and A′ carbonyl signals. As expected, the manganese CO signals (2027, 1922 and 1895 cm^{-1}) are found at slightly higher wavenumbers than those of the rhenium complex (2018, 1898 and 1867 cm^{-1}). These values are lower compared to those of the corresponding bis(pyrazol-1-yl)acetato complexes (*40*) and suggest that pyrazol-1-yl and 1-mcthylimidazol-2-yl groups may havc almost the same donor strength, but that the 1-methylimidazol-2-yl group is a weaker π acceptor. This observation is in good agreement with our DFT calculations. In conclusion, it can be said that bmip and similar ligands might be versatile ligands in different fields of coordination and bioinorganic chemistry. They may be applied in the synthesis of structural and possibly functional iron and zinc model complexes for enzymes mimicing the 2-Histidine-1-carboxylate motif. It has to be mentioned, that parallel to our work, this ligand has been investigated in the van Koten group (*175*). Very recently van Koten, Klein Gebbink and coworkers reported on sterically demanding analogous of bmip and ferric and ferrous model complexes thereof (*176*). Furthermore, the complexes **45** and **46** suggest, that bmip might be a suitable ligand in future rhenium and technetium complexes for radiopharmaceutical applications (*177–179*).

VII. Immobilization of *N,N,O* Complexes

Bioinorganic models for metalloenzymes are investigated for more than 20 years. Despite this long history and thousands of structural models, the amount of functional models especially in the field of oxygenases is rather disappointing. It has become obvious, that various problems are responsible for this: First of all often rather small sterically less hindered organic molecules are used to mimic the metal environment of the active site. These ligands tend to coordinate twice to metal ions and form 2:1 bisligand complexes which are often a dead end in reactivity. The complexes [Fe(bpza)$_2$] (**4a**) and [Fe(bdmpza)$_2$] (**4b**) which were discussed earlier are typical examples for this bisligand complex formation. A second problem occurs once the formation of an oxygenase model was successful. After activation of O_2 or H_2O_2 highly reactive intermediates such as Fe(IV)=O species are formed which can react instantaneously under autoxidation with another model complex which is still ferrous. This usually causes the formation of a dinuclear ferric μ-oxo species as a themodynamic dead end. Very recently we observed an example for such an autoxidation and μ-oxo complex formation for our ruthenium model complexes. Although the acetato complex [Ru(bdmpza)(O_2CCH_3)(PPh_3)] (**37a**) seems to be rather stable under aerobic conditions due to its low-spin state, it decomposes within weeks in acetone and forms a μ-oxo dimer [Ru_2(bdmpza)$_2$(μ-O)(μ-O_2CR)$_2$] (**47**) (Fig. 32). The two ruthenium(III) centers in **47** are antiferromagnetically coupled, so that [Ru_2(bdmpza)$_2$ (μ-O)(μ-O_2CR)$_2$] (**47**) is a diamagnetic compound, suitable for NMR analyses. Due to the simple ^{1}H and ^{13}C NMR spectra with one set of

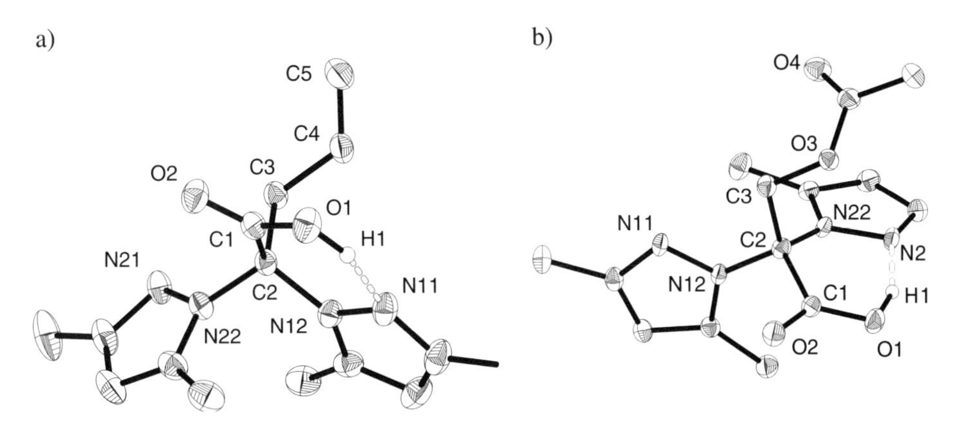

FIG. 32. Molecular structure of (a) 2,2-bis(3,5-dimethylpyrazol-1-yl)pent-4-enoic acid (**48**) and (b) of 2,2-bis(3,5-dimethylpyrazol-1-yl)-3-acetatopropionic acid (**50**); thermal ellipsoids are drawn at the 50% probability level (*68*).

signals for an acetate and a bdmpza ligand we deduced a C_2 symmetric structure of the complex.

To overcome these problems of bisligand complex formation and autooxidation often sterically more bulky ligands are applied in these models. The bd*t*bpza ligand which was discussed earlier is an example for this concept. The backdraw of these bulky ligands is that it restricts the access to the metal center and therefore changes the reactivity of the model complexes. Often these complexes and the metal environment are too crowded to allow any functional activity.

Fixation of model complexes on a solid phase would cause a separation of the metal centers without causing a change in reactivity and thus overcome these problems. Thus, solid phase fixation of the model complexes or ligands is of special interest, with regard to a potential use as catalysts with easy workup or to achieve dedicated structural environments with sterically less hindered ligands. Many methods used for immobilization of ligands or transition metal complexes introduce linker to the ligand which connects the ligand via covalent bonds to the solid material (*180*). Vinyl, allyl or acryl groups with their isolated double bonds open up a broad field of methods to immobilize the ligand (*181,182*). Common examples are copolymerization with MMA/EGDMA, metathesis reaction and especially radical-induced carbon–sulfur bond formation (*182*). Recent investigations on bis(oxazoline) ligands have shown the feasibility of immobilizing such ligands while maintaining catalytic activity of the resulting immobilized complexes (*182*). On the other hand, a hydroxymethyl linker allows solid phase fixation via esterification or ether formation.

We found bis(3,5-dimethylpyrazol-1-yl)acetic acid Hbdmpza (**3b**) to be a convenient starting material for linker-modified ligands. Deprotonation at the bridging CH in **3a** with excess LDA and subsequent alkylation with allyl bromide leads to 2,2-bis(3,5-dimethylpyrazol-1-yl)pent-4-enoic acid Hbdmpzpen (**48**) (Scheme 27; Pathway 1). Similar modifications of tris(pyrazolyl)methane ligands with various bases have been reported recently by D. L. Reger *et al.* (*183*). The molecular structure of **48** (Fig. 32a) reveals an intramolecular hydrogen bridge [d(O1-N11) = 2.504 (2) Å] between the carboxylate and the pyrazolyl group. The ^{1}H and ^{13}C NMR data of the resulting ligand **48** are almost identical to Hbdmpza (**3b**), except for the lack of a ^{1}H resonance of the bridging CH group and the additional allyl resonances. A similar reaction with paraformaldehyde yields 2,2-bis(3,5-dimethylpyrazol-1-yl)-3-hydroxypropionic acid Hbdmpzhp (**49**) (Scheme 27; Pathway 2). According to the NMR data, the molecular structure exhibits C_s symmetry in solution as well. Since the linker groups are introduced at the bridging carbon atom, this concept may be extended to the homochiral bis(pyrazol-1-yl)acetic acids, which were discussed earlier, without breaking the chirality or chiral induction.

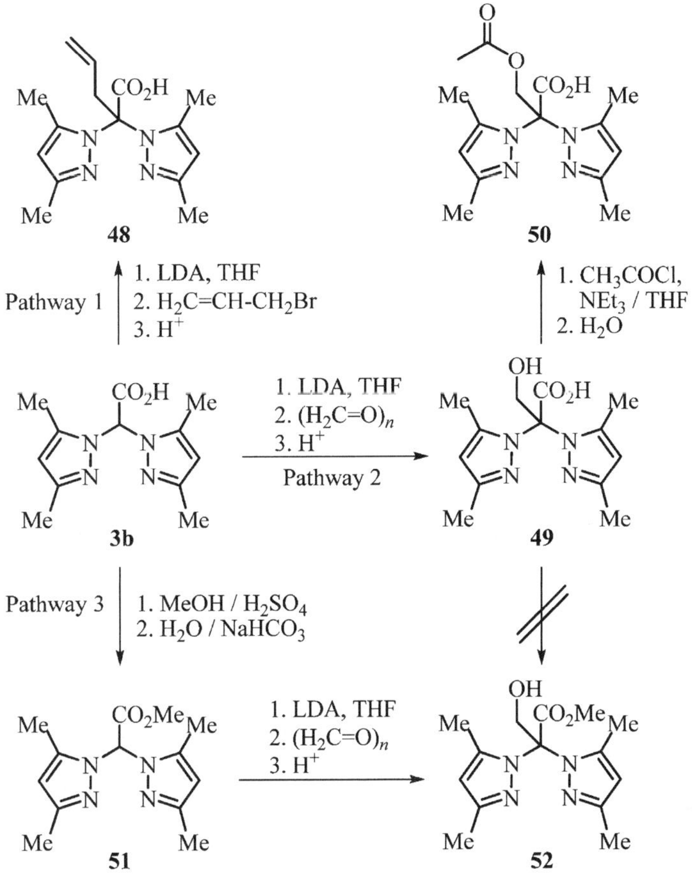

SCHEME 27. Synthesis of bis(pyrazol-1-yl)acetic acids with solid phase linkers (*68*).

Besides direct solid phase fixation, the functional OH-group of **49** may be used for further transformations. Esterification with acetyl chloride at the OH-linker (Scheme 27) leads to **50**, providing a heteroscorpionate ligand with the protected OH-linker. An intramolecular hydrogen bridge is found in the X-ray structure [d(O1-N21) = 2.478 (3) Å] between the carboxylic acid and the pyrazole nitrogen (Fig. 32b).

Very recently we obtained also acrylic and metacrylic esters of the bdmpza ligand via esterification of **49** (*184*). These ligands should allow copolymerization of ligands and model complexes in the near future.

On the other hand, ligand **3b** may be modified at the carboxylate donor instead of the OH-linker. This prevents the carboxylate donor from acting as linking group. Although the most obvious protecting group is the methyl ester **52** (Scheme 27), direct esterification of

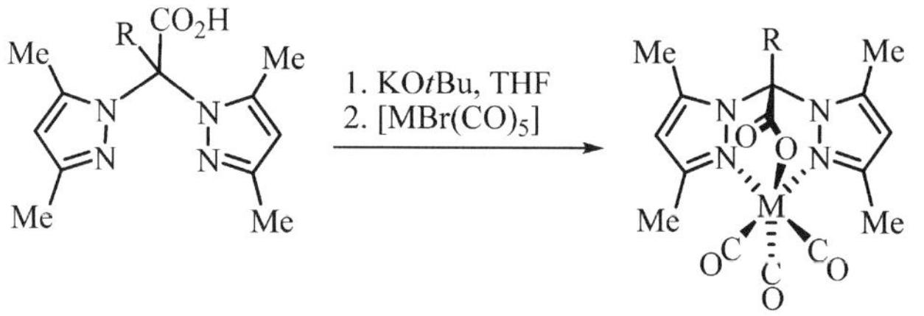

1. KO*t*Bu, THF
2. [MBr(CO)$_5$]

48 (R = allyl linker)
50 (R = $CH_2O_2CCH_3$)

M = Mn (**53**), Re (**54**)
M = Mn (**55**), Re (**56**)

SCHEME 28. Manganese and rhenium tricarbonyl complexes with solid phase linkers (*68*).

49 results in 2,2-bis(3,5-dimethylpyrazol-1-yl)-1-ethanol caused by decarboxylation. Instead, the methyl ester **52** is synthesized by esterification of **3b** and subsequent reaction of **51** with LDA and paraformaldehyde to **52** (Scheme 27; Pathway 3).

To examine the chemical behavior of these new ligands **48–50** in coordination chemistry, tricarbonyl complexes with manganese and rhenium were synthesized from [MnBr(CO)$_5$] and [ReBr(CO)$_5$] as described earlier for the other heteroscorpionate ligands (Scheme 28). The purpose of these tricarbonyl complexes was to verify tripodal binding of the solid-bound ligand. The protected OH linker in **50** acts as model for the solid phase bound ligand.

Ligand **50** reacts with the metal carbonyls as it is well known for the heteroscorpionate ligand Hbdmpza. This is also observed with the allyl-functionalized ligand **48** (Scheme 28).

The presence of the tricarbonyl complexes is indicated by three A′, A″ and A′ signals typical for C_s symmetrical ‘piano stool’ type carbonyl complexes (**53**: ν(CO) = 2033, 1939 and 1911 cm^{-1}; **54**: ν(CO) = 2023, 1915 and 1892 cm^{-1}; **55**: ν(CO) = 2034, 1940 and 1913 cm^{-1}; **56**: ν(CO) = 2024, 1918 and 1894 cm^{-1}). These IR absorptions are in good agreement with those of the complexes [M(bdmpza)(CO)$_3$] (M = Mn, Re) (**29b**, **30b**) and prove the κ^3*N,N,O* coordination of the ligands. IR bands at 1687 (**53**), 1697 (**54**), 1692 (**55**) and 1702 (**56**) cm^{-1} are assigned to the asymmetric carboxylate vibration $\nu_{asym}(CO_2^-)$. **55** and **56** show an additional IR signal at 1761 and 1762 cm^{-1} respectively which was assigned to the asymmetric carboxylate vibration in the ester residue. Two ^{13}C resonances are observed for the carbonyl ligands of the two complexes **55** and **56** (**55**: 220.6. ppm, 222.4 ppm, **56**: 195.7 ppm, 196.0 ppm). Finally, X-ray structure analyses (Fig. 33) doubtlessly reveal the κ^3*N,N,O* coordination of the tripodal *N,N,O*-ligands **53** and **55**.

Obviously, the coordination of the new ligands is not affected by the additional linker groups at the bridging carbon atom. No coordination of the linker to the metal center instead of the carboxylate donor takes place and the geometries of the allyl group in **53** (Fig. 33a) as well as of

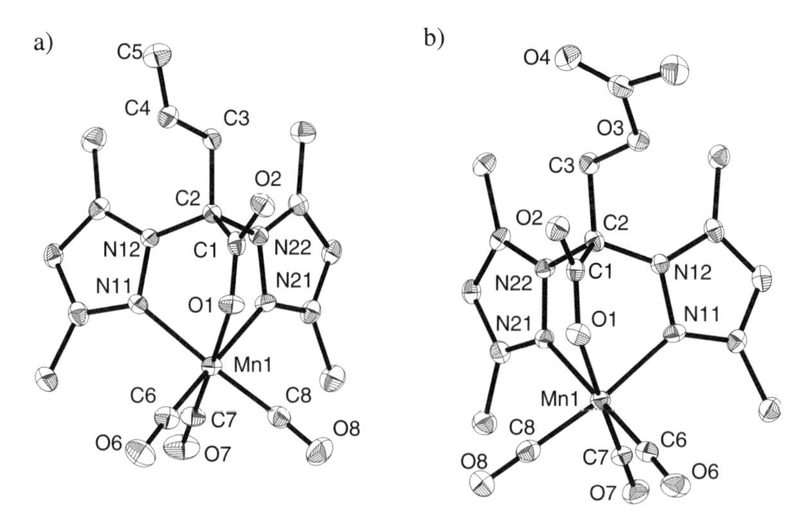

FIG. 33. Molecular structure of (a) [Mn(bdmpzpen)(CO)$_3$] (**53**) and (b) [Mn(bdmpzap)(CO)$_3$] (**55**); thermal ellipsoids are drawn at the 50% probability level (*68*).

the ester in **55** (Fig. 33b) prove their ability to act as a linking group to solid phase. The agreement of the bond distances and angles of **53** and **55** with those of [Mn(bdmpza)(CO)$_3$] implies similar properties of **48–50** in comparison to the Hbdmpza ligand **3b**.

A. SOLID PHASE BINDING

In our first attempt to bind linker-modified bis(pyrazol-1-yl)acetic acids to a solid phase we used Merrifield resin, which is one of the most popular solid phase supports. Since Merrifield polymer was designed to bind carboxylic acids, we used the ester methyl 2,2-bis(3,5-dimethylpyrazol-1-yl)-3-hydroxypropionate (**52**) instead of the 2,2-bis(3,5-dimethylpyrazol-1-yl)-3-hydroxypropionic acid (**49**).

Grafting on the resin was achieved via a nucleophilic substitution of the benzylic chlorine by the deprotonated OH-linker of **52** (Scheme 29) by using a mixture of KO*t*Bu, 18-crown-6 and CsBr. Determining the nitrogen content of solid phase samples by elemental analyses was accomplished, to verify the functionalization of the polymer. This enables calculation of the degree of functionalization. Usually, an occupancy of more than 20 percent of the theoretical sites was achieved. Saponification of the functionalized Merrifield resin **P-52** leads to the monoanionic *N,N,O* functionalized solid phase. Subsequent reaction with [ReBr(CO)$_5$] afforded the polymer mounted tricarbonyl rhenium complex **P-52-Re** (Scheme 29).

The IR spectra of **P-52-Re** exhibits the A′, A″ and A′ signals ($\tilde{\nu}$(CO) = 2022, 1915 and 1890 cm^{-1}). The location of these signals

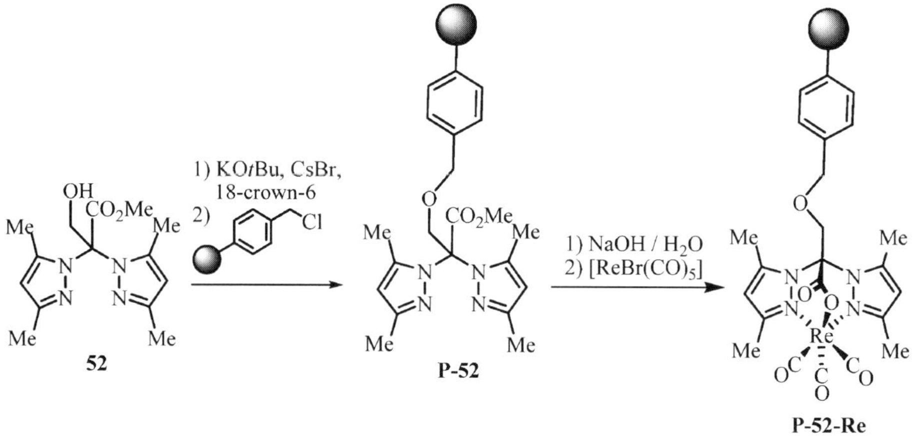

SCHEME 29. Grafting on Merriefield resin (*68*).

is almost identical to those of the complexes [Re(bdmpzap)$(CO)_3$] (**55**) which were described above (*68*).

A control experiment with the unfunctionalized methyl bis(3,5-dimethylpyrazol-1-yl)acetate (**51**) showed no occupancy of the polymer sites and subsequently no A′ and A″ signals but a typical IR spectrum of Merrifield polymer. Therefore, the results of our experiments prove a facial coordination of rhenium(I) by the monoanionic *N,N,O* tripod ligand as well as a solid phase fixation of the ligand and the resulting tricarbonyl complex (*68*).

Reactions with Merrifield polymer are quite limited regarding the solvents that can be used. Unfortunately, especially reactions in water are almost impossible. For future bioinorganic applications in e.g., aqueous solution, grafting on silica solid phase can be achieved via mercaptopropyl functionalized silica. Immobilization of chelating *N,N*-ligands bearing allyl linker groups on such functionalized silica has been recently reported by Mayoral *et al.* (*182*). Following a similar procedure, we immobilized the complexes [Mn(bdmpzpen)$(CO)_3$] (**53**) and [Re(bdmpzpen)$(CO)_3$] (**54**) on mercaptopropyl silica in the presence of AIBN (Scheme 30).

The deduced occupancy of mercaptopropyl sites by the tricarbonyl complexes added up to about 4% (**S–Mn** and **S–Re**). IR spectra of these silica phases **S–Mn** and **S–Re** show signals that can be assigned to the three A′, A″ and A′ signals of the grafted manganese and rhenium tricarbonyl complexes (**S–Mn**: $\tilde{\nu}(CO) = 2039$, 1948 and $1923\,cm^{-1}$; **S–Re**: $\tilde{\nu}(CO) = 2027$, 1920 and $1905\,cm^{-1}$) (*68*). A longer reaction time (48 h) resulted in higher occupancies, but caused also obscure IR spectra probably due to degradation of the carbonyl complexes or a further reaction with surrounding non innocent SH-groups. Our results prove an immobilization of manganese and

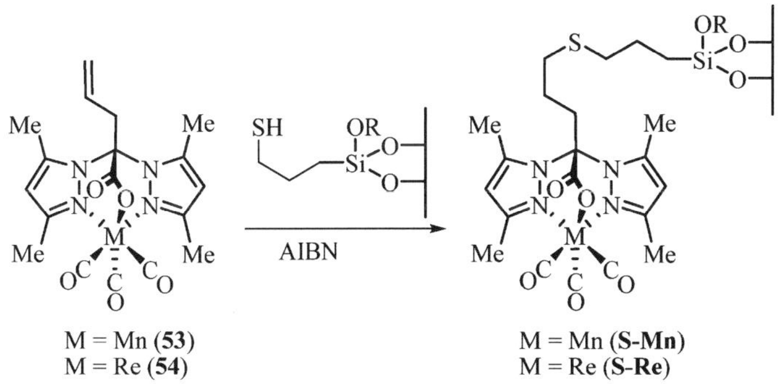

M = Mn (**53**)
M = Re (**54**)

AIBN

M = Mn (**S-Mn**)
M = Re (**S-Re**)

SCHEME 30. Grafting on mercaptopropyl silica (*68*).

rhenium tricarbonyl complexes on silica via a *N,N,O* tripod ligand with an allyl linker (*68*).

In conclusion, modifications of the ligand Hbdmpza (**3b**) lead to the new functionalized ligands 2,2-bis(3,5-dimethylpyrazol-1-yl)pent-4-enoic acid (**48**) and 2,2-bis(3,5-dimethylpyrazol-1-yl)-3-hydroxypropionic acid (**49**) bearing an allyl or an hydroxymethyl linker for solid phase fixation. These ligands or their protected analogues provide similar coordination properties as **3b**. Ligands **48**, **49** and complexes thereof have been grafted successfully on silica and on Merrifield polymer. This is proven by the characteristic IR spectra of immobilized tricarbonyl complexes.

VIII. Conclusion

The *N,N,O*-binding motif is found in many non-heme iron enzymes as well as in some zinc enzymes as metal-binding motif. Thus, to mimic this motif is the purpose of small organic κ^3 *N,N,O*-ligands such as various bis(pyrazol-1-yl)acetato ligands. These ligands can be tailored with bulky substituents to modify their steric hindrance and with solid phase linkers for solid phase fixation. With ferrous and ferric iron complexes bearing these ligands structural models for the iron oxygenases were obtained. Model complexes with ruthenium allow the coordination of substrate analogous. Zinc bis(pyrazol-1-yl)acetato complexes mimic the active sites of zinc proteases. The current interest in these zinc complexes focuses on the coordination of protease inhibitors to the zinc models. The new developed zinc protease models might serve as tools to develop and test new ZBGs. Zinc alkyl complexes as precursors allow an easy access to these model complexes. The anionic κ^3 *N,N,O*-ligands are also quite useful in organometallics and coordination chemistry and allow a chemistry comparable to that of cyclopentadienyl (Cp) or hydrotris(pyrazol-1-yl)borate ligands (Tp).

This includes future potential applications in radiopharmaceuticals. Furthermore, new chiral enantiopure *N,N,O* tripod ligands have been developed starting from cheap compounds of the chiral pool.

Acknowledgements

Generous financial support by the Fonds der Chemischen Industrie (Liebig-Stipendium to N.B.) and the Deutsche Forschungsgemeinschaft (BU 1223/2-1, BU 1223/4-1, BU 1223/5–1 and SFB 583) is gratefully acknowledged. Special thanks to Prof. Dr. H. Fischer for support and discussion. We acknowledge a generous gift of ruthenium trichloride hydrate by the Degussa AG.

References

1. Holm, R. H. *Chem. Rev.* **1987**, *87*, 1401–1449.
2. Feig, A. L.; Lippard, S. J. *Chem. Rev.* **1994**, *94*, 759–805.
3. Que, L., Jr.; Ho, R. Y. N. *Chem. Rev.* **1996**, *96*, 2607–2624.
4. Nivorozhkin, A. L.; Girerd, J.-J. *Angew. Chem Int. Ed. Engl.* **1996**, *35*, 609–611.
5. Bugg, T. D. H.; Winfield, C. J. *Nat. Prod. Rep.* **1998**, *15*, 513–530.
6. Lange, S. J.; Que, L., Jr. *Curr. Opin. Chem. Biol.* **1998**, *2*, 159–172.
7. Schofield, C. J.; Zhang, Z. *Curr. Opin. Struct. Biol.* **1999**, *9*, 722–731.
8. Solomon, E. I.; Brunold, T. C.; Davis, M. I.; Kemsley, J. N.; Lee, S.-K.; Lehnert, N.; Neese, F.; Skulan, A. J.; Yang, Y.-S.; Zhou, J. *Chem. Rev.* **2000**, *100*, 235–349.
9. Prescott, A. G.; Lloyd, M. D. *Nat. Prod. Rep.* **2000**, *17*, 367–383.
10. Que, L., Jr. *Nat. Struct. Biol.* **2000**, *7*, 182–184.
11. Bugg, T. D. H. *Curr. Opin. Chem. Biol.* **2001**, *5*, 550–555.
12. Messerschmidt, A.; Huber, R.; Poulos, T.; Wieghardt K. (Eds.) "*Handbook of Metalloproteins*"; Wiley: Weinheim, **2001**, pp. 613–682.
13. Bertini, I.; Sigel, A.; Sigel, H. (Eds.) "*Handbook on Metalloproteins*";, Dekker, Basel, **2001**, pp. 461–569.
14. Ryle, M. J.; Hausinger, R. P. *Curr. Opin. Chem. Biol.* **2002**, *6*, 193–201.
15. Bugg, T. D. H. *Tetrahedron* **2003**, *59*, 7075–7101.
16. Solomon, E. I.; Decker, A.; Lehnert, N. *Proc. Natl. Acad. Sci. USA* **2003**, *100*, 3589–3594.
17. Rohde, J.-U.; Bukowski, M. R.; Que, L., Jr. *Curr. Opin. Chem. Biol.* **2003**, *7*, 674–682.
18. Costas, M.; Mehn, M. P.; Jensen, M. P.; Que, L., Jr. *Chem. Rev.* **2004**, *104*, 939–986.
19. Hegg, E. L.; Que, L., Jr. *Eur. J. Biochem.* **1997**, *250*, 625–629.
20. Sami, M. Ph.D. Thesis, University of Oxford, 1998.
21. Burzlaff, N. I.; Rutledge, P. J.; Clifton, I. J.; Hensgens, C. M. H.; Pickford, M.; Adlington, R. M.; Roach, P. L.; Baldwin, J. E. *Nature* **1999**, *401*, 721–724.
22. Ogle, J. M.; Clifton, I. J.; Rutledge, P. J.; Elkins, J. M.; Burzlaff, N. I.; Adlington, R. M.; Roach, P. L.; Baldwin, J. E. *Chem. Biol.* **2001**, *8*, 1231–1237.
23. Elkins, J. M.; Rutledge, P. J.; Burzlaff, N. I.; Clifton, I. J.; Adlington, R. M.; Roach, P. L.; Baldwin, J. E. *Org. Biomol. Chem.* **2003**, *1*, 1455–1460.
24. Baldwin, J. E.; Bradley, M. *Chem. Rev.* **1990**, *90*, 1079–1088.
25. Baldwin, J. E.; Adlington, R. M.; Moroney, S. E.; Field, L. D.; Ting, H.-H. *Chem. Commun.* **1984**, 984–986.
26. Roach, P. L.; Clifton, I. J.; Fülöp, V.; Harlos, K.; Barton, G. J.; Hajdu, J.; Andersson, I.; Schofield, C. J.; Baldwin, J. E. *Nature* **1995**, *375*, 700–704.

27. Roach, P. L.; Schofield, C. J.; Baldwin, J. E.; Clifton, I. J.; Hajdu, J. *Protein Sci.* **1995**, *4*, 1007–1009.
28. Roach, P. L.; Clifton, I. J.; Hensgens, C. M. H.; Shibata, N.; Long, A. J.; Strange, R. W.; Hasnain, S. S.; Schofield, C. J.; Baldwin, J. E.; Hajdu, J. *Eur. J. Biochem.* **1996**, *242*, 736–740.
29. Roach, P. L.; Clifton, I. J.; Hensgens, C. M. H.; Shibata, N.; Schofield, C. J.; Hajdu, J.; Baldwin, J. E. *Nature* **1997**, *387*, 827–830.
30. Schenk, W. A. *Angew. Chem.* **2000**, *112*, 3551–3554. *Angew. Chem., Int. Ed. Engl.* **2000**, *39*, 3409–3411.
31. (a) Elkins, J. M.; Ryle, M. J.; Clifton, I. J.; Dunning Hotopp, J. C.; Lloyd, J. S.; Burzlaff, N. I.; Baldwin, J. E.; Hausinger, R. P.; Roach, P. L. *Biochemistry* **2002**, *41*, 5185–5519. (b) O'Brien, J. R.; Schuller, D. J.; Yang, V. S.; Dillard, B. D.; Lanzilotta, W. N. *Biochemistry* **2003**, *42*, 5547–5554.
32. Valegård, K.; Terwisscha van Scheltinga, A. C.; Lloyd, M. D.; Hara, T.; Ramaswamy, S.; Perrakis, A.; Thompson, A.; Lee, H.-J.; Baldwin, J. E.; Schofield, C. J.; Hajdu, J.; Andersson, I. *Nature* **1998**, *394*, 805–809.
33. Lee, H.-J.; Lloyd, M. D.; Clifton, I. J.; Harlos, K.; Dubus, A.; Baldwin, J. E.; Frere, J.-M.; Schofield, C. J. *J. Biol. Chem.* **2001**, *276*, 18290–18295.
34. Kivirikko, K. I.; Pihlajaniemi, T. *Adv. Enzym. Rel. Areas Mol. Biol.* **1998**, *72*, 325–398.
35. Myllyharju, J.; Kivirikko, K. I. *EMBO J.* **1997**, *16*, 1173–1180.
36. Franklin, T. *Chem. Ind.* **1992**, 54–57.
37. Otero, A.; Fernández-Baeza, J.; Tejeda, J.; Antiñolo, A.; Carrillo-Hermosilla, F.; Diez-Barra, E.; Lara-Sánchez, A.; Fernández-López, M.; Lanfranchi, M.; Pellinghelli, M. A. *J. Chem. Soc. Dalton Trans.* **1999**, 3537–3539.
38. Otero, A.; Fernández-Baeza, J.; Tejeda, J.; Antiñolo, A.; Carrillo-Hermosilla, F.; Díez-Barra, E.; Lara-Sánchez, A.; Fernández-López, M. *J. Chem. Soc. Dalton Trans.* **2000**, 2367–2374.
39. Otero, A.; Fernández-Baeza, J.; Antiñolo, A.; Carillo-Hermosilla, F.; Tejeda, J.; Díez-Barra, E.; Lara-Sanchez, A.; Sanchez-Barba, L.; López-Solera, I. *Organometallics* **2001**, *20*, 2428–2430.
40. Burzlaff, N.; Hegelmann, I.; Weibert, B. *J. Organomet. Chem.* **2001**, *626*, 16–23.
41. Beck, A.; Weibert, B.; Burzlaff, N. *Eur. J. Inorg. Chem.* **2001**, *5*, 521–527.
42. Otero, A.; Fernández-Baeza, J.; Antiñolo, A.; Carillo-Hermosilla, F.; Tejeda, J.; Lara-Sanchez, A.; Sanchez-Barba, L.; Fernández-López, M.; Rodríguez, A. M.; López-Solera, I. *Inorg. Chem.* **2002**, *41*, 5193–5202.
43. Burzlaff, N.; Hegelmann, I. *Inorg. Chim. Acta* **2002**, *329*, 147–150.
44. López Hernández, A.; Müller, R.; Kopf, H.; Burzlaff, N. *Eur. J. Inorg. Chem.* **2002**, *5*, 671–677.
45. Tang, L.-F.; Wang, Z.-H.; Chai, J.-F.; Wang, J.-T. *J. Chem. Cryst.* **2002**, *32*, 261–265.
46. Schwenzer, B.; Schleu, J.; Burzlaff, N.; Karl, C.; Fischer, H. *J. Organomet. Chem.* **2002**, *641*, 134–141.
47. Hegelmann, I.; Beck, A.; Eichhorn, C.; Weibert, B.; Burzlaff, N. *Eur. J. Inorg. Chem.* **2003**, 339–347.
48. Hegelmann, I.; Burzlaff, N. *Eur. J. Inorg. Chem.* **2003**, 409–411.
49. Beck, A.; Barth, A.; Hübner, E.; Burzlaff, N. *Inorg. Chem.* **2003**, *42*, 7182–7188.
50. Smith, J. N.; Shirin, Z.; Carrano, C. J. *J. Am. Chem. Soc.* **2003**, *125*, 868–869.
51. Hammes, B. S.; Kieber-Emmons, M. T.; Letizia, J. A.; Shirin, Z.; Carrano, C. J.; Zakharov, L. N.; Rheingold, A. L. *Inorg. Chim. Acta* **2003**, *346*, 227–238.
52. Kozlevčar, B.; Gamez, P.; de Gelder, R.; Driessen, W. L.; Reedijk, J. *Eur. J. Inorg. Chem.* **2003**, 47–51.
53. Otero, A.; Fernández-Baeza, J.; Antiñolo, A.; Tejeda, J.; Lara-Sanchez, A.; Sanchez-Barba, L.; Rodríguez, A. M. *Eur. J. Inorg. Chem.* **2004**, 260–266.

54. Otero, A.; Fernández-Baeza, J.; Antiñolo, A.; Tejeda, J.; Lara-Sanchez, A.; Sanchez-Barba, L.; Fernández-López, M.; López-Solera, I. *Inorg. Chem.* **2004**, *43*, 1350–1358.
55. Peters, L.; Burzlaff, N. *Polyhedron* **2004**, *23*, 245–251.
56. Ortiz, M.; Díaz, A.; Cao, R.; Otero, A.; Fernández-Baeza, J. *Inorg. Chim. Acta* **2004**, *357*, 19–24.
57. Ortiz, M.; Díaz, A.; Cao, R.; Suardíaz, R.; Otero, A.; Antiñolo, A.; Fernández-Baeza, J. *Eur. J. Inorg. Chem.* **2004**, 3353–3357.
58. Müller, R.; Hübner, E.; Burzlaff, N. *Eur. J. Inorg. Chem.* **2004**, 2151–2159.
59. Hammes, B. S.; Chohan, B. S.; Hoffman, J. T.; Einwachter, S.; Carrano, C. J. *Inorg. Chem.* **2004**, *43*, 7800–7806.
60. Marchetti, F.; Pellei, M.; Pettinari, C.; Pettinari, R.; Rivarola, E.; Santini, C.; Skelton, B. W.; White, A. H. *J. Organomet. Chem.* **2005**, *690*, 1878–1888.
61. Smith, J. N.; Hoffman, J. T.; Shirin, Z.; Carrano, C. J. *Inorg. Chem.* **2005**, *44*, 2012–2017.
62. Weng, Z.-H.; Song, H.-B.; Du, M.; Zhai, Y.-P.; Tang, L.-F. *Appl. Organomet. Chem.* **2005**, *19*, 1055–1059.
63. Porchia, M.; Papini, G.; Santini, C.; Lobbia, G. G.; Pellei, M.; Tisato, F.; Bandoli, G.; Dolmella, A. *Inorg. Chem.* **2005**, *44*, 4045–4054.
64. Porchia, M.; Papini, G.; Santini, C.; Lobbia, G. G.; Pellei, M.; Tisato, F.; Bandoli, G.; Dolmella, A. *Inorg. Chim. Acta* **2006**, *359*, 2501–2508.
65. Chandrasekhar, V.; Thilagar, P.; Sasikumar, P. *J. Organomet. Chem.* **2006**, *691*, 1681–1692.
66. Kitanovski, N.; Golobic, A.; Ceh, B. *Inorg. Chem. Commun.* **2006**, *9*, 296–299.
67. Kopf, H.; Pietraszuk, C.; Hübner, E.; Burzlaff, N. *Organometallics* **2006**, *25*, 2533–2546.
68. Hübner, E.; Haas, T.; Burzlaff, N. *Eur. J. Inorg. Chem.* **2006**, 4989–4997.
69. Manivannan, V.; Hoffman, J. T.; Dimayuga, V. L.; Dwight, T.; Carrano, C. J. *Inorg. Chim. Acta* **2007**, *360*, 529–534.
70. Otero, A.; Fernández-Baeza, J.; Antiñolo, A.; Tejeda, J.; Lara-Sánchez, A. *J. Chem. Soc. Dalton Trans.* **2004**, 1499–1510.
71. Pettinari, C.; Pettinari, R. *Coord. Chem. Rev.* **2005**, *249*, 663–691.
72. (a) Juliá, S.; Sala, P.; del Mazo, J.; Sancho, M.; Ochoa, C.; Elguero, J.; Fayet, J.-P.; Vertut, M.-C. *J. Heterocycl. Chem.* **1982**, *19*, 1141–1145. (b) Antiñolo, A.; Carrillo-Hermosilla, F.; Díez-Barra, E.; Fernández-Baeza, J.; Fernández-López, M.; Lara-Sánchez, A.; Moreno, A.; Otero, A. *J. Chem. Soc. Dalton Trans.* **1998**, *5*, 3737–3743.
73. Zhang, Z.; Ren, J.; Stammers, D. K.; Baldwin, J. E.; Harlos, K.; Schofield, C. J. *Nat. Struct. Biol.* **2000**, *7*, 127–133.
74. Ito, M.; Amagai, H.; Fukui, H.; Kitajima, N.; Moro-oka, Y. *Bull. Chem. Soc. Jpn.* **1996**, *69*, 1937–1945.
75. Hikichi, S.; Ogihara, T.; Fujisawa, K.; Kitajima, N.; Akita, M.; Moro-oka, Y. *Inorg. Chem.* **1997**, *36*, 4539–4547.
76. Kepert, D. L. *"Inorganic Stereochemistry"*; Springer: Berlin, **1982**; p. 36.
77. Kepert, D. L. In: *"Comprehensive Coordination Chemistry"*; vol. 1; Eds. Wilkinson, G.; Gillard, R. D.; McCleverty, J. A. Pergamon: Oxford, **1987**, p. 31.
78. Holm, R. H.; Kennepohl, P.; Solomon, E. I. *Chem. Rev.* **1996**, *96*, 2239–2314.
79. Lainé, P.; Gourdon, A.; Launay, J.-P.; Tuchagues, J.-P. *Inorg. Chem.* **1995**, *34*, 5150–5155.
80. Tommasi, L.; Shechter-Barloy, L.; Varech, D.; Battioni, J.-P.; Donadieu, B.; Verelst, M.; Bousseksou, A.; Mansuy, D.; Tuchagues, J.-P. *Inorg. Chem.* **1995**, *34*, 1514–1523.
81. Ogihara, T.; Hikichi, S.; Akita, M.; Uchida, T.; Kitagawa, T.; Moro-oka, Y. *Inorg. Chim. Acta* **2000**, *297*, 162–170.
82. Chiou, Y.-M.; Que, L., Jr. *J. Am. Chem. Soc.* **1992**, *114*, 7567–7568.

83. Chiou, Y.-M.; Que, L., Jr. *Inorg. Chem.* **1995**, *34*, 3270–3278.
84. Chiou, Y.-M.; Que, L., Jr. *J. Am. Chem. Soc.* **1995**, *117*, 3999–4013.
85. Ha, E. H.; Ho, R. Y. N.; Kisiel, J. F.; Valentine, J. S. *Inorg. Chem.* **1995**, *34*, 2265–2266.
86. Hegg, E. L.; Ho, R. Y. N.; Que, L., Jr. *J. Am. Chem. Soc.* **1999**, *121*, 1972–1973.
87. Ho, R. Y. N.; Mehn, M. P.; Hegg, E. L.; Liu, A.; Ryle, M. J.; Hausinger, R. P.; Que, L., Jr. *J. Am. Chem. Soc.* **2001**, *123*, 5022–5029.
88. Mehn, M. P.; Fujisawa, K.; Hegg, E. L.; Que, L., Jr. *J. Am. Chem. Soc.* **2003**, *125*, 7828–7842.
89. Scott, A. P.; Radom, L. *J. Phys. Chem.* **1996**, *100*, 16502–16513.
90. Wiberg, K. B.; Wendoloski, J. J. *J. Phys. Chem.* **1984**, *88*, 586–593.
91. (a) Deacon, G. B.; Phillips, R. J. *Coord. Chem. Rev.* **1980**, *33*, 227–250. (b) Nakamoto, K. *"Infrared and Raman Spectra of Inorganic and Coordination Compounds"*; Wiley-VCH: New York, **1997**; pp. 59–62.
92. (a) Oliver, J. D.; Mullica, D. F.; Hutchinson, B. B.; Milligan, W. O. *Inorg. Chem.* **1980**, *19*, 165–169. (b) Briois, V.; Sainctavit, P.; Long, G. J.; Grandjean, F. *Inorg. Chem.* **2001**, *40*, 912–918. (c) Calogero, S.; Gioia Lobbia, G.; Cecchi, P.; Valle, G.; Friedl, J. *Polyhedron* **1994**, *13*, 87–97. (d) Cecchi, P.; Berrettoni, M.; Giorgetti, M.; Gioia Lobbia, G.; Calogero, S.; Stievano, L. *Inorg. Chim. Acta* **2001**, *318*, 67–76.
93. Armstrong, W. H.; Lippard, S. J. *Inorg. Chem.* **1985**, *24*, 981–982.
94. (a) Fukui, H.; Ito, M.; Moro-oka, Y.; Kitajima, N. *Inorg. Chem.* **1990**, *29*, 2868–2870. (b) Arulsamy, N.; Bohle, D. S.; Hansert, B.; Powell, A. K.; Thomson, A. J.; Wocaldo, S. *Inorg. Chem.* **1998**, *37*, 746–750.
95. Cho, S.-H.; Whang, D.; Han, K.-N.; Kim, K. *Inorg. Chem.* **1992**, *31*, 519–522.
96. Baldwin, J. E. Private communication.
97. Andersen, O. A.; Flatmark, T.; Hough, E. *J. Mol. Biol.* **2001**, *314*, 266–291.
98. Holland, D. R.; Barclay, P. L.; Danilewicz, J. C.; Matthews, B. W.; James, K. *Biochemistry* **1994**, *33*, 51–56.
99. Lipscomb, W. N.; Sträter, N. *Chem. Rev.* **1996**, *96*, 2375–2433.
100. Gomis-Rüth, F. X. *Mol. Biotech.* **2003**, *24*, 157–202.
101. Natesh, R.; Schwager, S. L. U.; Sturrock, E. D.; Acharya, K. R. *Nature* **2003**, *421*, 551–554.
102. Kim, H. M.; Shin, D. R.; Yoo, O. J.; Lee, H.; Lee, J.-O. *FEBS Lett.* **2003**, *538*, 65–70.
103. Pannifer, A. D.; Wong, T. Y.; Schwarzenbacher, R.; Renatus, M.; Petosa, C.; Bienkowska, J.; Lacy, D. B.; Collier, R. J.; Park, S.; Leppla, S. H.; Hanna, P.; Liddington, R. C. *Nature* **2001**, *414*, 229–233.
104. Weissmann, G. Annual meeting of the American College of Rheumatology, San Francisco, 2001.
105. Parkin, G. *Chem. Rev.* **2004**, *104*, 699–767.
106. (a) Hammes, B. S.; Carrano, C. J. *Inorg. Chem.* **1999**, *38*, 3562–3568. (b) Hammes, B. S.; Carrano, C. J. *Inorg. Chem.* **1999**, *38*, 4593–4600.
107. (a) Dowling, C.; Parkin, G. *Polyhedron* **1996**, *15*, 2463–2465. (b) Ghosh, P.; Parkin, G. *J. Chem. Soc. Dalton Trans.* **1998**, *5*, 2281–2283.
108. Puerta, D. T.; Cohen, S. M. *Inorg. Chem.* **2002**, *41*, 5075–5082.
109. (a) Puerta, D. T.; Schames, J. R.; Henchman, R. H.; McCammon, J. A.; Cohen, S. M. *Angew. Chem.* **2003**, *115*, 3902–3904. *Angew. Chem. Int. Ed. Engl.* **2003**, *42*, 3772–3774.
110. Puerta, D. T.; Cohen, S. M. *Inorg. Chem.* **2003**, *42*, 3423–3430.
111. (a) Alsfasser, R.; Powell, A. K.; Vahrenkamp, H. *Angew. Chem.* **1990**, *102*, 939–941. *Angew. Chem., Int. Ed.* **1990**, *29*, 898–899. (b) Alsfasser, R.; Powell, A. K.; Trofimenko, S.; Vahrenkamp, H. *Chem. Ber.* **1993**, *126*, 685–694.
112. (a) Gorrell, I. B.; Looney, A.; Parkin, G. *J. Chem. Soc. Chem. Commun.* **1990**, *5*, 220–222. (b) Looney, A.; Han, R.; Gorrell, I. B.; Cornebise, M.; Yoon, K.; Parkin, G.; Rheingold, A. L. *Organometallics* **1995**, *14*, 274–288.

113. Niklas, N.; Walter, O.; Alsfasser, R. *Eur. J. Inorg. Chem.* **2000**, 1723–1731.
114. Gockel, P.; Gelinsky, M.; Vogler, R.; Vahrenkamp, H. *Inorg. Chim. Acta* **1998**, *272*, 115–124.
115. LeCloux, D. D.; Tokar, C. J.; Osawa, M.; Houser, R. P.; Keyes, M. C.; Tolman, W. B. *Organometallics* **1994**, *13*, 2855–2866.
116. Trofimenko, S. "*Scorpionates – The Coordination Chemistry of Polypyrazolylborate Ligands*"; Imperial College Press: London, **1999**, pp. 5–9.
117. Moberg, C. *Angew. Chem.* **1998**, *110*, 260–281. *Angew. Chem., Int. Ed.* **1998**, *37*, 249–268.
118. (a) Tokar, C. J.; Kettler, P. B.; Tolman, W. B. *Organometallics* **1992**, *11*, 2737–2739. (b) LeCloux, D. D.; Tokar, C. J.; Osawa, M.; Houser, R. P.; Keyes, M. C.; Tolman, W. B. *Organometallics* **1994**, *13*, 2855–2866.
119. Review on C_2 symmetry: J. K. Whitesell, J. K. *Chem. Rev.* **1989**, *89*, 1581–1590.
120. Hirschmann, H.; Hanson, K. R. *Tetrahedron* **1974**, *30*, 3649–3656.
121. House, D. A.; Steel, P. J.; Watson, A. A. *Aust. J. Chem.* **1986**, *39*, 1525–1536.
122. Herrmann, W. A.; Serrano, R.; Bock, H. *Angew. Chem.* **1984**, *96*, 364–365. *Angew. Chem. Int. Ed. Engl.* **1984**, *23*, 383–384.
123. Wieghardt, K.; Pomp, C.; Nuber, B.; Weiss, J. *Inorg. Chem.* **1986**, *25*, 1659–1661.
124. Okuda, J.; Herdtweck, E.; Herrmann, W. A. *Inorg. Chem.* **1988**, *27*, 1254–1257.
125. Herrmann, W. A.; Ladwig, M.; Kiprof, P.; Riede, J. *J. Organomet. Chem.* **1989**, *371*, C13–C17.
126. Herrmann, W. A.; Taillefer, M.; de Méric de Bellefon, C.; Behm, J. *Inorg. Chem.* **1991**, *30*, 3247–3248.
127. Herrmann, W. A.; Roesky, P. W.; Kühn, F. E.; Scherer, W.; Kleine, M. *Angew. Chem.* **1993**, *105*, 1768–1770. *Angew. Chem. Int. Ed. Engl.* **1993**, *32*, 1714–1716.
128. Kühn, F. E.; Herrmann, W. A.; Hahn, R.; Elison, M.; Blümel, J.; Herdtweck, E. *Organometallics* **1994**, *13*, 1601–1606.
129. Walther, D.; Gebhardt, P.; Fischer, R.; Kreher, U.; Görls, H. *Inorg. Chim. Acta* **1998**, *281*, 181–189.
130. Kühn, F. E.; Haider, J. J.; Herdtweck, E.; Herrmann, W. A.; Lopez, A. D.; Pillinger, M.; Romão, C. C. *Inorg. Chim. Acta* **1998**, *279*, 44–50.
131. Schlecht, S.; Deubel, D. V.; Frenking, G.; Geiseler, G.; Harms, K.; Magull, J.; Dehnicke, K. *Z. Anorg. Allg. Chem.* **1999**, *625*, 887–891.
132. Degnan, I. A.; Herrmann, W. A.; Herdtweck, E. *Chem. Ber.* **1990**, *123*, 1347–1349.
133. Domingos, Â.; Marçalo, J.; Paulo, A.; Pires de Matos, A.; Santos, I. *Inorg. Chem.* **1993**, *32*, 5114–5118.
134. Herrmann, W. A.; Thiel, W. R.; Kühn, F. E.; Fischer, R. W.; Kleine, M.; Herdtweck, E.; Scherer, W. *Inorg. Chem.* **1993**, *32*, 5188–5194.
135. (a) Slugovc, C.; Schmid, R.; Kirchner, K. *Coord. Chem. Rev.* **1999**, *185–186*, 109–126. (b) Cadierno, V.; Díez, J.; Pilar Gamasa, M.; Gimeno, J.; Lastra, E. *Coord. Chem. Rev.* **1999**, *193–195*, 147–205. (c) Gemel, C.; Wiede, P.; Mereiter, K.; Sapunov, V. N.; Schmid, R.; Kirchner, K. *J. Chem. Soc. Dalton Trans.* **1996**, 4071–4076. (d) Corrochano, A. E.; Jalón, F. A.; Otero, A.; Kubicki, M. M.; Richard, P. *Organometallics* **1997**, *16*, 145–148. (e) Chen, Y. Z.; Chan, W. C.; Lau, C. P.; Chu, H. S.; Lee, H. L. *Organometallics* **1997**, *16*, 1241–1246. (f) Trimmel, G.; Slugovc, C.; Wiede, P.; Mereiter, K.; Sapunov, V. N.; Schmid, R.; Kirchner, K. *Inorg. Chem.* **1997**, *36*, 1076–1083. (g) Gemel, C.; Kickelbick, G.; Schmid, R.; Kirchner, K. *J. Chem. Soc. Dalton Trans.* **1997**, 2113–2117. (h) Slugovc, C.; Mauthner, K.; Kacetl, M.; Mereiter, K.; Schmid, R.; Kirchner, K. *Chem. Eur. J.* **1998**, *4*, 2043–2050. (i) Slugovc, C.; Mereiter, K.; Schmid, R.; Kirchner, K. *J. Am. Chem. Soc.* **1998**, *120*, 6175–6176. (j) Lo, Y.-H.; Lin, Y.-C.; Lee, G.-H.; Wang, Y. Y. *Organometallics* **1999**, *18*, 982–988. (k) Slugovc, C.; Gemel, C.; Shen, J.-Y.; Doberer, D.; Schmid, R.; Kirchner, K.; Mereiter, K.

Monatsh. Chem. **1999**, *130*, 363–375. (l) Slugovc, C.; Mereiter, K.; Schmid, R.; Kirchner, K. *Eur. J. Inorg. Chem.* **1999**, 1141–1149.
136. Slugovc, C.; Mereiter, K.; Zobetz, E.; Schmid, R.; Kirchner, K. *Organometallics* **1996**, *15*, 5275–5277.
137. Slugovc, C.; Doberer, D.; Gemel, C.; Schmid, R.; Kirchner, K.; Winkler, B.; Stelzer, F. *Monatsh. Chem.* **1998**, *129*, 221–233.
138. Buriez, B.; Burns, I. D.; Hill, A. F.; White, A. J. P.; Williams, D. J.; Wilton-Ely, J. D. E. T. *Organometallics* **1999**, *18*, 1504–1516.
139. Pavlik, S.; Gemel, C.; Slugovc, C.; Mereiter, K.; Schmid, R.; Kirchner, K. *J. Organomet. Chem.* **2001**, *617–618*, 301–310.
140. (a) Bruce, M. I. *Chem. Rev.* **1991**, *91*, 197–257. (b) Bruce, M. I. *Chem. Rev.* **1998**, *98*, 2797–2858. (c) Touchard, D.; Dixneuf, P. H. *Coord. Chem. Rev.* **1998**, *178–180*, 409–429. (d) Puerta, M. C.; Valerga, P. *Coord. Chem. Rev.* **1999**, *193–195*, 977–1025. (e) Cadierno, V.; Pilar Gamasa, M.; Gimeno, J. *Eur. J. Inorg. Chem.* **2001** 571–591. (f) Rigaut, S.; Touchard, D.; Dixneuf, P. H. *Coord. Chem. Rev.* **2004**, *248*, 1585–1601. (g) Cadierno, V.; Pilar Gamasa, M.; Gimeno, J. *Coord. Chem. Rev.* **2004**, *248*, 1627–1657.
141. (a) Guerchias, V. *Eur. J. Inorg. Chem.* **2002**, *5*, 783–796. (b) Slugovc, C.; Mereiter, K.; Schmid, R.; Kirchner, K. *Organometallics* **1998**, *17*, 827–831. (c) Rüba, E.; Gemel, C.; Slugovc, C.; Mereiter, K.; Schmid, R.; Kirchner, K. *Organometallics* **1999**, *18*, 2275–2280. (d) Sanford, M. S.; Valdez, M. R.; Grubbs, R. H. *Organometallics* **2001**, *20*, 5455–5463.
142. Slugovc, C.; Sapunov, V. N.; Wiede, P.; Mereiter, K.; Schmid, R.; Kirchner, K. *J. Chem. Soc. Dalton Trans.* **1997**, 4209–4216.
143. Buriez, B.; Cook, D. J.; Harlow, K. J.; Hill, A. F.; Welton, T.; White, A. J. P.; Williams, D. J.; Wilton-Ely, J. D. E. T. *J. Organomet. Chem.* **1999**, *578*, 264–267.
144. Harlow, K. J.; Hill, A. F.; Wilton-Ely, J. D. E. T. *J. Chem. Soc. Dalton Trans.* **1999** 285–291.
145. Jiménez-Tenorio, M. A.; Jiménez-Tenorio, M.; Puerta, M. C.; Valerga, P. *Organometallics* **2000**, *19*, 1333–1342.
146. Sanford, M. S.; Henling, L. M.; Grubbs, R. H. *Organometallics* **1998**, *17*, 5384–5389.
147. Pavlik, S.; Mereiter, K.; Puchberger, M.; Kirchner, K. *J. Organomet. Chem.* **2005**, *690*, 5497–5507.
148. Bruneau, C.; Dixneuf, P. H. *Acc. Chem. Res.* **1999**, *32*, 311–323.
149. Fürstner, A. *Angew. Chem. Int. Ed.* **2000**, *39*, 3012–3043.
150. Katayama, H.; Yoshida, T.; Ozawa, F. *J. Organomet. Chem.* **1998**, *562*, 203–206.
151. Sanford, M. S.; Love, J. A.; Grubbs, R. H. *Organometallics* **2001**, *20*, 5314–5318.
152. Bruce, M. I.; Hall, B. C.; Zaitseva, N. N.; Skelton, B. W.; White, A. H. *J. Chem. Soc. Dalton Trans.* **1998**, 1793–1803.
153. Bianchini, C.; Innocenti, P.; Peruzzini, M.; Romerosa, A.; Zanobini, F. *Organometallics* **1996**, *15*, 272–285.
154. Pavlik, S.; Schmid, R.; Kirchner, K.; Mereiter, K. *Monatsh. Chem.* **2004**, *135*, 1349–1357.
155. Barthel-Rosa, L. P.; Maitra, K.; Fischer, J.; Nelson, J. H. *Organometallics* **1997**, *16*, 1714–1723.
156. Bianchini, C.; Casares, J. A.; Peruzzini, M.; Romerosa, A.; Zanobini, F. *J. Am. Chem. Soc.* **1996**, *118*, 4585–4594.
157. Jiménez-Tenorio, M.; Palacios, M. D.; Puerta, M. C.; Valerga, P. *J. Organomet. Chem.* **2004**, *689*, 2853–2859.
158. Hansen, H. D.; Nelson, J. H. *Organometallics* **2000**, *19*, 4740–4755.
159. Dötz, K. H.; Sturm, W.; Alt, H. G. *Organometallics* **1987**, *6*, 1424–1427.
160. Le Bozec, H.; Ouzzine, K.; Dixneuf, P. H. *Organometallics* **1991**, *10*, 2768–2772.
161. Beddoes, R. L.; Grime, R. W.; Hussain, Z. I.; Whiteley, M. W. *J. Organomet. Chem.* **1996**, *526*, 371–378.

162. Leung, W. H.; Chan, E. Y. Y.; Williams, I. D.; Wong, W.-T. *Organometallics* **1997**, *16*, 3234–3240.
163. Barthel-Rosa, L. P.; Maitra, K.; Nelson, J. H. *Inorg. Chem.* **1998**, *37*, 633–639.
164. Keller, A.; Jasionka, B.; Glowiak, T.; Ershov, A.; Matusiak, R. *Inorg. Chim. Acta* **2003**, *344*, 49–60.
165. Fürstner, A.; Guth, O.; Düffels, A.; Seidel, G.; Liebl, M.; Gabor, B. *Chem. Eur. J.* **2001**, *7*, 4811–4820.
166. Schanz, H.-J.; Jafarpour, L.; Stevens, E. D.; Nolan, S. P. *Organometallics* **1999**, *18*, 5187–5190.
167. (a) Niazi, M. S. K. *Bull. Chem. Soc. Jpn.* **1989**, *62*, 1253–1257. (b) Fleury, M. B.; Dufresne, J. C. *Bull. Soc. Chim. Fr.* **1972**, 844–850. (c) Kunchev, K. V.; Tur'yan, Y. I.; Dinkov, Kh. A. *Russ. J. Gen. Chem.* **1992**, *62*, 311–315. (d) Pedersen, K. J. *Acta Chem. Scand.* **1952**, *6*, 243–256. (e) Kawabata, N.; Higuchi, I.; Yoshida, J.-I. *Bull. Chem. Soc. Jpn.* **1981**, *54*, 3253–3258.
168. Sandford, M. S.; Valdez, M. R.; Grubbs, R. H. *Organometallics* **2001**, *20*, 5455–5463.
169. Pavel, E. G.; Zhou, J.; Busby, R. W.; Gunsior, M.; Townsend, C. A.; Solomon, E. I. *J. Am. Chem. Soc.* **1998**, *120*, 743–753.
170. (a) Joseph, M.; Leigh, T.; Swain, M. L. *Synthesis* **1977**, 459–461. (b) Gimeno, B.; Sancho, A.; Soto, L.; Legros, J.-P. *Acta Cryst. C* **1996**, *52*, 1226–1228. (c) Gimeno, B.; Soto, L.; Sancho, A.; Dahan, F.; Legros, J.-P. *Acta Cryst. C* **1992**, *48*, 1671–1673.
171. (a) Akhriff, Y.; Server-Carrió, J.; Sancho, A.; García-Lozano, J.; Escrivá, E.; Folgado, J. V.; Soto, L. *Inorg. Chem.* **1999**, *38*, 1174–1185. (b) Akhriff, Y.; Server-Carrió, J.; Sancho, A.; García-Lozano, J.; Escrivá, E.; Soto, L. *Inorg. Chem.* **2001**, *40*, 6832–6840. (c) Sancho, A.; Gimeno, B.; Amigó, J.-M.; Ochando, L.-E.; Debaerdemaeker, T.; Folgado, J.-V.; Soto, L. *Inorg. Chim. Acta* **1996**, *248*, 153–158. (d) Núñez, H.; Escrivà, E.; Server-Carrió, J.; Sancho, A.; García-Lozano, J.; Soto, L. *Inorg. Chim. Acta* **2001**, *324*, 117–122.
172. (a) Likó, Z.; Süli-Vargha, H. *Tetrahedron Lett.* **1993**, *34*, 1673–1676. (b) Ősz, K.; Várnagy, K.; Süli-Vargha, H.; Sanna, D.; Micera, G.; Sóvágó, I. *Dalton Trans.* **2003**, 2009–2016. (c) Várnagy, K.; Sóvágó, I.; Ágoston, K.; Likó, Z.; Süli-Vargha, H.; Sanna, D.; Micera, G. *J. Chem. Soc. Dalton Trans.* **1994**, 2939–2945. (d) Ősz, K.; Várnagy, K.; Süli-Vargha, H.; Csámpay, A.; Sanna, D.; Micera, G.; Sóvágó, I. *J. Inorg. Biochem.* **2004**, *98*, 24–32. (e) Várnagy, K.; Sóvágó, I.; Goll, W.; Süli-Vargha, H.; Micera, G.; Sanna, D. *Inorg. Chim. Acta* **1998**, *283*, 233–242.
173. Braussaud, N.; Rüther, T.; Cavell, K. J.; Skelton, B. W.; White, A. H. *Synthesis* **2001** 626–632.
174. Peters, L.; Hübner, E.; Burzlaff, N. *J. Organomet. Chem.* **2005**, *690*, 2009–2016.
175. Bruijnincx, P. C. A.; Lutz, M.; Spek, A. L.; Hagen, W. R.; Weckhuysen, B. M.; van Koten, G.; Klein Gebbink, R. J. M. *Eur. J. Inorg. Chem.* **2005**, 779–787.
176. Bruijnincx, P. C. A.; Lutz, M.; Spek, A. L.; van Faasen, E. E.; Weckhuysen, B. M.; van Koten, G.; Klein Gebbink, R. J. M. *J. Am. Chem. Soc.* **2007**, *129*, 2275–2286.
177. Bernard, J.; Ortner, K.; Spingler, B.; Pietzsch, H.-J.; Alberto, R. *Inorg. Chem.* **2003**, *42*, 1014–1022.
178. Metzler-Nolte, N. *Angew. Chem., Int. Ed. Engl.* **2001**, *113*, 1072–1076. *Angew. Chem., Int. Ed. Engl.* **2001**, *40*, 1040–1044.
179. Le Bideau, F.; Salmain, M.; Top, S.; Jaouen, G. *Chem. Eur. J.* **2001**, *7*, 2289–2294.
180. Recent reviews: (a) Tada, M.; Iwasawa, Y. *Chem. Commun.* **2006**, *5*, 2833–2844. (b) Dioos, B. M. L.; Vankelecom, I. F. J.; Jacobs, P. A. *Adv. Synth. Catal.* **2006**, *348*, 1413–1446.
181. Findeis, R. A.; Gade, L. H. *J. Chem. Soc. Dalton Trans.* **2003**, 249–254.

182. (a) Fraile, J. M.; García, J. I.; Mayoral, J. A.; Royo, A. J. *Tetrahedron: Asymmetry* **1996**, *7*, 2263–2276. (b) Burguete, M. I.; Fraile, J. M.; García, J. I.; García-Verdugo, E.; Herrerías, C. I.; Luis, S. V.; Mayoral, J. A. *J. Org. Chem.* **2001**, *66*, 8893–8901.
183. Reger, D. L.; Grattan, T. C. *Synthesis* **2003**, 350–356.
184. Hübner, E.; Türkoglu, G.; Wolf, M.; Zenneck, U.; Burzlaff, N. *Eur. J. Inorg. Chem.* **2008**, 1226–1235.

HYDROXYPYRANONES, HYDROXYPYRIDINONES, AND THEIR COMPLEXES

JOHN BURGESS[a] and MARIA RANGEL[b]

[a]Department of Chemistry, University of Leicester, Leicester, LE1 7RH, UK
[b]Requimte, Instituto de Ciências Biomédicas de Abel Salazar, Universidade do Porto, 4099-033 Porto, Portugal

I. Introduction

Hydroxypyranones, of which a number occur in the plant kingdom, and hydroxypyridinones, of which a great variety can be synthesized from hydroxypyranones, form complexes with the majority of metal cations. These ligands and complexes are of considerable interest not only in their own right but also in relation to numerous current and potential uses and applications. Thus the hydroxypyranones are of long-term importance as natural products, in food, and as analytical reagents for colorimetric and spectrophotometric determination of a wide range of elements. They are also important, especially in the present context, as precursors of hydroxypyridinones. Hydroxypyridinones and their complexes have been, and are, of importance in solvent extraction, analysis, and several areas of diagnosis and therapy. In many applications, particularly in solvent extraction and

ADVANCES IN INORGANIC CHEMISTRY
VOLUME 60 ISSN 0898-8838 / DOI: 10.1016/S0898-8838(08)00005-6

chelation therapy, the properties of the ligands are almost as important as those of their complexes.

This chapter reviews several areas of the chemistry of these ligands and complexes, including synthetic methods, structures, and some aspects of their solution chemistry – stability constants, solvation, partition, redox potentials, and kinetics and mechanism. It also provides an introduction to their applications in solvent extraction and analysis, to their roles in medical diagnosis and therapy, and to a variety of minor uses. It is not possible to provide anything approaching comprehensive coverage even of our selected areas, but we have tried to provide an overview of a field which extends from natural products through synthetic organic chemistry and fundamental physical inorganic chemistry to medically relevant biochemistry. We have also tried to show how this area has developed from the isolation and characterization of maltol, meconic acid, and kojic acid in the last decade of the 19th century and the first decade of the 20th century, up to the first years of the 21st century. A few historic and general references to hydroxypyranones and hydroxypyridinones provide the starting point for our chronicle of successive advances in the chemistry of this area.

II. Ligands and Complexes – Synthesis and Structure

A. Ligands

A.1. General

Hydroxypyranones and hydroxypyridinones with hydroxylic and ketonic oxygen atoms at appropriate positions on the heterocyclic rings are potentially chelating ligands, their anions forming stable complexes with a range of metal ions. This group of compounds[1] includes 3-hydroxy-2-pyranones (**1**), 3-hydroxy-4-pyranones (**2**), and the closely related 3-hydroxy-2-pyridinones (**3**), 3-hydroxy-4-pyridinones (**4**), and 1-hydroxy-2-pyridinones (**5**). Compounds **1–4** are intermediate in structure between the exhaustively studied β-diketones and catechols (1,2-dihydroxybenzenes), while **5** is a close relative of hydroxamic acid. The general formulae of these ligands are shown in Fig. 1, while Fig. 2 shows the formulae of several simple hydroxypyranones (**6–11**) and hydroxypyridinones (**12**, **13**) which will appear in this review, with their trivial names. Tetra-, hexa-, and octadentate analogues can be prepared by the incorporation of two, three, or four of the chelating units. Hexadentate hydroxypyridinones have been of particular interest in recent years in view of the high stability and substitution-inertness of their metal complexes.

[1]Nomenclature and abbreviations are outlined at the end of this chapter.

1 2 3 4 5

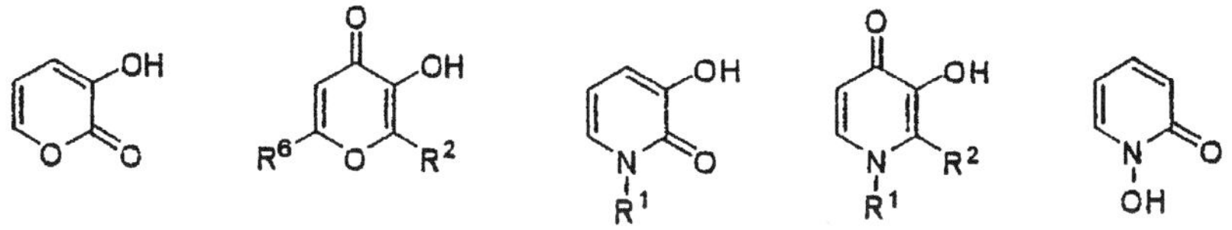

FIG. 1. General formulae for hydroxypyranones and hydroxypyridinones: **1**, 3-hydroxy-2-pyranone; **2**, 3-hydroxy-4-pyranones; **3**, 3-hydroxy-2-pyridinones; **4**, 3-hydroxy-4-pyridinones; **5**, 1-hydroxy-2-pyridinone. In each case the ring atoms are numbered anticlockwise, starting with the ring-oxygen or ring-nitrogen atom.

maltol

6

kojic acid

7

meconic acid

8

pyromeconic acid

9

comenic acid

10

allomaltol

11

mimosine

12

L1; Deferiprone

13

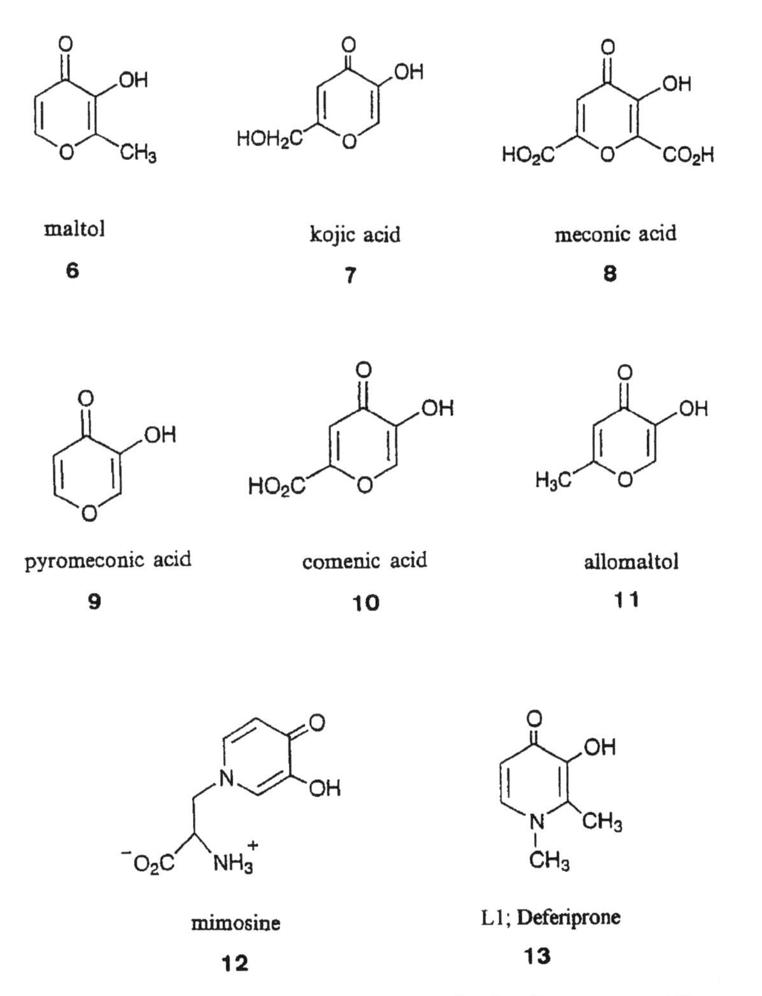

FIG. 2. Named 3-hydroxy-4-pyranones and 3-hydroxy-4-pyridinones.

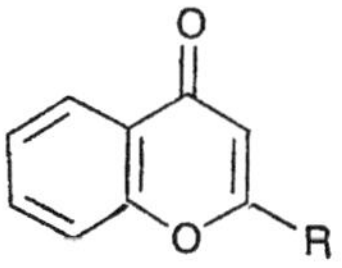

14 R = H : chromone

15 R = Ph : flavone

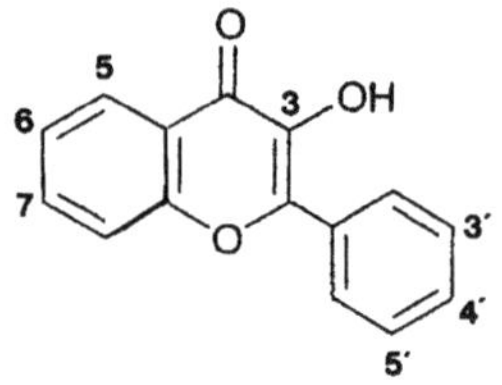

16 flavanol (3-hydroxyflavone)

17 quercetin (3,5,7,3′,4′-pentahydroxyflavone)

18 myricetin (3,5,7,3′,4′,5′-hexahydroxyflavone)

19 quercetagetin (3,5,6,7,3′,4′-hexahydroxyflavone)

20 morin (3,5,7,2′,4′-pentahydroxyflavone)

FIG. 3. Chromone, flavone, and flavanols.

Chromones (**14**) and flavones (phenylchromones, **15**) are pyranones; when they have a hydroxyl group substituted in such a position that they can act as bidentate *O,O*-donors then they and their complexes are relevant to the present review. Fig. 3 shows formulae for chromone, flavone, and the polyhydroxyflavones (**16–20**) which appear in this review.

A.2. Occurrence

Several 3-hydroxy-4-pyranones are natural products. Maltol (maltH; **6**) was first isolated (*1*) from larch bark (*Larix decidua*, hence a rarely encountered alternative name of larixinic acid); it is also found in pine needles. Maltol is a key component of the odor of baking – it is present in many foodstuffs, such as bread and coffee, whose preparation involves roasting (*2*) – and indeed is a permitted food additive. Maltol, and a number of its derivatives and analogues, have been identified in tobacco smoke condensate (*3*). Kojic acid (kojH; **7**) is produced by various bacterial or fungal strains of *Aspergillus*, *Penicillium*, or *Acetobacter* species. It was first isolated from mould of *Aspergillus oryzae* (*4*) (from which it is still commercially produced). It is present in Japanese rice beer (sake) and is a permitted food additive. Meconic acid (mecH; **8**) provides the anion for certain opium poppy alkaloids, and pyromeconic acid (pyromecH; **9**) can be isolated from the flowers and leaves of *Erigeron annus*. The structure of meconic acid was deduced as early as 1884, from its reduction to pimelic acid, $HO_2C(CH_2)_5CO_2H$ (*5*). The 3-hydroxy-4-pyridinone mimosine (leucenol; leucaenol; leucaenine; **12**) is a naturally occurring alkaloid (in *Mimosa pudica*, *Leucaena glauca*, and *Leucaena leucocephala*), the elucidation of whose structure provides an illuminating example of the laboriously indirect classical methods still required in the middle of the 20th century (*6*). Mimosine is a pyridoxal antagonist, inhibiting protein synthesis and growth (*7*) through its chelating effects on the relevant enzymes (*8*). In particular it has depilatory properties, inducing alopecia in animals and in humans (*9*). Observations on the development of alopecia in sheep which had grazed on *L. leucocephala*, a leguminous shrub common in northern Australia, led to (unsuccessful) trials of mimosine as a wool growth inhibitor and defleecing agent (*10*).

A large number of flavone (**15**) derivatives occur in a wide range of plants – in flowers, in wood, and in roots (*11*). Most of these contain a 3-hydroxy-4-pyranone unit, generally with several more hydroxy substituents. Many are used as dyes or colorants, while a few have modest therapeutic value (Section IV.A.2). Quercetin (3,5,7,3′,4′-pentahydroxyflavone; **17**) is the most important and widespread plant pigment of this group. It takes its name from the American oak, *Quercus velutina*, in whose bark it occurs; it is also found in, e.g., roses, wallflowers, pansies, hops, and tea. Myricetin (3,5,7,3′,4′,5′-hexahydroxyflavone; **18**) occurs widely in leaves, wood, and bark; quercetagetin (3,5,6,7,3′,4′-hexahydroxyflavone; **19**) colors the African marigold (*Tagetes patula*). Morin (3,5,7,2′,4′-pentahydroxyflavone; **20**) which occurs in, and is extracted from, mulberry wood (*Morus tinctoria*) is the most relevant to the present review.

A.3. Availability and synthesis

(a) Availability. Maltol and kojic acid have been fairly readily available from natural sources since soon after their isolation, but it was many years before they were synthesized in the laboratory – kojic acid, from glucose via 2,3,4,6-tetraacetyl glucosone hydrate, in 1946 (*12*), maltol (in 17% yield) the following year (*13*). Around this time it was shown that maltol was produced by alkaline hydrolysis of streptomycin (*14*), that it could be isolated, in very small yield, from milk which had undergone prolonged heat treatment (*15*), and that it could be obtained, again in only small amounts, by heating maltose or lactose with glycine (*16*). A high-yield (70%) synthesis for maltol followed some years later, described in a paper which provides a terse review of the various syntheses of maltol and ethylmaltol then extant (*17*). An industrial process for the production of kojic acid was patented in 1959. Thus maltol and kojic acid have been commercially available for many years; ethylmaltol has more recently become readily available, being a commercially used flavorant since the mid-1970s. Flavanol (**16**) has also been commercially available for many years – it is a natural colorant, albeit rather readily bleached by peroxide (*18*).

Comenic acid (**10**) – not to be confused with comanic acid, a 2-pyranone carboxylic acid – can be obtained by decarboxylation (by boiling in 50% hydrochloric acid (*19*)) of meconic acid (**8**), or from kojic acid (**7**) via comenaldehyde (*20*). Allomaltol (**11**), a reduction product of kojic acid, can be prepared by a straightforward method from chlorokojic acid (*21*). Other 3-hydroxy-4-pyranones can be synthesized (*22*), albeit often with difficulty, including the parent unsubstituted 3-hydroxy-4-pyranone and a range of variously alkyl-substituted derivatives (*3*). There are several ring-closure reactions which produce 4-pyranones (*23*), e.g., from MeOCH=CH–C≡C–CH(OEt)$_2$ or from acid anhydrides RCH$_2$COOCOCH$_2$R (*24*) – the latter approach offers possibilities for producing multisubstituted and multifunctional 4-pyranones. However there may well be problems in obtaining starting materials containing the correctly positioned hydroxyl group for the synthesis of a specific required 3-hydroxy-4-pyranone derivative.

(b) 3-Hydroxy-4-pyridinones. Hydroxypyranones such as maltol, kojic acid, and comenic acid can be converted into hydroxypyridinones by reaction with ammonia, with primary amines, or with bifunctional amines such as amino acids. In such reactions the pyranone ring –O– is replaced by –NR–, thus providing a route to a great variety of potential ligands. An early example involved the production of 3-hydroxy-4-pyridinones from kojic acid by reaction of its mono- or dimethyl ester with ammonia or methylamine (*25*). Subsequently a "total synthesis" of mimosine through the appropriate 3-hydroxy-4-pyranone and 3-methoxy-4-pyridinone, based on a 19th century conversion of 4-pyranones into 4-pyridinones, was detailed (*26*). Further examples of conversions of hydroxypyranones into hydroxypyridinones followed (*27–30*), including another synthesis of mimosine, through reaction of the benzyl or methyl ester of pyromeconic acid (**9**) with HO$_2$C(TosHN)CHCH$_2$NH$_2$ (*31*). In principle these conversions are straightforward, sometimes involving simply a one-pot synthesis, as in the reactions of kojic acid or comenic acid with ammonia. Several 1-aryl derivatives were synthesized by heating aromatic amines with maltol in a sealed tube at 150°C (*32,33*); subsequently it was established that the 1,2-dimethyl (*34,35*) and many 1-(4′-X-phenyl) (X = e.g., H (*36*) or *n*-alkyl (*37*) and 1-(4′-X-benzyl) ligands (X = e.g., Me, F, or CF$_3$ (*38*)) could be synthesized simply by refluxing a solution of the reactants for many hours. However for some amines reaction with a hydroxypyranone is a long, tedious, and low yield process. It is often necessary to protect the hydroxyl group (*39–43*), which leads to a time-consuming sequence of methylation or benzylation, reaction with amine, extraction, deprotection by hydrogenation, and purification. Sometimes the use of a dipolar aprotic solvent such as dimethyl sulfoxide or acetonitrile can encourage the amine to react, as e.g., with trifluoroethylamine, but this approach can fail in that a solvent of this type may encourage Schiff base formation at the expense of the desired Michael reaction (*35*). Purification is usually by recrystallization, but occasionally, as for instance for 1-*n*-hexyl-2-methyl-3-hydroxy-4-pyridinone (*42*), sublimation has been preferred.

The fact that almost any amine can be reacted with a hydroxypyranone, given the right conditions, means that a great variety of derivatized hydroxypyridinones are accessible. The hydrophilic–lipophilic balance (HLB) (*44*) of ligands, and indeed of complexes, is crucial to transport across biological membranes. It is usually assessed through determination of partition or distribution coefficients (see Section III.C.3 later). The HLB of ligands can be tailored, with small or negligible changes in their chelating properties (*45,46*), to optimize this property for delivery to the sites where the ligand is required (*47*). This can be achieved simply by introducing appropriate substituents on the endocyclic nitrogen atom (*48*). Thus it is possible to vary the alkyl groups on the ring-nitrogen, with e.g., increasing size of alkyl or aryl substituent giving increasing lipophilicity. Substituted benzyl groups provide a means of varying the ligand's HLB while having an insignificant effect on its donor properties and thus on stabilities of its complexes (*49*). Lipophilicity can also be tuned by F-for-H replacement, an approach which may be relevant for pharmaceutical applications. Indeed 1-CH_2CF_3 and 1-$CH_2CH_2CF_3$ derivatives of 3-hydroxy-2-pyridinones have been claimed to be of particular interest (*50*). Maltol and ethylmaltol react, albeit reluctantly, with amines such as $CF_3CH_2NH_2$. Maltol and ethylmaltol react much more readily with aromatic amines such as 4F- or 4-CF_3-aniline (*51*), and even react with long-chain fluorinated amines such as n-$C_6F_{13}CH_2CH_2NH_2$ (*52*). However direct fluorination, attempted for several ligands, was unsuccessful (*53*). Hydrophilicity can be enhanced by the introduction of appropriate groups such as hydroxyl, carboxylate, or sulfonate. This can be achieved by starting from kojic acid (*54*) rather than maltol, or by reacting the starting hydroxypyranone with amines containing groups such as carboxy, sulfonate, or imidazole. 1-Carboxymethyl-2-methyl-3-hydroxy-4-pyridinone can readily, if slowly, be prepared by reacting maltol with glycine (*55*), or from the reaction of isomaltol with glycine (*29*). Sulfonate has been introduced by reacting taurine ($HO_3SCH_2CH_2NH_2$) with kojic acid (*35*). The incorporation of an imidazole group (*56,57*) also provides a second center for substrate binding. The presence of groups such as imidazole, carboxy, sulfonate, or amino permits ligand variation, through pH control of protonation equilibria, of charge and of the HLB. The HLB of these ligands, and of course of their complexes, may also be tuned by varying substituted phenyl groups on the ring-nitrogen (*58*). Indeed it is possible to probe the extent of substituent effects across a $-CH_2-$ barrier by synthesizing ranges of ligands with substituted phenyl and analogous ring-substituted benzyl groups on the ring-nitrogen (*59*). The attachment of a glucose moiety or a ribonucleoside to a hydroxypyridinone may facilitate transmembrane transport through the sugar-transport pathway (*60*). 1-(2′-Hydroxyethoxy)methyl-2-alkyl-3-hydroxy-4-pyridinones form a link between these sugar derivatives and simple 1-alkyl ligands, in that the $-CH_2OCH_2CH_2OH$ approximates to half a monosaccharide molecule (*61*).

It may be advantageous to be able to target specific iron-binding sites or reservoirs. Thus ester prodrugs have been developed which target the liver (*62,63*), while the incorporation of an imidazole group, as mentioned in an earlier paragraph, offers the possibility of targeting lysosome iron (*57*). Particularly small lipophilic chelators which can make their way through narrow hydrophobic channels can be used to target active sites of lipoxygenase (*64*) – bulky or hydrophilic chelators are feeble inhibitors of lipoxygenase (*65*).

Chelating resins have been prepared in which hydroxypyridinones have been incorporated into sepharose gels (*66*). These materials were developed, along with desferrioxamine analogues (*67*), with a particular view to immobilizing iron(III). Other specialty hydroxypyridinones have included several containing alkene units, incorporated for their lipophicity, their potential for use in hydroboration (*68*), or for their intrinsic interest (*69*).

The fact that pyromeconic acid and allomaltol were only available with difficulty meant that direct synthesis of certain 3-hydroxy-4-pyridinones was not possible. However the demonstration that some of these compounds were accessible from maltol or ethylmaltol by functionalizing the position adjacent to the ring-oxygen by an aldol condensation and N-oxide intermediates led to the preparation of 2-(1′-hydroxyalkyl) and 2-amido derivatives with usefully high affinities for Fe^{3+} (*70*).

2-Alkyl-3-hydroxy-4-pyridinones can be converted into analogues containing, e.g., anilino-, phenylthio-, or 2-hydroxyethylthio-substituents by silver(I) oxidation (Ag_2O in ethanol) followed by Michael addition (*71*). In aminomethylation of 3-hydroxy-4- and -2-pyridinones under Mannich conditions the position of substitution can be tailored, by reaction conditions to position C4 or C6, or by converting the OH into OMe, which directs substitution to C5 (*72*).

Many years ago (*29*), and recently (*73*), it has been reported that 1-alkyl-2-methyl-3-hydroxy-4-pyridinones can be prepared from carbohydrates. This was achieved, though only in low yields, simply by heating maltose or lactose with primary aliphatic amines in neutral aqueous solution. The reaction of maltose with methylammonium acetate yields a score of products, including 1,2-dimethyl-3-hydroxy-4-pyridinone in less than 2% yield (*29*). 1-Alkyl-2-methyl-3-hydroxy-4-pyridinones can be synthesized in reasonable yields in a three-stage sequence from carbohydrate precursors (*74*). This involves dehydration of, e.g., α-D-lactose using piperidine acetate to give *O*-galactosylisomaltol, which can be converted into isomaltol enzymatically or by steam distillation from aqueous phosphoric acid (*75*). Isomaltol is then converted into a 1-alkyl-2-methyl-3-hydroxy-4-pyridinone by heating with methylammonium acetate (*29*). We digress here to point out that isomaltol is not a hydroxypyranone but a hydroxyfuranone. However it can act as a bidentate *O,O*-donor ligand, through its adjacent hydroxyl and –COMe substituents, though it has a considerably lower affinity for metal ions than maltol (*cf.* footnote d to Table VII in Section III.A.3.a) (*76,77*).

(c) 3-Hydroxy-2-pyridinones. Bidentate 3-hydroxy-2-pyridinones may be prepared by amidation. The methyl ester of 2,5-dimethoxytetrahydrofuran-2-carboxylic acid, obtained by electrolytic oxidation of methyl 2-furoate, is reacted with a primary alkylamine, followed by acid-catalyzed rearrangement. This gives a 1-alkyl-3-hydroxy-2-pyridinone. The rearrangement is promoted by metal ions, such as Fe^{3+}, which form stable complexes with the product hydroxypyridinone (*78*).

(d) Hexadentate hydroxypyridinones. By appropriate choice of reagents it is possible to synthesize tetradentate, hexadentate, and octadentate hydroxypyridinonates. The chelating hydroxypyridinone units can be attached to a central linking or capping unit by a range of synthetic strategies (*79*). Figs. 4(a) and 4(b) shows a range of capping units (scaffolds) that have been used for the assembly of tripodal (semi-encapsulating) hexadentate ligands, Fig. 4(c) shows how a hopo unit may be linked to the scaffold through bonding to ring-C or to ring-N to give the hexadentate tris-hydroxypyridinonate tren(3,2-hopo)$_3$, **21** (Fig. 4(d)). A few examples of tetradentate, hexadentate, and octadentate hydroxypyridinones are shown in Fig. 5. A template synthesis of the triserine trilactone scaffold of enterobactin permits the preparation of hydroxypyridinone analogues of this powerful siderophore (*80*). Linear as well as tripodal hexadentate and octadentate ligands have been synthesized and assessed as metal-binding chelators (see e.g., **23** in Fig. 5 and Ref. (*81*)).

A rather different route to hexadentate ligands containing 3-hydroxy-2-pyridinone units is based on the reaction of an appropriate pyrazinone with dimethyl acetylene dicarboxylate, which replaces one of the ring-nitrogen atoms by carbon (*82*).

A.4. *Structures*

The structures of many hydroxypyranones and hydroxypyridinones have been established by X-ray diffraction techniques, although it has sometimes proved difficult to grow suitable crystals. 3-Hydroxy-4-pyridinones crystallize in the form of hydrogen-bonded centrosymmetric dimers, as established for the 1,2-dimethyl (*42,83,84*), 1-ethyl-2-methyl (*85*), and 1,2-diethyl compounds (*43,86*), and for several 1-aryl (*58*) and 1-benzyl analogues (*59*). The situation is sometimes complicated by the existence of more than one polymorph, as is the case for, for instance, maltol (two polymorphs, one a chain structure, the other consisting of hydrogen-bonded dimers) (*87*), ethylmaltol (two of whose three polymorphs are analogous to those of maltol, the third consisting of spiral chains) (*88*), and 1-carboxymethyl-2-methyl-3-hydroxy-4-pyridinone (*89*). Much of the interest in the structural chemistry of this group of ligands lies in

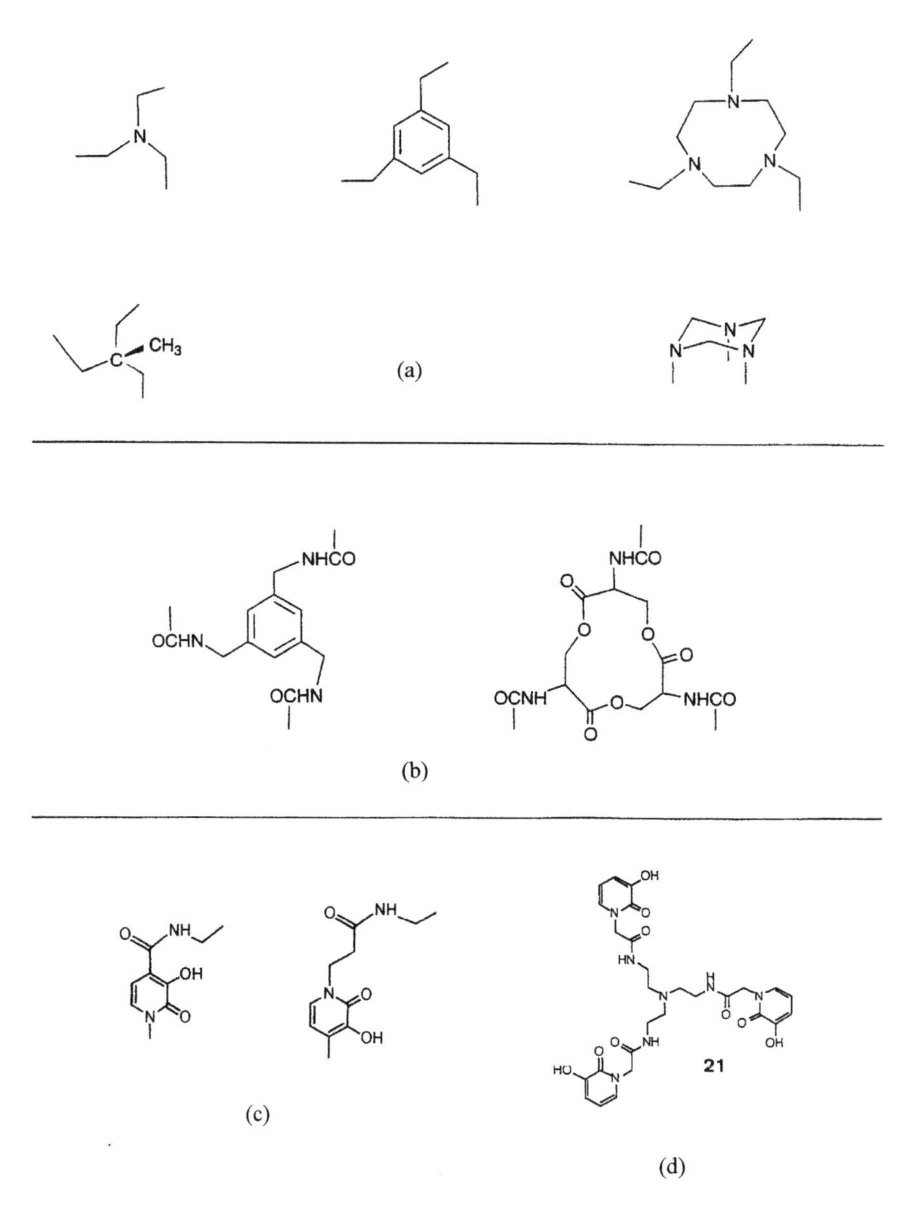

FIG. 4. Capping units (a) bonding through peptide links, (b) to ring-carbon or ring-nitrogen, (c) to give hexadentate hydroxypyridinones, (d) such as **21**, tren(3,2-hopo)$_3$.

the information on hydrogen-bonding, both within and between ligand molecules, and on interactions with and within the surrounding water molecules in the case of hydrates. There is a further aspect of some fundamental interest, and that is the use of established bond distances within the heterocyclic rings to throw light on the question of

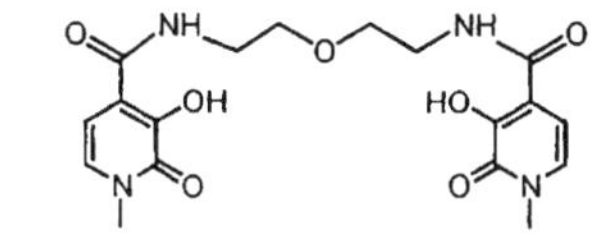

22 5LIO(Me-3,2-hopo)$_2$

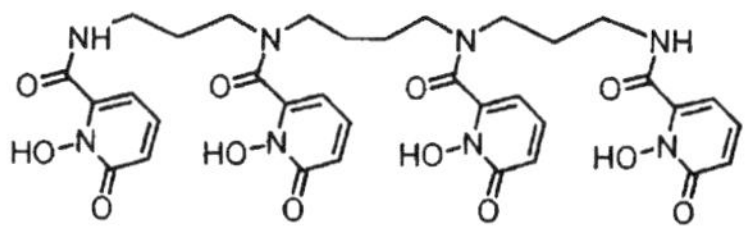

23 3,4,3-LI(1,2-hopo)$_4$

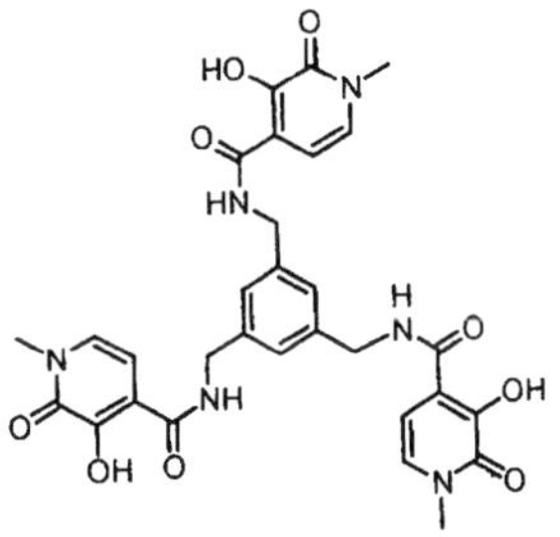

24 mes(3,2-hopo)$_3$

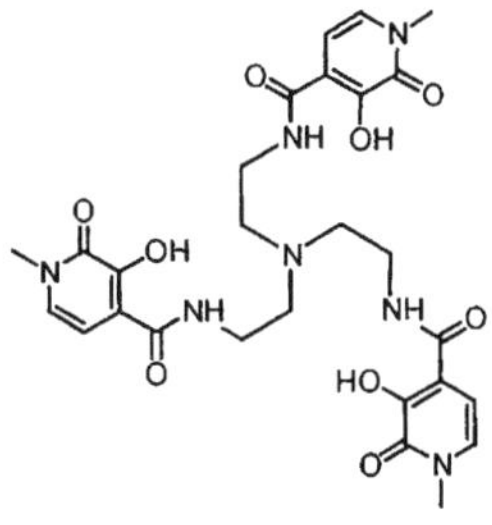

25 tren(Me-3,2-hopo)$_3$

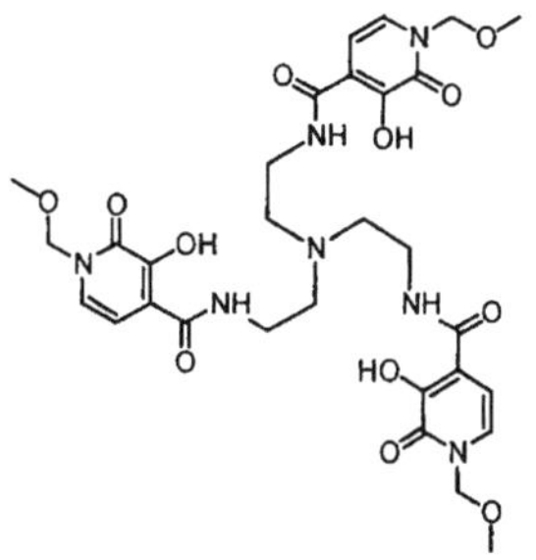

26 tren(MOE-3,2-hopo)$_3$

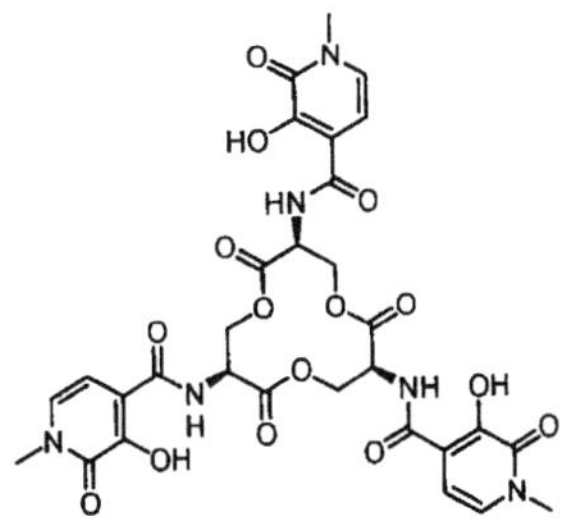

27 entero(3,2-hopo)$_3$ = hopobactin

FIG. 5. Tetra-, octa-, and hexa-dentate hydroxypyridinones.

the tautomeric/mesomeric equilibrium between the delocalized but non-aromatic forms shown in the representations **1–6** earlier and their zwitterionic forms. The report of the structure of a second monoclinic polymorph of 1-carboxymethyl-2-methyl-3-hydroxy-4-pyridinone illustrates these points well (*89*).

Many crystal structures have also been reported for hydrochlorides or hydrobromides of hydroxypyranones and hydroxypyridinones. Such salts, unlike their parent bases, generally prove much more amenable to growing crystals of good quality for X-ray structure determination. Waters of crystallization in the halide salts are hydrogen-bonded to the halide anions, often bridging the halides to hydroxyl groups of the organic cation. This is the case in, e.g., the hydrated hydrochloride of 1-*n*-hexyl-2-ethyl-3-hydroxy-4-pyridinone, where hydrated chlorides form a chain along the *c* axis, the chain being linked to the organic cations by hydrogen-bonding to both the 3- and the 4-hydroxy groups (*86*). In contrast, in the hydrobromide hydrate of 1,2-dimethyl-3-hydroxy-4-pyridinone there is no significant hydrogen-bonding between the hydrated bromide and the organic cation (*83*). The formate and acetate salts of 1,2-dimethyl-3-hydroxy-4-pyridinone are probably better considered as formic acid and acetic acid solvates. The unit cell of the latter contains two hydroxypyridinone and two acetic acid molecules alternating in an almost planar hydrogen-bonded ring of 2+2 stoichiometry; the formic acid solvate has a similar molecular arrangement, but in the form of a near-planar four-bladed propellor (*84*).

B. Complexes

Hydroxypyranones and hydroxypyridinones give stable complexes with a variety of 2+, 3+, 4+, 5+ and 6+ metal centers. The earliest reported complexes were of M^{2+} cations (see Section II.B.2 later); the most studied complexes have been those of M^{3+} cations. Of these, Fe^{3+}, the Group 13 cations Al^{3+}, Ga^{3+}, and In^{3+}, and the lanthanide Gd^{3+} have been especially fully investigated. These cations normally form tris-ligand complexes ML_3 and hexadentate ligand analogues ML, while f-block cations also form complexes with tetradentate and octadentate hydroxypyridinones. Metal 4+ cations often form ternary complexes, such as the metal-ligand-(pseudo)halide complexes ML_2X_2, where M may be Ti, Zr, Hf, or Sn and X = halide or thiocyanate. In the cases of vanadium(IV), metal(V), and metal(VI) centers, bidentate, tetradentate, and hexadentate hydroxypyridinones form oxometal complexes, e.g., VOL_2, UO_2L_2, or MoO_3L. Many metal(II) cations, both from the sp-block and the d-block, form complexes of the expected formula ML_2, though sometimes such complexes have unexpected structures.

B.1. Preparations

Once the required ligands have been obtained, the formation of complexes is usually straightforward. Metal complexes can often be prepared by direct reaction in solution between the ligand and a metal salt, generally at pHs above seven so that the hydroxypyranone or hydroxypyridinone is in its anionic form. There can be difficulties with purification, as solubility characteristics of ligands and their respective complexes may be inconveniently similar, but recrystallization is usually effective. In cases of difficulty sublimation may successfully separate unreacted ligand from the complex.

It is quite often possible to prepare hydroxypyridinone complexes directly by one-pot synthesis from the appropriate hydroxypyranone, amine, and metal salt (*90–92*). They can also be prepared by reacting complexes such as β-diketonates with hydroxypyridinones (see e.g., Ce^{IV}, Mo^{VI} later). Several maltolate complexes, of stoichiometry ML_2, ML_3, ML_4, or MOL_2, have been prepared by electrochemical oxidation of the appropriate metal anode, M = a first-row d-block metal (Ti, V, Cr, Mn, Fe, Co, Ni), In, Zr, or Hf, in a solution of maltol in organic solvent mixtures (*92*). Preparations of, e.g., manganese(III), vanadium(III), or vanadium(V) complexes generally involve oxidation or reduction in solution, while reduction is unavoidable for technetium since pertechnetate is the inevitable starting material.

B.2. Metal(II) complexes

Many 2+ cations form uncharged bis-ligand complexes $M(malt)_2$, anhydrous or hydrated. The earliest characterized examples were $Zn(malt)_2$ ($\frac{1}{2}H_2O$; $3H_2O$), $Cu(malt)_2$ (anhydrous), and $Ca(malt)_2$ (anhydrous; $5H_2O$) (*93*). $M(malt)_2(H_2O)_2$, M = Co, Ni, crystallize from aqueous solution (*94*); anhydrous $M(malt)_2$, M = Co, Ni, or Cu, have been generated electrochemically in non-aqueous media (*92*). Kojates of Mn^{2+}, Co^{2+}, Ni^{2+}, and Cu^{2+} (*95*), and maltolate, ethylmaltolate, and kojate complexes of Zn^{2+} (*96*), have been characterized, as have 1,2-dimethyl- and 1,2-diethyl-3-hydroxy-4-pyridinonates of Cu^{2+} (*97*). $Pd(malt)_2$ (*98*), $Pd(dmpp)_2$, and $Pd(3,2\text{-hopo})_2$ (*99*) have been prepared, as have ternary salts $[M(malt)(PPh_3)_2]BPh_4$ with M = Pd or Pt (*98*) and $[PdL(PPh_3)_2]BPh_4$ with L = dmpp or 3,2-hopo (*99*). The stability of $Pd(kojate)_2$ in solution has been assessed (see Section III.A.2 later). Tin(II) maltolate, ethylmaltolate, and kojate complexes have been characterized (*96*); $Sn(malt)_2$ has been prepared electrolytically, in an acetonitrile–toluene solvent mixture. $Sn(malt)_2$ is readily oxidized by iodine to the ternary tin(IV) complex $Sn(malt)_2I_2$ (*92*). Cadmium forms ternary complexes such as tetramethylethylenediamine-bis(3-hydroxy-4-pyranonate), $Cd(tmen)(malt)_2$ (*92*). 3-Hydroxy-4-pyranonate complexes of mercury(II) in the form RHgL, where R = e.g., phenyl or 2-furyl, L = malt (*100*), and R = 4-XC_6H_4, X = Me, OMe, NO_2 for L = malt, koj, have been described (*101*).

Ternary 3-hydroxy-4-pyronate complexes of ruthenium(II) include $Ru(malt)_2(dmso)_2$, $Ru(malt)_2(PPh_3)_2$, $Ru(malt)_2$(cycloocta-1,5-diene) (*102*), and $Ru(3,2\text{-hopo})_2(PPh_3)_2$ (*103*). There are several areneruthenium(II) complexes containing ligands of these types, e.g., Ru(mesitylene)(etmalt)Cl (*104*), its maltol and 3-hydroxy-4-pyridinone analogues (*105*), and Ru(4-cymene)LCl with L^- = 1-alkyl- or 1-aryl-2-methyl-3-hydroxy-4-pyridinonate (*106*). These areneruthenium compounds are soluble in water and in dichloromethane, a considerable advantage in several ways. Ru(4-cymene)LCl with L^- = 2-methyl-3-hydroxy-4-pyridinonate undergoes a self-assembly reaction in the presence of sodium methoxide to give a triangular trimer (*107*).

Some bis-ligand complexes have non-simple structures – the unit cell of $Zn(malt)_2 \cdot 1\frac{1}{2}H_2O$ contains both five- and six-coordinated zinc (*108*). Bis-maltolatotin(II) is monomeric, containing five-coordinated tin in a distorted trigonal bipyramidal environment with a stereoactive lone pair in an equatorial position (*96*). $Pb(dmpp)_2 \cdot 7H_2O$ is dimeric, containing bridging dmpp (*108*); $Mn(etmalt)_2(H_2O) \cdot 2H_2O$ is also dimeric, with pairs of ethylmaltolate bridges (*109*). Tetrahedral Zn(OH)(htpzb), where htpzb = hydrotris(3,5-phenylmethylpyrazolyl)borate, reacts with hydroxypyridinones to give ternary complexes Zn(htpzb)(hopo) in which the Zn^{2+} is five-coordinated. The stereochemistry approximates to trigonal bipyramidal, but the extent of distortion varies considerable with the nature of the ligand (*110*).

B.3. Metal(III) complexes

Early examples include $M(malt)_3$, M = Cr, Mn, Fe (*111*), and maltolate (*112*) and kojate (*112,113*) complexes of the lanthanides. As indicated earlier, metal(III) complexes of hydroxypyranones and hydroxypyridinones are generally prepared by simple metal ion plus ligand reactions, sometimes in mixed aqueous or non-aqueous media, especially for the more lipophilic hydroxypyridinones. Occasionally a redox-based preparation has been used – $Cr(malt)_3$ and $Cr(etmalt)_3$ can be prepared either by the direct reaction of Cr^{3+}(aq) with the potential ligand or by reaction of dichromate with an excess of the potential ligand (*114*), while vanadium(III) complexes of hydroxypyranones and hydroxypyridinones are readily prepared by dithionite reduction of oxovanadium(IV) precursors (*v.i.*). Aqueous solutions of these vanadium(III) complexes are surprisingly reluctant to undergo aerial oxidation to vanadium(IV).

The coordination geometry around the central metal ion in tris-bidentate ligand complexes, established by X-ray diffraction or EXAFS (*38,115*) methods, is close to octahedral, for 3-hydroxy-4-pyranonate and 3-hydroxy-4-pyridinonate complexes and for complexes of both p- and d-block elements. Examples include $M(malt)_3$, M = Fe (*116*) or Al (*117*), $V(dmpp)_3$ (*118*), $Fe(dmpp)_3$ (*119*), and $M(empp)_3$, M = Fe, Al, or Ga (*85*). The difference between *mer* geometry for $M(malt)_3$, but *fac* for $M(etmalt)_3$, has been ascribed to the greater bulk of the 2-ethyl group (*120*). $Bi(etmalt)_3$ has a stereochemically active lone pair, so the BiO_6 unit is considerably distorted from octahedral (*121*), indeed the bismuth coordination unit in this compound, as in other examples such as the recently established structure for tris(*N,N′*-dimethyldithiocarbamate)bismuth (*122*), can be described as a pentagonal pyramid. 1,2-Dimethyl-3-hydroxy-4-pyridinonate stabilizes manganese(III) in relation to manganese(II) (*109*); $Mn(dmpp)_2Cl \cdot \frac{1}{2}H_2O$ provides a rare example of five-coordinated high-spin Mn^{3+} (*123*).

Apart from the metals which appear in the summary Tables I and II, hydroxypyranonate and hydroxypyridinonate complexes have been prepared and characterized for several other metal(III) centers, including rhodium, iridium, and ruthenium. For rhodium(III) tris-hydroxypyranonate and tris-hydroxypyridinonate complexes the ligands include maltolate (*98*), dmpp (*99*), and 3-hydroxy-2-pyranonate (*99*). Organometallic compounds include the pentamethylcyclopentadienyl complexes $Rh(C_5Me_5)LCl$, L = malt, koj, or 2-ethyl-3-hydroxy-4-pyridinonate, which are significantly water-soluble (*124*), and series of 1-alkyl-2-methyl-3-hydroxy-4-pyridinonates $M(C_5Me_5)LCl$ with M = Rh or Ir (*106*). $Ir(C_5Me_5)LCl$, where L is the parent (H-on-N)

TABLE I

STEREOCHEMISTRY AND DEGREE OF HYDRATION OF 3-HYDROXY-4-PYRANONATE COMPLEXES OF 3+ METAL IONS

	R^6	R^2	Metal(III)	Stereo	Hydration	Reference
Pyromeconate	H	H	Fe	*fac*	anhydrous	(*139*)
Maltolate	H	Me	Al	*mer*	anhydrous	(*117*)
			Fe	*mer*	anhydrous	(*116*)
Ethylmaltolate	H	Et	Al	*fac*	anhydrous	(*134*)
			Fe	*fac*	anhydrous	(*120*)
			In	*fac*	anhydrous	(*21*)
			Bi	[a]	anhydrous	(*140*)
	CH_2O^nPr	H	In	*fac*	anhydrous	(*21*)
Kojate	CII_2OII	II	Fe	*fuc*	$\frac{1}{2}H_2O$	(*141*)

[a]The lone pair on Bi is stereochemically active; the geometry is closer to *mer* than to *fac*.

TABLE II

STEREOCHEMISTRY AND DEGREE OF HYDRATION OF 3-HYDROXY-4-PYRIDINONATE COMPLEXES OF 3+ METAL IONS

R^1	R^2	Metal(III)	Stereo	Hydration	Reference
Me	Me	Al; Ga; In; V; Fe	*fac*	$12H_2O$	[a]
		Co	*fac*	$6H_2O$	(*144*)
Et	Me	Al; Ga	*fac*	$12H_2O$	(*133*)
		Fe; Al; Ga	*fac*	$3H_2O$	(*85*)
^nPr	Me	Al; Ga	*fac*	$3H_2O$	(*91*)
^nBu	Me	Al	*fac*	$3H_2O$	(*91*)
$4\text{-}MeC_6H_4$	Me	Al; Ga	*fac*	$5\frac{1}{2}H_2O$	(*36*)
Me	Et	Al	*mer*	$6H_2O$	(*134*)

[a]Source: Al, Ga from Refs. (*132,142*); In, Ref. (*143*); V, Ref. (*118*); and Fe, Ref. (*119*).

2-methyl-3-hydroxy-4-pyridinonate, trimerizes under the influence of sodium methoxide to give a chiral triangular metallomacrocycle (*107*). Iridium(III) also forms $Ir(dmpp)_3$ (*99*), while ruthenium(III) maltolate(*98*), ethylmaltolate (*125*), kojate (*126*), and dmpp (*98,99,125*) complexes have all been described. Several of the ruthenium complexes were prepared in an attempt to find suitable substitution-inert pairs of complexes for the examination of the application of Marcus's theory (*127*) to hydroxypyridinonate complexes, to avoid problems of substitution-lability presented by many iron(II) complexes, which with O_6 donor sets are generally high-spin.

Bond distances in the chelate rings in hydroxypyranonate and hydroxypyridinonate complexes reflect the extent of bond delocalization within the chelating O–C–C–O moiety, suggesting greater delocalization in the more stable 3-hydroxy-4-pyridinonate complexes (*128,129*) than in analogous 3-hydroxy-4-pyranonate complexes (*130*). The difference in bond lengths between the two C–O bonds, Δ(C–O), in

Al^{3+} and Fe^{3+} complexes of maltol and ethylmaltol is within the range 0.05–0.08 Å, whereas Δ(C–O) lies between 0.03 and 0.04 for 3-hydroxy-4-pyridinonate complexes of these cations. The geometry of the chelate ring in the boron(III) compound B(1-(4′-tolyl)-2-methyl-3-hydroxy-4-pyridinonate)Cl_2 is very similar to that of iron(III) or aluminum(III) tris-ligand analogues (*131*), the O–C–C–O bond lengths in the boron compound suggesting a similar degree of delocalization in the chelate ring to that in the analogous Fe^{3+} or Al^{3+} complexes.

Whereas 3-hydroxy-4-pyranonate complexes are usually anhydrous, the analogous 3-hydroxy-4-pyridinonate complexes are often hydrated, sometimes heavily as in the case of the $ML_3 \cdot 12H_2O$ series with M = Al or Ga, L = dmpp (*132*) or empp (*133*). The water molecules in these dodecahydrates form chains filling channels through the structure, with the molecules of complex residing in holes in a three-dimensional clathrate water matrix. The structure of this matrix closely resembles that of ice I, and it is the hydrogen-bonding in this structure which constrains the complex to adopt *fac* geometry (*117*). In $Al(mepp)_3 \cdot 6H_2O$ the waters form layers perpendicular to the *b* axis (*134*); in some of the lower hydrates the waters appear to represent what remains of the dodecahydrate matrix after bulky groups such as phenyl substituents have crowded out some of the 12 waters (*36*).

Structures of 1-hydroxy-2-pyridinonate and 3-hydroxy-2-pyridinonate complexes may be compared with those of the more extensively studied 3-hydroxy-4-pyridinonates through the published crystal structures of their tris-ligand iron(III) complexes. The iron(III) tris-ligand complex of 1-*n*-butyl-3-hydroxy-2-pyridinone has *fac* geometry; it crystallizes as a trihydrate (*135*).

The donor atoms of the hexadentate ligand **21** (Fig. 4(d)) form an approximately octahedral coordination sphere around Fe^{3+} (*136*); the *N*-methyl derivative of **21** forms a stable complex with Gd^{3+} which is eight-coordinate, with two water ligands (*82*). Luminescence properties of 1-hydroxy-2-pyridinonate complexes of Eu^{3+} indicate that these too have two water molecules in the coordination sphere of the cation (*137*). An example of a structure determination of a complex of hexadentate ligand with one arm different from the others – in this case two hopo legs and one catecholate leg – is provided by a Ga^{3+} complex of a tren scaffold ligand (*138*). Summaries of the structures and states of hydration of tris-bidentate-ligand 3-hydroxy-4-pyranonate and 3-hydroxy-4-pyridinonate complexes of metal(III) cations are presented in Table I (*21,116,117,120,134,139–141*) and Table II (*36,85,91,118,119,132–134,142–144*), respectively.

B.4. *Metal(IV) complexes*

The importance of hydroxypyridinones in connection with the decorporation of plutonium (Section IV.C.7) and the similarities between the coordination chemistries of early actinides and lanthanides

(*145*) led to the investigation of complexes of cerium(IV). Indeed Ce(IV) provides a good model for the solution chemistry of Pu(IV) (*146*). Cerium(IV) forms complexes CeL_4 and CeL_2 with several bi- and tetradentate 3-hydroxy-2-pyridinonate ligands, which may be prepared from $Ce(acac)_4$. The Ce^{4+} is eight-coordinated in these complexes, but the detailed geometry of its coordination shell depends not only on the nature of the ligand but also on the solvent of crystallization. Thus the complex of the tetradentate ligand containing two 3-hydroxy-2-pyridinonate groups linked through $-CONH(CH_2)_2O(CH_2)_2NHCO-$ is a distorted square antiprism as $CeL_2 \cdot 4H_2O$, but a trigonal dodecahedron when crystallized as $CeL_2.2MeOH$ from methanol (*147*). Thorium(IV) forms homoleptic complexes ThL_4 with several bidentate hydroxypyridinones. The complex with L^- = 1-methyl-4-CONH^{n}Pr-3-hydroxy-2-pyridinonate crystallizes as a monohydrate in which the thorium is nine-coordinated, by the four ligands plus an oxygen from an adjacent amide group, giving a chain structure (*148*). These cerium(IV) and thorium(IV) complexes have been investigated not just for their intrinsic interest, but also as safer substitutes for assessing the chelating properties towards plutonium(IV).

There are many examples of tris-bidentate complexes ML_3^+. An early example is provided by the germanium complex $[Ge(malt)_3]^+$, whose formation, along with $[Ge(OH)_2(malt)_2]$, was documented in 1966 (*149*). Better-characterized examples are provided by the series of tin(IV) compounds $[SnL_3]X \cdot nH_2O$, with L^- = 1,2-dimethyl-3-hydroxy-4-pyridinonate or its 1H or 1-Et analogues and X = Cl, Br, or I (*150*). For M = Si the *mer* and *fac* forms coexist in solution (*151*); for M = V such complexes are best prepared in a non-aqueous medium such as acetic acid, though they are stable over a short pH range in acidic aqueous media (*152,153*). Technetium(IV) complexes $[TcL_3]^+$ with L^- = 1-ethyl- or 1-(4′-methoxyphenyl)-2-methyl-3-hydroxy-4-pyridinonate have been prepared, by dithionite reduction of pertechnetate in the presence of the potential ligand, characterized, and their electrochemical behavior examined. This indicates accessible oxidation states of +3, +4, and +5 for the technetium. This contrasts with $[Tc(acac)_3]^{n+}$, where the accessible oxidation states are +2, +3, and +4 (*154*).

As stated earlier, the M^{4+} cations of Sn, Ti, Zr, Hf generally form ternary complexes ML_2X_2, e.g., the tin(IV) series with L^- = kojate, maltolate, ethylmaltolate, or hydroxypyridinonate, X = halide or thiocyanate (*155*). The structures of the complexes SnL_2X_2, with L = malt or etmalt and X = F or Cl, have been reported (*156*). The compounds $M(etmalt)_2Cl_2$, M = Sn or Ti (*157*), are, as expected, octahedral and $VO(malt)_2$ square-pyramidal with respect to the central metal; the $M(etmalt)_2Cl_2$ complexes are *cis*, but the compounds $Sn(kojate)_2X_2$ were, on the basis of their infrared and NMR spectra, deduced to be the *trans* forms (*158*).

For vanadium(IV), the normal form is VOL_2. The EPR spectrum of bis-maltolato-oxovanadium(IV), $VO(malt)_2$, sometimes known as

BMOV, was reported in 1972 (*159*) and in 1987 (*160*). Its electrochemical preparation was described in 1978 (*92a*), and EPR monitoring of its redox behavior, in chloroform, in 1987 (*160*). However this now-important compound seems not to have been properly characterized until 1992 (*161*). Since then complexes of several 3-hydroxy-4-pyridinones (*162–164*), and of 1-hydroxy-2-pyridinone (*165*), have been synthesized and characterized, especially by EPR (*164*). $VO(malt)_2$ exists as a *cis* ⇆ *trans* equilibrium mixture in aqueous solution, and generally crystallizes as a mixture of the two isomers. However the crystal structure of the *trans* structure was eventually solved, confirming the expected square-pyramidal stereochemistry (*166*). The relative stabilities of the *cis* and *trans* forms of $V^{IV}OL_2$ complexes depend on the nature of the bidentate ligand L^-, with the *cis* configuration favored by $VO(malt)_2$ and $VO(koj)_2$ (*167*), but the *trans* by 3-hydroxy-4-pyridinonate ligands (*164*).

BMOV and its hydroxypyridinone analogues have a rich redox chemistry, which is now fairly well established. Kinetic parameters (k, $\Delta H^{\ddagger}$, $\Delta S^{\ddagger}$) have been established for the oxidation of BMOV by dioxygen in aqueous solution (*168*).

B.5. *Metal(V) complexes*

Rhenium(V) forms the complex salt $[Re(malt)_2(NPh)(PPh_3)][BPh_4]$ and the uncharged benzoyldiazenido^{3-} complex $[ReCl(malt)(N_2COPh)(PPh_3)_2]$. The maltolate ligands are *cis* to each other in the former (*169*). The difference in bond lengths between the two C–O bonds, Δ(C–O), in this complex is 0.050 Å, which is similar to Δ(C–O) for Al^{3+} and Fe^{3+} complexes of maltol and ethylmaltol, suggesting a similar degree of bond delocalization around the chelate rings. Three related rhenium(V) and technetium(V) maltolate-halide complexes have been structurally characterized, *viz.* $^{n}Bu_4N[TcCl_3(malt)]$, $^{n}Bu_4N[ReBr_3(malt)]$, and *cis*-$[ReOBr(malt)_2]$; several hydroxypyridinone analogues have been prepared (*170*).

Vanadium(V) forms oxo complexes $[VO_2L_2]^-$, e.g., for $L^- =$ maltolate; the coordination environment of the vanadium in $K[VO_2(malt)_2] \cdot H_2O$ is approximately octahedral, the two oxo ligands being in *cis* positions. $[K(H_2O)_6]^+$ units link adjacent vanadium(V) complex anions to give a chain structure (*166*). The main products of aerobic oxidation of $[V^{IV}O(dmpp)_2]$ in aqueous solution are $[VO_2(dmpp)]$ and $[VO_2(dmpp)_2]^-$. High pH favors these V^V products, whereas at low pH V^{IV} species predominate (*171*). Vanadium(V) also forms a $VO(OR)(malt)_2$ series, readily prepared from ammonium vanadate, maltol, and the appropriate alcohol in a water–alcohol–dichloromethane medium (*172*), and 3-hydroxy-4-pyridinonate analogues $VO(OR)L_2$ on oxidation of their oxovanadium(IV) precursors in solution in the appropriate alcohol ROH (*168*).

B.6. *Metal(VI) complexes*

Molybdenum(VI) forms a range of hydroxypyranone and hydroxypyridinone complexes *cis*-MoO_2L_2; the maltolate (*98*), kojate, and 1-phenyl-2-methyl-3-hydroxy-4-pyridinonate complexes have been structurally characterized by X-ray diffraction (*173*). The water-solubility and HLB of complexes MoO_2L_2 can be tuned by varying the HLB of hydroxypyridinones LH, as has been demonstrated by the incorporation of hydrophilic groups such as hydroxyl and morpholine (*174*) or lipophilic groups such as alkene or ether moieties (*68*) in the ligand substituents. An X-ray diffraction determination of the structure of a propyl vinyl ether derivative confirmed *cis* geometry. Molybdenum(VI) forms *cis*-dioxo complexes not only with bidentate hydroxypyridinones but also with tetradentate bis-3-hydroxy-4-pyridinone derivatives of iminodiacetate and of edta (*175*). The latter group of 3-hydroxy-4-pyridinones also form complexes containing the MoO_3 moiety (*175*). Complexes *cis*-$MoOL_2$ have also been prepared, for L^- = 3-hydroxy-2-pyridinonate with groups –CH_2CO_2R (R = Me, Et, or Bu) on the ring-nitrogen, by reacting $MoO_2(acac)_2$ with the respective hydroxypyridinones (*176*).

Osmium(VI) forms *trans*-$OsO_2(malt)_2$ (*98*), while dioxouranium(VI) forms *trans*-$UO_2(malt)_2$ (*98*) and many hydroxypyridinonate complexes, with bidentate and with tetradentate (*177*) ligands – *trans*-UO_2L_2 and UO_2L, respectively. Several actinide elements form complexes with hexa- and octa-dentate hydroxypyridinonates (see Section IV.C.7 later).

III. Solution Properties

A. Stability Constants

A.1. *Context*

To put hydroxypyranonate and hydroxypyridinonate complexes in context, stability constants[2] for kojate and 1,2-dimethyl-3-hydroxy-4-pyridinonate complexes of Mg^{2+}, Al^{3+}, Fe^{3+}, and Gd^{3+} are compared with stability constants for complexes of these cations with a few other ligands in Table III. That these hydroxypyranonate and hydroxypyridinonate ligands form stable complexes is immediately apparent. In this section we shall present and discuss a generous, but far from

[2]The values for stability (formation) constants given in the Tables should be approached with some degree of caution, especially as they have been obtained at a variety of ionic strengths. Where two or more values have been reported we have sometimes quoted a mean value, sometimes the most recent, depending on circumstances. All data refer to aqueous solution, at 298.2 K, unless stated otherwise; all are based on concentration units of mol dm^{-3}; all log K or log β values are $\log_{10}$.

TABLE III

COMPARISONS OF STABILITY CONSTANTS (LOG K_1)[a] FOR SELECTED M^{n+} COMPLEXES OF KOJATE AND OF 1,2-DIMETHYL-3-HYDROXY-4-PYRIDINONATE WITH THOSE FOR A RANGE OF OTHER LIGANDS

	Mg^{2+}	Al^{3+}	Fe^{3+}	Gd^{3+}
dtpa	9	18	27	23
edta	9	17	25	17
dmpp	4	12	15	
acac	3	8	11	6
koj	3	10	10	6
citrate	3	8	11	8
oxalate	3	6	7	5
phen	1	6	7	
acetate	1	2	3	2

[a]Values are given only to the nearest whole number, as there is considerable uncertainty over some values (especially for less stable complexes); the purpose of this table is merely to give a general idea of the stability of hydroxypyranonate and hydroxypyridinonate complexes in the context of stabilities of complexes of other common ligands. Stability constants will be given to a precision of an order of magnitude higher in subsequent tables, where sources will be cited.

comprehensive, selection of stability constant data for hydroxypyranonate and hydroxypyridinonate complexes.

A.2. Complexes of 2+ cations

Stepwise stability constants decrease regularly in the normal manner (*178*), log K_1 > log K_2 > log K_3, as may be exemplified by the values for the Ni^{2+}-maltolate system, *viz.* 5.5, 4.3, and 2.7, respectively (*179*). Stability constants (log K_1) for a selection of 3-hydroxy-4-pyranonate and 3-hydroxy-4-pyridinonate complexes of some first-row transition metal 2+ cations are listed in Table IV (*10,128,180–184*). The values for the 3d transition metal cations conform to the long-established Irving–Williams order

$$Mn^{2+} < (Fe^{2+}) < Co^{2+} < Ni^{2+} < Cu^{2+} > Zn^{2+}$$

with the expected marked maxima at Cu^{2+} (and, as so often, no values for Fe^{2+}). The stability of the kojate complex of palladium(II) (log $K_1 = 5.2$) (*185*) is comparable with that of complexes of the first-row transition metal 2+ cations.

There are very few stability constants published for complexes of alkaline earth cations with members of this group of ligands; an early estimate of log $K_1 = 2.9$ for the monokojate complex of Mg^{2+} (*186*) is, as expected, lower than analogous values for the transition metal 2+ cations. Stability constants determined for three 3-hydroxy-4-pyridinonate ligands and one 3-hydroxy-2-pyridinonate ligand complexing Mg^{2+} and Ca^{2+}, although determined at 310 K rather than 298 K,

TABLE IV

STABILITY CONSTANTS (LOG K_1) FOR HYDROXYPYRONATE AND HYDROXYPYRIDINONATE COMPLEXES OF 2+ TRANSITION METAL CATIONS

		Mn^{2+}	Co^{2+}	Ni^{2+}	Cu^{2+}	Zn^{2+}
3-Hydroxy-4-pyranones[a]						
R^6	R^2					
H	Me	4.2	5.1	5.4	7.7	5.6
CH_2OH	H	3.7	4.6	4.9	6.6	5.0
CO_2H	H					4.9[b]
CO_2H	CO_2H					7.3[b]
3-Hydroxy-4-pyridinones						
R^1	R^2					
H	Me		6.4[c]	6.8[c]	9.9[c]	7.5[c]
Me	Me		6.6[d]	6.9[d]	10.6[e]	7.2[e]
Et	Et		6.8[e]	7.1[e]	10.7[e]	7.7[e]
3-Hydroxy-2-pyridinone[f]						
$R^1 = R^4 = R^5 = R^6 = H$					7.7	5.6
1-Hydroxy-2-pyridinone						
$R^3 = R^4 = R^5 = R^6 = H$			4.9[g]	5.2[g]	7.3[g]	5.0[h]

[a]From Ref. (*179*) unless otherwise noted.
[b]From Ref. (*180*).
[c]Gameiro, P.; Rangel, M., *unpublished determinations*.
[d]From Ref. (*128*).
[e]From Ref. (*182*).
[f]From Ref. (*10*).
[g]From Ref. (*183*).
[h]From Ref. (*184*).

confirm that these cations form relatively weak complexes with this group of ligands (*10*).

Stability constants (log K_1) for Zn^{2+} complexes of pyromeconate (pyromecH = **9**) and comenate (comH = **10**), and for chlorokojate (anion from **2** with $R^6 = CH_2Cl$, $R^2 = H$) and iodokojate (anion from **2** with $R^6 = CH_2I$, $R^2 = H$) are all, at 4.9 or 5.0, essentially equal to that for kojate (kojH = **7**) – electronic effects of the Cl and I are insulated by $-CH_2-$ from the metal–ligand bonds. However meconate (mecH = **8**) binds considerably more strongly (log $K_1 = 7.3$) (*180*). The stability order

$$\text{kojate} < \text{maltolate} < \text{meconate}$$

has also been observed for Fe^{3+} and UO_2^{2+} (see later); it parallels the order of pK_a values for the parent 3-hydroxy-4-pyranones.

Stability constants are also available for several M^{2+} complexes of mimosine (**12**, i.e., **4** with $R^1 = CH_2CH(NH_2)CO_2H$, $R^2 = H$), of its methyl ether, of isomimosine (**4**, $R^1 = H$, $R^2 = CH_2CH(NH_2)CO_2H$), and of its isomer **3** with $R^1 = H$, $R^2 = CH_2CH(NH_2)CO_2H$, in water at 310 K

(*8,10*). Log K_1 values for mimosine complexes of Cd^{2+}, Pb^{2+} are 5.9, 8.5 (at 310 K; *cf.* Zn^{2+}, log $K_1 = 6.5$) (*8*).

Measurements in aqueous dioxan confirm and extend some of the trends established in aqueous media. Thus values for log K at 303 K in 50% dioxan for maltolate (*187*) and kojate (*188*) complexes and in 75% dioxan for kojate and chlorokojate complexes (*189*) of many of the cations in Table IV provide further examples of conformance with the Irving–Williams stability order. They also confirm that stability constants for Ca^{2+} complexes (*187,188*) are, as one would expect, considerably lower than for analogous complexes of first-row transition metal 2+ cations. Kojate and chlorokojate complexes of Be^{2+} have been claimed to have similar stability constants to those for their first-row d-block analogues (*189*). Chlorokojates are, in 75% dioxan, of slightly lower stability than the respective kojates (*189*) – the difference in log K is between 0.3 and 0.7. For these, as for most ligands, stability constants increase markedly on going from aqueous media to aqueous dioxan (Table V) (*187–189*).

In contrast to stability constants, there are very few data for enthalpies and entropies of complex formation for hydroxypyranonate and hydroxypyridinonate complexes. Early studies on zinc-maltolate (*190*) and first-row transition metal(II) complexes of kojate (*191*) gave estimates of enthalpies and entropies of formation from temperature variation of stability constants, though as accurate stability constant measurements are only possible over a rather short temperature range the ΔH and ΔS values obtained cannot be of high precision.

A.3. Complexes of 3+ cations

(a) Bidentate ligands. Stability constants for a selection of maltolate and kojate complexes of 3+ cations from the p-, d-, and f-block (Table VI) (*192–195*) can be used to illustrate that stabilities of hydroxypyranonates follow the usual trend of K_n decreasing as n increases (*178*). For the iron(III)-maltolate (*196*) and iron(III)-kojate (*194*) systems, and probably for many others, it is necessary to make allowance for small concentrations of hydroxo-species in equilibrium with the main mono-, bis-, and tris-ligand complexes of Fe^{3+}.

Further examples of stability constants for complexes of Al^{3+} and Fe^{3+} are given in Table VII (*43,64,76,182,183,192,194,195,197,198*).

TABLE V

STABILITY CONSTANTS FOR KOJATE COMPLEXES IN WATER–DIOXAN MIXTURES

	Water (298 K)	50% dioxan (303 K)	75% dioxan (303 K)
Co^{2+}	4.6	6.8	9.5
Ni^{2+}	4.9	7.1	9.7
Zn^{2+}	5.0	7.4	10.4

TABLE VI

STEPWISE STABILITY CONSTANTS FOR THE FORMATION OF 3-HYDROXY-4-PYRANONATE COMPLEXES OF SELECTED 3+ CATIONS

	Al^{3+} (malt)[a]	Al^{3+} (koj)[b]	Fe^{3+} (koj)[c]	Gd^{3+} (koj)[d]
Log K_1	8.3	7.6	6.7	5.5
Log K_2	7.5	6.5	5.4	5.2
Log K_3	6.7	5.4	7.5	

[a]From Ref. (*192*).
[b]From Ref. (*193*).
[c]From Ref. (*194*).
[d]From Ref. (*195*).

This Table serves to illustrate how stability constants vary as the nature of the hydroxypyranone or hydroxypyridinone ligands varies. Table VIII (*36,43,128–130,182,199*) shows how stability constants for complexes of several hydroxypyridinone ligands vary with the nature of the 3+ cation. A selection of the available stability constants for the five 3+ cations whose hydroxypyranone and hydroxypyridinone complexes have been most thoroughly investigated is given in this Table. The cations involved, of iron, aluminum, gallium, indium, and gadolinium, span the p-, d-, and f-blocks of the Periodic Table. The order of stabilities is

$$Fe^{3+} \sim Ga^{3+} > In^{3+} \sim Al^{3+} \gg Gd^{3+}$$

with very similar values for Fe^{3+} and Ga^{3+} reflecting the close similarity of their ionic radii, and the relatively low values for Gd^{3+} consistent with its larger size.

Tables VI–VIII show that hydroxypyranonate complexes are generally of lower stability than their hydroxypyridinonate analogues, for 2+ and for 3+ cations. The stability order for complexes of hydroxypyridinonates is (*43,135,148*):

1-hydroxy-2-pyridinonates < 3-hydroxy-2-pyridinonates
< 3-hydroxy-4-pyridinonates

Within the series of 3-hydroxy-4-pyridinonates, it is difficult to discern which alkyl substituents at positions 1 and 2 maximize stability constants. Table IV suggests that, from log K_1 values for 2+ cations, 1,2-diethyl complexes $M(depp)^+$ are marginally more stable than 1,2-dimethyl complexes $M(dmpp)^+$. However Table VIII shows log β_3 for $M(depp)_3$ to be smaller than log β_3 for $M(depp)_3$, for M = Fe, Ga, and In, but the opposite for M = Gd. Unfortunately differences between experimental conditions, particularly in relation to the nature and concentration of salts added to maintain ionic strength, often

TABLE VII

STABILITY CONSTANTS (LOG β_3) FOR HYDROXYPYRONATE AND HYDROXYPYRIDINONATE COMPLEXES OF THE Fe^{3+} AND Al^{3+} CATIONS[a]

		Fe^{3+}	Al^{3+}
3-Hydroxy-4-pyranones			
R^6	R^2		
H	Me	29[b]	22.5[c,d]
H	Et	28[e]	
CH_2OH	H	26[f]	19.3[c]
3-Hydroxy-4-pyridinones			
R^1	R^2		
H	H	36.9[g]	
H	Me	37.2[g]	32.1
Me	Me	37.2[g]	32.3
Et	Et	36.8[h]	
R^i	H	34.7	29.2
1-Hydroxy-2-pyridinone			
$R^3 = R^4 = R^5 = R^6 = H$		27.3[j]	21.6
$R^4 = OMe$; $R^3 = R^5 = R^6 = H$		29.3[k]	
3-Hydroxy-2-pyridinone			
$R^1 = R^4 = R^5 = R^6 = H$		32.3	23.1

[a]From Ref. (*197*) unless otherwise indicated.

[b]Ref. (*195*) quotes 28.5; Ref. (*198*) quotes 29.5. There is also an early value of 29.7 at 18–22°C (Stefanovi, A.; Havel, J.; Sommer, L. *Coll. Czech. Chem. Commun.* **1968**, *33*, 4198–4214).

[c]From Ref. (*192*).

[d]Contrast this value with that for the furanone analogue isomaltol, for which log $\beta_3 = 14.5$ (Ref. (*76*)).

[e]From Ref. (*64*).

[f]Averaged from Ref. (*194*), which gives log $\beta_3 = 26.5$ and McBryde, W. A.; Atkinson, G. F. *Can. J. Chem.* **1961**, *39*, 510–525 and references therein (mean log $\beta_3 = 25.5$).

[g]From Ref. (*43*).

[h]From Ref. (*182*).

[i]$R = CH_2CH(NH_2)CO_2H$ (mimosine).

[j]From Ref. (*183*).

[k]From Ref. (*198*).

preclude meaningful direct comparisons between values which may differ by little more than experimental uncertainties.

Stability constants (log K_n and log β_3) have been determined for a number of iron(III) complexes of 3-hydroxy-4-pyridinones bearing hydrophilic substituents, including those with $R^1 = CH_2CH_2OH$ and $CH_2CH_2CO_2H$, $R^2 = Et$ (*200*) and with the secondary alcohol group $CH(OH)CH_3$ (*201*). Log K_n and log β_3 have also been determined for the iron(III) complexes of meconate (*202*), and for the sequence of iron(III) complexes with $R^1 = H$, Me, Et, ^{n}Pr at constant $R^2 = Me$; all four log β_3 lie between 37.2 and 37.7 (*43*).

Stability constants for maltolate and 2-methyl-3-hydroxy-4-pyridinonate complexes have been discussed in the context of a score of

TABLE VIII

STABILITY CONSTANTS (LOG β_3) FOR SELECTED HYDROXYPYRIDINONATE COMPLEXES OF SOME 3+ CATIONS

		Fe^{3+}	Al^{3+}	Ga^{3+}	In^{3+}	Gd^{3+}
3-Hydroxy-4-pyridinones						
R^1	R^2					
H	Me	37.2[a]	32.3[b]	38.0[c]	32.8[c]	
Me	Me	37.2[a]	32.3[b]	38.4[c]	32.9[c]	17.3[d]
Et	Et	36.8[e]		36.1[e]	32.4[e]	19.8[e]
Ph	Me		30.7[f,g]	36.3[f]	31.1[f,h]	
3-Hydroxy-2-pyridinone[i]						
R^1 – Me; R^4 – R^5 – R^6 – II		30.0	25.1	29.7	24.4	15.7

[a] From Ref. (*43*).
[b] From Ref. (*130*).
[c] From Ref. (*199*).
[d] From Ref. (*128*).
[e] From Ref. (*182*).
[f] From Ref. (*36*).
[g] At 310 K log $\beta_3 = 32.4$.
[h] Ishii, H.; Numao, S.; Odashima, T. *Bull. Chem. Soc. Jpn.* **1989**, *62*, 1817–1821 give a value of 32.6 for log β_3.
[i] From Ref. (*129*).

"hard", mainly oxygen-donor, ligands (*203*), while stability constants – log β_1, log β_2, and log β_3 – for aluminum complexes of several 3-hydroxy-4-pyranonate and 3-hydroxy-4-pyridinonate ligands have been tabulated in two reviews which very usefully place these log β_n values in the context of many other ligands (*193,204*). Speciation in biofluids has been modeled for Al^{3+} and Ga^{3+} in an aqueous citrate-transferrin medium containing maltol, kojic acid, or 1,2-dimethyl-3-hydroxy-4-pyridinone (*205*).

(b) Tetradentate ligands. Tetradentate ligands can lead to complications with cations which favor octahedral stereochemistry, often forming binuclear species M_2L_3 as well as ML. However oxocations such as VO^{2+} and MO_2^{2+} form a number of mononuclear complexes VOL and MO_2L. Tetradentate ligands are of particular relevance to f-block cations, since these often favor coordination numbers of eight or more and can thus readily form complexes ML_2 or ML_2X. Stability constants have been determined, e.g., for cerium(III) complexes of two linear bis-3-hydroxy-2-pyridinonate ligands with the two hopo units joined by $-CH_2CH_2XCH_2CH_2-$ with $X = CH_2$ or O. Values of log β_2 are 21.6 and 20.9, respectively (*147*). These values may be compared with log $\beta_4 = 19.7$ for bidentate 4-^{n}PrNHCO-3-hydroxy-2-pyridinonate, and log $\beta_1 = 11.4$ for the $-CH_2CH_2OCH_2CH_2-$ linked complex with log $\beta_2 = 11.1$ for the same bidentate ligand (*147*). These comparisons indicate a rather small chelate effect for these linear

TABLE IX

STABILITY CONSTANTS FOR IRON(III) COMPLEXES (LOG_{10} K_1 FOR HEXADENTATE LIGANDS, LOG_{10} β_3 FOR BIDENTATE LIGANDS)

	Log_{10} K_1 or Log_{10} β_3		Log_{10} K_1
Synthetic tris-catecholates[a]	41–49	enterobactin	~52
Tris-3-hydroxy-4-pyridinonates	32–38	desferrioxamine	31
Hexadentate 3-hydroxy-2-pyridinonates	27–38	dtpa	28
Tris-3-hydroxy-2-pyridinonates	29–32	edta	25
Tris-3-hydroxy-4-pyranonates	28–30	transferrin	~23[b]

[a]This range is for tripodal ligands – a log K_1 value of about 59 has been estimated for the Fe^{3+} complex of an encapsulating tris-catecholate ligand (Vögtle, F. "*Supramolecular Chemistry*"; Wiley: Chichester, **1991**, Section 2.3.7.).

[b]This value should be treated with caution, as it is a conditional stability constant determined in a medium modeling blood plasma (Martin, R. B.; Savory, J.; Brown, S.; Bertholf, R. L.; Wills, M. R. *Clin. Chem.* **1987**, *33*, 405–407).

tetradentate hydroxypyridinonates, perhaps reflecting some degree of difficulty in wrapping them around the Ce^{3+} cation.

(c) Hexadentate ligands. A selection of stability constants for hexadentate complexes is given in Table IX; these may be compared with values for their tris-bidentate analogues in Tables VII and VIII earlier in this section. Values of K_1 for hexadentate ligand complexes are, as expected, usually significantly greater than β_3 for tris complexes of analogous bidentate ligands. There are, however, some anomalies. For the complexes of ligands with $–CH_2NRCOCH_2–$ spacers linking N-bonded 3,2-hopo units to the 1-, 3-, 5- positions of a benzene cap, log K_1 values of 28.2 and 28.7 (for R = H, Me) are almost the same as log β_3 of 29.1 for the bidentate analogue (*79*). Further, K_1 for the iron(III) complex of hexadentate tren-(3,2-hopo)$_3$ (**21**) is actually smaller than β_3 for its tris-bidentate analogue (*206*) – log K_1 and log β_3 are 28.8 and 32.3, respectively. This reversal has been ascribed to strain in wrapping the ligand around the Fe^{3+}; computations of relative free energies of complexation by two conformations of the ligand have cast some light on this apparent anomaly (*207*). In contrast, the Fe^{3+} complex of MeC-tris(3,2-hopo)$_3$, which ligand comprises three hopos attached via ring-N to $MeC(CH_2OCH_2CH_2CH_2–)_3$, has a high stability constant (log K_1 = 37.6). The longer links between the bridgehead MeC$\leqslant$atom and the hydroxypyridinone groups presumably provide greater flexibility (*208*). Replacing $–CH_2CH_2–$ spacers in the tren cap of tren-hopo$_3$ ligands by $–CH_2CH_2CH_2–$ reduces the log K values, but replacing successive hopo arms by 2,3-dihydroxyterephthalamide (tam) arms (Fig. 6) leads to increasing stability constants. At the extreme of three replaced arms, log K has risen to 45.2 (*138*). Stabilities of iron(III) complexes of hydroxypyranones and hydroxypyridinones, hexadentate or tris-bidentate, are put into context with some other iron(III) complexes in Table IX. The incorporation of a second nitrogen into the heteronuclear pyridinone ring gives a pyrimidinone or a

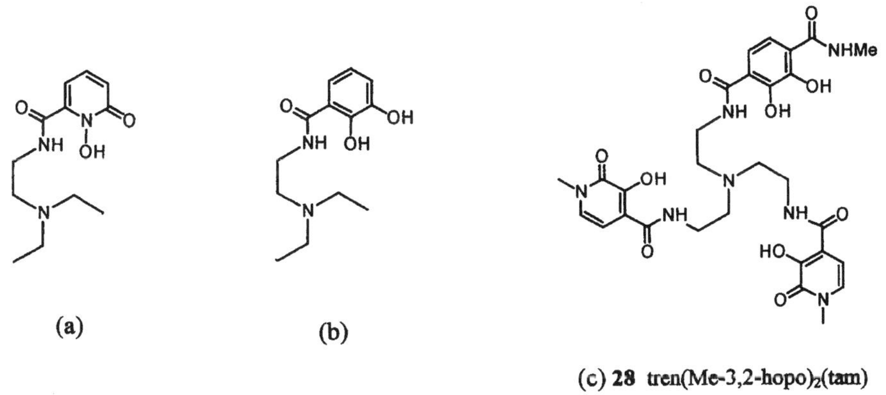

(a) (b) (c) **28** tren(Me-3,2-hopo)$_2$(tam)

FIG. 6. (a) → (b) Replacement of a peptide-bound 1-hydroxy-2-pyridinone unit by a catechol unit on a tris(aminoethyl)amine (tren) cap. (c) A tren-capped hexadentate ligand with two 3-hydroxy-2-pyridinone (3,2-hopo) and one terephthalamide (tam) chelating units.

pyrazinone; three 1-hydroxy-2-pyrimidinonate or 1-hydroxy-2-pyrazinonate units can be incorporated into tripodal ligands which are effective chelators of Fe^{3+}. Stability constants for these Fe^{3+} complexes are within the range $20 < \log K_1 < 27$ (*209*).

(d) Log K and pM. The stoichiometry and concentration of species present in a metal ion – ligand – complex(es) system depends on the initial concentrations of metal and of ligand and on pH. The parameters defining the system are the pK values for ligand and aqua-metal ion and the stability constants for the complexes involved. The distribution of species depends on the interrelation of these parameters and will, except in the case of ligands and aqua-cations with no significant acid–base properties within the pH range of interest, depend strongly on pH. Tabulated stability constant data usually refer to fully deprotonated ligands, so may not be directly applicable to solutions with pH ~ 7. Thus dtpa^{5-}, diethylenetriaminepentaacetate, $(^-O_2CCH_2)_2NCH_2CH_2N(CH_2CO_2^-)$ $CH_2CH_2N(CH_2CO_2^-)_2$, will at the physiological pH of 7.4 be present in protonated forms – pK values for successive protonation of dtpa^{5-} are 10.5, 8.6, 4.3, 2.7, and 2.0. Thus the relative effectiveness of ligands as chelating agents in various situations may not follow directly from tabulated stability constants. Conditional stability constants, β_{eff}, were therefore introduced to provide a quantitative measure of stability in neutral solution which allows for ligand and aqua-cation (de)protonation. This has been usefully and informatively illustrated for aluminum complexes of a range of ligands (*210*). For the tris(hydroxypyridonate) complex of a cation such as Gd^{3+}, which is in the aqua-cation form at pHs of biological and pharmacological interest, $\log \beta_{eff} = \log \beta_3 - 3(\log K_2 - \text{pH})$, where K_2 is the second protonation constant of the hydroxypyridinone.

Another complication in comparing the effectiveness of chelators is the difficulty in comparing ligands of different denticity, a factor arising from the same cause as the well-known and much debated chelate effect. As a result of different dependences on concentrations, hexadentate ligands, for instance, become more effective chelators relative to bidentate ligands as concentrations decrease. To provide a reasonably relevant basis for inter-ligand comparisons at an appropriate pH it has become customary to calculate the concentration of free metal ions present under given conditions, pM = –log ([M^{n+}]free) for a total ligand concentration of 10^{-5} mol dm^{-3} and a total metal ion concentration of 10^{-6} mol dm^{-3}, at the physiological pH of 7.4 (though at 298 K rather than the physiological temperature of 310 K). These conditions, though now used almost universally, were unfortunately not used in some early systems. Indeed in the useful illustration of the effect of the difference in behavior between bi- and hexa-dentate ligands as a function of metal ion concentration, set out in Table X, pM values refer to conditions where the concentration of excess free ligand is equivalent to the total metal ion concentration (*211*). The advantages of hexadentate ligands over bidentate under physiological conditions can also be illuminatingly illustrated by comparing the complexes of ligands with –$CH_2NRCOCH_2$– spacers linking N-bonded 3,2-hopo units to the 1-, 3-, and 5-positions of a benzene cap with their bidentate analogues. Whereas stability constants for the hexadentate and tris-bidentate complexes are very similar – log K_1 = 28.2 and 28.7 (for R = H, Me) and log β_3 = 29.1 – pFe values for the hexadentate complexes are several orders of magnitude higher for the hexadentates – 24.8 and 25.1 versus 18.3 for the bidentate analogue (*79*).

Table XI lists a selection of log K and pM values for a range of iron(III) complexes (*197*), to illustrate the range of maximum free iron concentrations obtainable (down to pM = 32.6, i.e., less than 10^{-32} mol dm^{-3} for tris-hopo complexes) as well as the pM-log K

TABLE X

VALUES OF pM (25°C; I = 0.10 MOL DM^{-3}) IN DIFFERENT CONCENTRATION REGIMES FOR Fe^{3+} IN SOLUTIONS OF BI- AND HEXADENTATE LIGANDS

Ligand	Log β_3 or log K_1	pM (10^{-3} M)	pM (10^{-6} M)
Bidentate ligands			
1,2-Dimethyl-3-hydroxy-4-pyridinonate, dmpp	36	24	18
Acethydroxamate	28	18	13
Hexadentate ligands			
Diethylenetriaminepentaacetate, dtpa	28	25	25
Desferrioxamine	31	26	26
Enterobactin	~52	38	38

TABLE XI

VALUES OF STABILITY CONSTANTS AND pM FOR Fe^{3+} COMPLEXES WITH HYDROXYPYRIDINONATE LIGANDS

		Log β_3	pFe
3-Hydroxy-2-pyridinonate			
R^1 = Et; $R^4 = R^5 = R^6$ = H		32.3	18.3
3-Hydroxy-4-pyridinonates			
R^1	R^2		
Et	Et	36.8	19.7
H	H	35.1	20.8
Me	Me	37.2	21.0[a]
$CH_2CH_2CO_2H$	Me	38.0	21.5
		Log K_1	pFe
Hexadentate 3-hydroxy-2-pyridinonates			
mes(3,2-hopo)$_3$, **24**		28.2	24.8
tren(Me-3,2-hopo)$_3$, **21**		26.7	26.7
enter(Me-3,2-hopo)$_3$, **27**		26.7	27.4
5LIO(hopo)$_2$(tam), (*81*)		32.1	30.4
MeC-tris(3,2-hopo)$_3$, (*208*)		37.6	32.6
Hexadentate tris-catecholates			
tren(tam)$_3$		45.2	34.2
5-LIO(tammeg)$_2$(tam)		45.0	34.4
enterobactin		~52	35.5

[a]A range of values for the pFe of this most-studied 3-hydroxy-4-pyridinone will be found in the literature, down to 19.4 (see e.g. Ref. (*212*)).

difference outlined in this paragraph. The entries in this Table are arranged in order of increasing pFe values. Note that there is no overlap between bi- and hexa-dentate ligands – all the latter are more effective than the former at sequestering Fe^{3+} under physiological conditions. Substituents containing hydrophilic and hydrogen-bonding groups such as carboxylate (Table XI), amido, or hydroxyalkyl (*70,201*) increase pFe values somewhat beyond that for dppm. Thus 2-amido derivatives have pFe values up to 2.3 units higher than dppm, i.e. they enhance the chelating properties of the ligand by up to 200-fold (*212*). The polyaminocarboxylate ligands edta and dtpa have pFe values of 22.2 and 24.7; transferrin has pFe = 23.6. These values fit neatly between the bidentate and hexadentate hydroxypyridinones in Table. The final three entries in the Table, for enterobactin, and for a tripodal (*138*) and a linear hexadentate tris-catecholate ligand (*213*) – tammeg stands for the 2,3-dihydroxy-terephthalamide (tam) moiety with a solubilizing $-CH_2CH_2OMe$ group attached to its free amide end – have been included to show that even the most strongly binding hopo ligands do not have quite such a high affinity for Fe^{3+} as the most strongly binding tris-catecholates. A recent approach to maximizing affinity and selectivity has involved the synthesis of iron-binding dendrimers with hydroxypyranone or hydroxypyridinone terminating groups (*214*).

The relation of pM values for bidentate hydroxypyranones and hydroxypyridinones to those for some other ligands can be illustrated by reference to aluminum. Table XII shows values from two compatible sources (*197,205*). Further pAl values (*215*), which refer to somewhat different conditions (ligand concentration 50 times higher than that of aluminum) can be related approximately to Table XII data through the respective desferrioxamine values. The entries in Table XII are arranged according to decreasing free Al^{3+} concentration on going from citrate through transferrin and the two most common polyaminocarboxylates to 3-hydroxy-4-pyranones and 3-hydroxy-4-pyridinones. The Table also documents the much lower effectiveness of 1,2- and 3,2-hydroxypyridinones than of 3,4-hydroxypyridinones in sequestering Al^{3+}.

(e) Hexadentate complexes of gadolinium. Apart from Fe^{3+}, the only other 3+ cation for which many stability constant data relating to hexadentate ligands are available is that of gadolinium. It is necessary to maximize stability, selectivity, relaxivity, and solubility in the search for Gd^{3+} complexes suitable for use as contrast agents in magnetic resonance imaging (MRI). The first two properties concern us here. The stability constant, or rather pGd, needs to be as high as possible to ensure negligible dissociation, as $Gd^{3+}(aq)$ is toxic. It competes successfully with Ca^{2+} for binding sites (*216*) – it can cause serious damage to the kidneys, especially when these are not functioning well. High selectivity with respect to Zn^{2+} as well as to Ca^{2+} is desirable, hence the interest in comparing stability constants, and pM values, for Gd^{3+}, Ca^{2+}, and Zn^{2+}.

The first report of the preparation and characterization of a tripodal hexadentate ligand complex of Gd^{3+}, that of tren-Me-3,2-hopo (**25** in

TABLE XII

pAl Values[a] for Selected Ligands

	pAl
3-Hydroxy-4-pyridinonate	12.3
3-Hydroxy-4-pyridinonate	12.4
Nitrilotriacetate	12.6
Maltolate	12.6
Kojate	12.6
Citrate	13.3
Transferrin	14.5
Ethylenediaminetetraacetate	14.7
Diethylenetriaminepentaacetate	15.3
1,2-Dimethyl-3-hydroxy-4-pyridinonate	15.7
Desferrioxamine	20.4

[a] At $[Al^{3+}] = 1\,\mu M$, $[L] = 10\,\mu M$; pH 7.4; 298 K.

Fig. 4(d)), reported a stability constant log $K_1 = 20.3$, corresponding to pGd = 20.3. The stability constant for this complex may be compared with values for the well-known polyaminocarboxylate ligands nitrilotriacetate (nta), the linear edta and dtpa, and the macrocyclic dota, which are log $K_1 = 11.5$, 17.3, 22.5, and 25.0, respectively. The pGd value for the complex represents a small increase beyond the pGd values for e.g., dtpa (pGd = 19.1) and dota (pGd = 18.9). Comparison of the value of log K_1 for Gd^{3+} with values of 7.6 and 13.1 for Ca^{2+} and Zn^{2+}, respectively suggests acceptable selectivity (*217*); for tren-1,2-hopo conditional stability constants are given by pCa = 8.8, pZn = 15.2, pGd = 19.3 (*218*). Replacement of the Me groups on the pyridinone ring-nitrogens by $-CH_2OMe$ results in a minor reduction in stability (log K_1 = pGd = 19.8) (*82*). Increasing the distance between the bridgehead and the hopo units by going from tren to trpn, or by introducing spacers, reduced stability constants, and pGd values (*219*). Overall, log K_1 values for Gd^{3+} complexes of hexadentate ligands range from 15 to 24 – considerably less than values for analogous complexes of Fe^{3+}, Al^{3+}, Ga^{3+}, and In^{3+} (*220*), as is the situation for bidentate complexes of these cations.

Variation of one of the three hopo-containing arms of tren-Me-3,2-hopo results in considerably bigger changes in stability. Three of the variants tested gave lower stability, but replacement by a 2,3-dihydroxyterephthalamide moiety – giving an arm 2,3-$(OH)_2$-4-CONHMe-C_6H_2-$CONHCH_2CH_2$– and an abbreviation tren-(Me-3,2-hopo)$_2$(tam) – gave a Gd^{3+} complex with a high log K_1 of 24.1, though pGd, at 20.1, is no higher than for the parent tren-Me-3,2-hopo (*220*). It is interesting to compare the effects of further changes of arms from hopo to tam on pGd and on pFe (Table XIII). Whereas pFe increases steadily from [Fe(tren(Me-hopo)$_3$)] to [Fe(tren(tam)$_3$)], reflecting the particularly high affinity of Fe^{3+} for catechol-binding units, pGd decreases from [Gd(tren(Me-hopo)$_3$)] to [Gd(tren)(tam)$_3$] (*221*). Adding the hydrophilic group $-CH_2CH_2OCH_2CH_2OCH_2CH_2OEt$ to a tam ligand arm increases the solubility of the complex usefully without making a significant change to its pGd value (*222*). Adding an ethanolamine unit to the tam arm of [Gd(tren(6-Me-3,2-hopo)$_2$(tam))] increases solubility, pGd,

TABLE XIII

pGd and pFe Values for Selected Tripodal Hexadentate Ligands[a]

	tren(Me-hopo)$_3$	tren(Me-hopo)$_2$(tam)	tren(Me-hopo)(tam)$_2$	tren(tam)$_3$
pGd[b]	19.5, 20.3	20.1	15.2	~13
pFe[c]	26.8	30.9	33.6	34.2

[a] hopo = 3,2-hopo, with Me variously in the 1- or 6-positions – the tren(Me-hopo)$_3$ entry for Gd shows how small a difference this makes.
[b] From Ref. (*221*).
[c] From Ref. (*138*).

and selectivity for Gd^{3+} over Ca^{2+} and Zn^{2+} – compare pCa = 6.0, pZn = 12.3, and pGd = 20.3 with the pM values for tren-1,2-hopo in the earlier paragraph (*223*). The incorporation of various amino-, carboxylate-, and hydroxyl-containing groups in tripodal ligands of this type permits variation of both hydrophilicity and charge; pGd is maximal when the charge on the complex is zero (*224*).

A.4. Complexes of 4+, 5+, and 6+ cations

(a) General. Very few stability constants have been determined for complexes of 4+, 5+, or 6+ metal ions, which is hardly surprising in view of the rarity of M^{4+}(aq) cations (f-block only) and lack of M^{5+}(aq) and M^{6+}(aq) species. However there are some values for Ce^{4+} and Th^{4+}, and for oxocations such as VO^{2+}, VO_2^{+}, and UO_2^{2+}.

(b) Ce^{4+} and Th^{4+}. Stability constants for Ce^{4+} were calculated from the respective values for Ce^{3+} and appropriate redox potentials – see Section III.B.3 later. The stability constants for the 1-CHCONH^{n}Pr-2-methyl-3-hydroxy-4-pyridinonate complexes of Th^{4+} decrease regularly in the normal manner:

$$\log K_1\ (13.5) > \log K_2(11.5) > \log K_3(10.8) > \log K_4(6.0)$$

Log β_4 for the 3-hydroxy-4-pyridinonate complex (41.8) is, again as normal, greater than that for its 3-hydroxy-2-pyridinonate (38.3) and 1-hydroxy-2-pyridinonate (36.0) analogues (*148*).

There are a few values in the literature relating to the M^{4+} cations Zr^{4+} and Ge^{4+}. In the zirconium case log K_1 and log K_2 values are reported for formation of quercetin complexes, but as these were obtained in 0.5 M HCl these values may well apply to ternary chloride-containing species (*225*). Equilibrium constants have been determined for reaction of two or of three maltol ligands with $Ge(OH)_4$ (*149*), but these values are not comparable with values in the Tables elsewhere in this section, since these all refer to complex formation from aqua-cations or oxoaqua-cations.

(c) Oxocations. Stability constants are available for several oxo- or hydroxo-metal derivatives of 4+, 5+, or 6+ metal ions, the important cases being VO^{2+}, VO_2^{+}, and UO_2^{2+} (Table XIV). VO^{2+} complexes are slightly more strongly bonded than Cu^{2+} (*167,226*); the highest stability constants in this group are for the UO_2^{2+} complexes (*227,228*), reflecting the formal 6+ charge on the central metal. Stability constants for formation of three bis-3-hydroxy-4-pyranonate complex of the oxovanadium(IV) and dioxovanadium(V) cations are shown in Table XV (*166,229*). Both the number and position of the methyl groups in the ligands have small but significant effects. The rather small differences between values for the VO^{2+} and VO_2^{+} cations, and the fact that the value for VO^{2+} is higher than that for VO_2^{+} in one case, but lower in the other two, reflects a fine balance between the higher charge on the VO^{2+} cation and the higher charge on the central

TABLE XIV

STABILITY CONSTANTS (LOG K_1) FOR HYDROXYPYRONATE AND HYDROXY-PYRIDINONATE COMPLEXES OF OXOVANADIUM(IV), DIOXOVANADIUM(V), AND DIOXOURANIUM(VI), WITH VALUES FOR Cu^{2+} AND Zn^{2+} FOR COMPARISON

			VO^{2+}	UO_2^{2+}	VO_2^+	Cu^{2+}	Zn^{2+}
3-Hydroxy-4-pyranones							
R^6	R^2						
H	Me	malt	8.7[a]	8.3[b]	7.5[c]	7.7	5.6
Me	Me		9.2[d]				
CH_2OH	H	koj	7.6[a]	7.1[e]		6.6	5.0
CO_2H	CO_2H	mec		11.8[e]			7.3
3-Hydroxy-4-pyridinones							
R^1	R^2						
H	Me		12.1[f]			9.9	7.5
Me	Me	dmpp	12.2[g]	9.1[h]	10.5[i]	10.1	6.3
Et	Et	depp	12.3[f]			10.2	7.1
R[j]	H	mimosine	10.2[k]			9.5	6.5
1-Hydroxy-2-pyridinone							
$R^3 = R^4 = R^5 = R^6 = H$			8.3[l]			7.3	5.0

[a]From Ref. (*167*).
[b]From Ref. (*227*).
[c]From Ref. (*166*).
[d]From Ref. (*184*).
[e]From Ref. (*230*); log K_1 for the kojate complex in 75% dioxan at 303 K is 10.2 (Kido, H.; Fernelius, W. C.; Haas, C. G. *Anal. Chim. Acta* **1960**, *23*, 116–123).
[f]From Ref. (*164*).
[g]From Ref. (*153*).
[h]From Ref. (*231*).
[i]From Ref. (*171*).
[j]$R = CH_2CH(CO_2^-)NH_3^+$.
[k]From Ref. (226).
[l]From Ref. (165).

TABLE XV

STABILITY CONSTANTS (LOG β_2) FOR HYDROXYPYRONATE COMPLEXES OF OXOVANADIUM(IV) AND DIOXOVANADIUM(V)[a]

Ring substituents in **1**:	Maltolate R^2 = Me, R^6 = H	Allomaltolate R^2 = H, R^6 = Me	Methylmaltolate $R^2 = R^6$ = Me
Oxovanadium(IV), VO^{2+}	15.5	14.8	17.1
Dioxovanadium(V), VO_2^+	13.7[b]	16.8	18.0

[a]From Ref. (*229*) except where indicated otherwise.
[b]From Ref. (*166*).

metal atom in VO_2^+. Stability constants for ternary VO^{2+}/dmpp/L systems with L = oxalate, lactate, citrate have been established as a prelude to modeling and quantifying speciation in model biofluids containing other ligands which may be present in such media (*153*).

Log $K_1 = 10.2$ for the VO^{2+} complex of mimosine (*226*) shows that this mimosine complex is, like the Cu^{2+} and Zn^{2+} complexes of this ligand, significantly less stable than 3-hydroxy-4-pyridinonates with H, Me, or Et at positions 1 and 2 of the ring. The VO^{2+} complexes of the thio-analogue of 1-hydroxy-2-pyridinonate (1,2-topo) are considerably less stable than the mono- and bis-1-hydroxy-2-pyridinonate (1,2-hopo) complexes – stability constants are log K_1, log $\beta_2 = 6.9$, 13.3 for $[VO(topo)]^+$, $[VO(topo)_2]$; log K_1, log $\beta_2 = 8.3$, 16.0 for $[VO(hopo)]^+$, $[VO(hopo)_2]$ (*165*).

As shown in Table XIV, the stability constant for the meconate complex of UO_2^{2+} is much higher than that for the corresponding kojate complex (log K_1(mec) = 11.8, log K_1(koj) = 7.1) (*230*). This behavior parallels that for the Zn^{2+} complexes of this pair of ligands (log K_1(mec) = 7.3, log K_1(koj) = 5.0; *cf.* Section III.A.2 earlier), and in turn reflects a large difference in pK_a values for meconic and kojic acids.

Direct comparisons of the stability constants for formation of the dmpp complexes of molybdenum(VI) with those for uranium(VI) are not possible. The mono-ligand complexes have different stoichiometries, MoO_3(dmpp) versus UO_2(dmpp), while although log β_2 is available for $MoO_2(dmpp)_2$ (*175*) the value of 40.2 refers to formation from MoO_4^{2-} rather than from MoO_2^{2+}(aq), and $UO_2(dmpp)_2$ is too sparingly soluble for its formation constant to be determined (*231*).

B. Redox and Stability

B.1. General

Oxidation–reduction potentials for complexes in solution are determined by the relative stabilities of the complexes of the metal ion in the lower and higher oxidation states. The thermodynamic cycle connecting redox potentials and stability constants is shown in Fig. 7. This cycle can be useful both in rationalizing aspects of aqueous solution chemistry of complexes and in predicting or estimating values for stability constants or redox potentials for systems which are difficult or impossible to access experimentally. Thus knowledge of stability

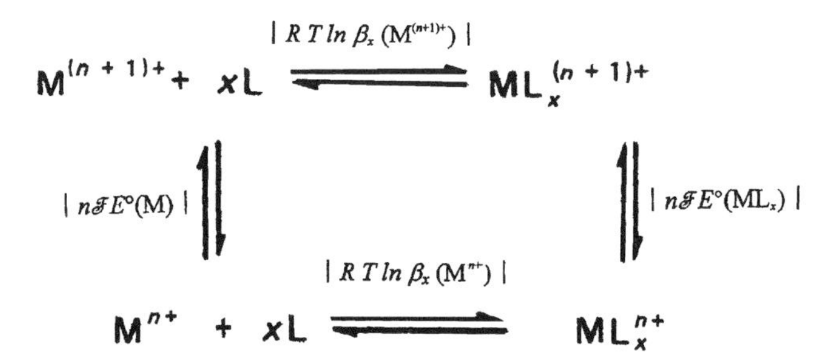

Fig. 7. Thermodynamic cycle linking stability constants and redox potentials for metal ions and their complexes in two oxidation states.

constants for both lower and higher oxidation states permits estimation of the redox potential for the complex couple. Complementarily, the stability constant for, e.g., an iron(II) complex where aerial oxidation causes difficulties of measurement can be calculated from the stability constant for the iron(III) complex and the redox potentials for $Fe^{3+/2+}$ (aq) and for the complex couple (*232*). The usefulness of cycles of this type will be illustrated in the next section for discussing iron complexes of hydroxypyranones and hydroxypyridinones in the context of other ligands, and in the following section for estimating stability constants for cerium(IV) complexes of these ligands.

B.2. Iron complexes

Table XVI shows a selection of stability constants and redox potentials for iron(II) and iron(III) complexes. This Table covers a wide range of the latter, showing how the relative stabilities of the iron(II) and iron(III) complexes are reflected in $E^{o}(Fe^{3+}/Fe^{2+})$ values. A more detailed illustration is provided by the complexes of a series of linear hexadentate hydroxypyridinonate and catecholate ligands, where again high stabilities for the respective iron(III) complexes are reflected in markedly negative redox potentials (*213*). The combination of the high stabilities of iron(III) complexes of hydroxypyridinones, as of hydroxamates, catecholates, and siderophores, and the low stabilities of their iron(II) analogues is also apparent in Fig. 8. Here redox potentials for hydroxypyranonate and hydroxypyridinonate complexes of iron are placed in the overall context of redox potentials for iron(III)/iron(II) couples. The $E^{o}(Fe^{3+}/Fe^{2+})$ range for e.g., water, cyanide, edta, 2,2′-bipyridyl, and (substituted) 1,10-phenanthrolines is

TABLE XVI

STABILITY CONSTANTS AND REDOX POTENTIALS FOR IRON(II) AND IRON(III) COMPLEXES[a]

	Log K_1 ($Fe^{III}L$)	Log K_1 ($Fe^{II}L$)	E°/V
1,10-phenanthroline[b]	15	21	1.07
cyanide[c]	44	35	0.36
aspartate	11	4	0.33
edta	25	14	0.10
hopobactin	26	8	−0.34
tren-hopo	27	6	−0.44
desferrioxamine	31	10	−0.47
enterobactin	52	22	−0.99

[a]The values for log K and E° in this table are published data from various standard reference sources, hence thermodynamic cycles of the type shown in Fig. 7 may not sum exactly to zero when applied to these data due to differences in experimental conditions.

[b]Values apply to the tris-ligand complexes.

[c]Values apply to the hexakis-ligand complexes, i.e., the hexacyanoferrates (Watt, G. D.; Christensen, J. J.; Izatt, R. M. *Inorg. Chem.* **1965**, *4*, 220–222).

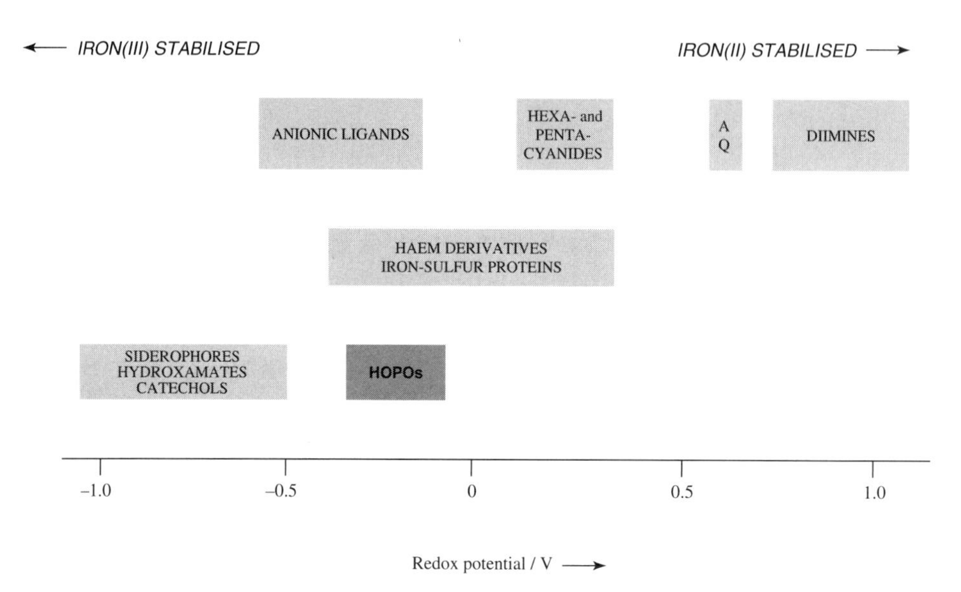

FIG. 8. Ranges of redox potentials for iron(III)/(II) couples; AQ indicates the position of the redox potential for the hexa-aqua iron(III)/(II) couple; diimines include 2,2′-bipyridyl, 1,10-phenanthroline, and their substituted derivatives.

from +1.2 V down to about −0.2 V; the addition of hydroxypyridinones, hydroxamates, catecholates, and siderophores to this list increases the range considerably (Table XVI), the enterobactin couple having a value as low as −0.99 V.

B.3. Cerium complexes

An opportunity to use the thermodynamic cycle shown in Fig. 7 was provided by the requirement to estimate stability constants for cerium(IV) complexes of a series of hydroxypyridinones. As stability constants for their cerium(III) analogues had been measured and $E^\circ(Ce^{4+}/Ce^{3+})$ values established, stability constants for one bidentate and two tetradentate 3-hydroxy-2-pyridinones could be obtained. Log β_4 for the former was calculated to be 40.9, log β_2 for the complexes of the tetradentate ligands 40.6 and 41.9. These very high values, expected for a 4+ cation, are paralleled by high pCe values between 37 and 38 for the tetradentate ligands (*147*).

C. SOLUBILITIES, SOLVATION, AND PARTITION

C.1. General

Solubility, solvation, and partition properties of ligands and complexes are of central importance to many applications. Most simple hydroxypyranones and hydroxypyridinones are sparingly or moderately soluble in water, and in methanol or ethanol, as are their metal(II) and metal(III) complexes. However the incorporation of suitably hydrophilic or lipophilic substituents can have significant, sometimes dramatic, effects on solubilities. The solubility properties of ligands and complexes can thus be tailored to specific requirements although, as will be seen later in relation to gadolinium complexes of hexadentate hydroxypyridinones, the synthesis of the required ligands may provide a challenge. Efficient solvent extraction requires a very much higher solubility in the extractant than in water, but solubility requirements for biomedical applications differ greatly from those for solvent extraction and associated analysis in that they generally require moderate to high solubility in water. Ligands administered for chelation therapy and complexes administered for treating deficiencies or for the introduction of appropriate species for diagnosis or therapy have to enter the body, reach the site required, and eventually be excreted, a trajectory that requires the crossing of several membranes. As the central layer of body membranes consists of fatty hydrocarbon chains, any administered ligand or complex needs to be sufficiently soluble in water to remain dissolved in body fluids, but also sufficiently fat-soluble to be able to cross membranes. There are rarely channels or pumps to assist their crossing, so this must take place by diffusion, through layers of high lipophilicity and very low dielectric constant. For the membrane crossing required in the metabolism of pharmaceuticals, partition coefficients of the order of unity are preferred. Solubility requirements for the measurement of partition (distribution) coefficients are not so demanding, though very high or very low solubility in either phase causes difficulties.

C.2. Solubilities

Despite their importance in many pharmacological uses and potential applications, solubilities of hydroxypyranones, hydroxypyridinones, and their complexes have not been extensively and systematically investigated and established. This situation contrasts sharply with that for partition coefficients, as will become apparent in the following section. The solubility of maltol in water is approximately $0.1\,\mathrm{mol\,dm^{-3}}$, of ethyl maltol $0.13\,\mathrm{mol\,dm^{-3}}$, at 298 K. 1,2-Alkyl-3-hydroxy-4-pyridinones show the expected decrease in water-solubility as the sizes of the alkyl groups increase; solubilities – in water at 298 K – of 1-aryl-2-methyl-3-hydroxy-4-pyridinones decrease from $9 \times 10^{-3}\,\mathrm{mol\,dm^{-3}}$ for the 1-phenyl compound through $6 \times 10^{-4}\,\mathrm{mol\,dm^{-3}}$ for 1-(4′-tolyl) to $6 \times 10^{-5}\,\mathrm{mol\,dm^{-3}}$ for 1-(4′-*n*-hexyl-phenyl) (*37*).

We present a selection of solubilities of 3-hydroxy-4-pyridinonate complexes of M^{3+} cations in Table XVII (*114,143,144,233–241*) and Table XVIII (*237*), to give some impression of the effects of their variation with the nature of the complex and of the solvent. It is very difficult

TABLE XVII

SOLUBILITIES (mM) OF TRIS(3-HYDROXY-4-PYRANONATE) AND TRIS(3-HYDROXY-4-PYRIDINONATE)METAL(III) COMPLEXES IN WATER AT 298 K

			FeL_3	AlL_3	GaL_3	InL_3	GdL_3	Other ML_3
3-Hydroxy-2-pyranones								
R^6	R^2							
H	H	pyromec		2.8	<1			
H	Me	malt	20	50	31	2.5		
H	Et	etmalt	0.3	1.3		0.11	2.1[a]	CrL_3: 0.3
CH_2OH	H	koj	13	5.5	2.0	1.9		
3-Hydroxy-4-pyridinones								
R^1	R^2							
H	Me	mpp	11	43		1.0		
Me	Me	dmpp	0.5	3.1		1.1	1.8	CoL_3: 1.3
1-hexyl	Me	hmpp	2.4					

[a] La^{3+}, Dy^{3+}, and Y^{3+} analogues have solubilities within the range of 1–5 mM.

TABLE XVIII

SOLUBILITIES OF ETHYLMALTOL AND OF TRIS(ETHYLMALTOLATO)IRON(III) IN WATER AND IN ALCOHOLS, AT 298 K

		Solubility EtmaltH ($mol\,dm^{-3}$)	10^3 Solubility $Fe(etmalt)_3$ ($mol\,dm^{-3}$)		Solubility etmaltH ($mol\,dm^{-3}$)	10^3 Solubility $Fe(etmalt)_3$ ($mol\,dm^{-3}$)
C_0	water	0.13	0.30			
C_1	methyl	0.94	1.2			
C_2	ethyl	1.5	4.1			
C_3	1-propyl	1.2	2.8	2-propyl	0.67	0.80
C_4				*t*-butyl	0.19	0.67
C_6	1-hexyl	0.31	0.44			
C_8	1-octyl	0.20	0.28			
C_{10}	1-decyl	0.16	0.23			

to discern an overall pattern from the scattered data available at the present, though the hydroxypyranonate complexes decrease in solubility

$$M(malt)_3 > M(koj)_3 > M(etmalt)_3$$

and for hydroxypyranonate/hydroxypyridinonate comparison,

$$M(dmpp)_3 < M(malt)_3, \text{ but } M(dmpp)_3 \approx M(etmalt)_3$$

The situation is further confused by qualitative observations, such as the conflicting views as to the solubilizing effect or otherwise of the CH_2OH group in kojates and derived hydroxypyridinonates (*238*). In relation to Table XVIII, the trend of increasing solubility from water to ethanol followed by a steady decrease to 1-decanol, common to ethylmaltol and its iron complex, resembles that established for $Fe(bipy)_2(CN)_2$ and $Fe(phen)_2(CN)_2$, though in the case of these complexes the maximum occurs at methanol rather than at ethanol (*239*). Ethylmaltol, its iron complex, $Fe(bipy)_2(CN)_2$, and $Fe(phen)_2(CN)_2$ all have both hydrophilic and lipophilic areas on their surfaces and can thus exhibit synergic solvation with maximum solubility reflecting favorable solvation of both hydrophilic and lipophilic regions. Comparison of indium complexes with those of gallium, aluminum, or iron is complicated by the possibility that the large size of In^{3+} may permit sufficiently close approach of water between the bidentate ligands towards the cation for it to hydrogen-bond with donor oxygen atoms of the bidentate ligands, as has been proposed for some β-diketonate complexes (*240*). For hydroxypyranones, hydroxypyridinones, and their complexes, as for all other solutes, solubilities are determined by the resultant of several factors, including solvation effects such as the magnitude of, and balance between, hydrophilic and hydrophobic effects, molecular volume and shape, and solvent structure (*241*).

There is a large temperature variation of solubility for the tris-dmpp complex of indium (*143*) but a much smaller variation for tris(maltolato)aluminum (*242*) in aqueous solution. The solubility of the former increases by a factor of 3.5, of the latter by only 1.3 times, on raising the temperature from 298 to 310 K, i.e., from the standard 25°C to the physiological 37°C. The enthalpy of solution of $Al(malt)_3$ in water is $23\,kJ\,mol^{-1}$, but is medium-dependent, rising to $56\,kJ\,mol^{-1}$ in 80% methanol.

Solubilities of hydroxypyranonate and hydroxypyridinonate complexes in aqueous salt solutions are also of some relevance to biological and medical applications. The solubility of the ethylmaltol complex $Fe(etmalt)_3$ is reduced fourfold on going from water to molar NaCl solution, that of $Fe(dmpp)_3$ is reduced by 30%. Solubilities of these complexes in a range of salt solutions have been measured (*243*), yielding Setschenow coefficients (*244*) to provide context and to supplement the remarkably few values known for inorganic complexes (*245*) and link to the rather greater body of information on organic solutes (*246*).

C.3. *Partition coefficients*

Partition (distribution) coefficients,[3] *P* (*D*) – often reported as $\log_{10} P$ or $\log_{10} D$, have been measured for a large number of

[3]The terms "partition coefficient" and "distribution coefficient" are often regarded as interchangeable. We shall generally use the term employed in the original report. Use of the term "distribution coefficient" implies that measurements were carried out in thermostatted apparatus with an aqueous layer buffered at a pH of, or very close to, 7.4, and that due regard was taken of any acid–base or complex dissociation reactions that may be involved. The p*K* values for the great majority of hydroxypyranones, hydroxypyridinones, and their complexes as isolated are such that they are in an uncharged form at pH ~7 and thus partition and distribution coefficients are equal. In the present review the term "partition coefficient" may refer to results that were obtained under less rigorous conditions, especially in early determinations (*cf.* text).

hydroxypyranones, hydroxypyridinones, and their complexes, especially of iron(III). The organic phase is almost invariably 1-octanol for the ligands and complexes under discussion here. In the early days, as in Hansch's pioneering study of the additivity of substituent effects (*247*), partition coefficients were estimated by simple flask-shaking methods without thermostatting, though apparatus for efficient shaking and *in situ* spectrophotometric monitoring was soon developed (*248*). Sometimes there was no pH control of the aqueous layer and often the possibility of the coexistence of species other than the complex under investigation was not considered. Values for partition coefficients of complexes of 2+ metal cations, and for hydroxypyranone complexes of some 3+ ions, may be concentration-dependent due to significant dissociation and hydrogen-bonding effects (*132*). It is therefore not surprising that there is sometimes a considerable lack of agreement between published values from various sources. Thus, for an extreme example, estimates for P for $Fe(dmpp)_3$ range from 0.0009 (*249*) up to 0.24 in an early patent application, with a value of 0.0025 (*203*) recently favored. The development of much improved measurement techniques such as the filter-probe extractor (*250*) and modified automated continuous flow methods (*21*), and due consideration of speciation have lead to much more reliable values for distribution coefficients. These are now usually determined for partition between 1-octanol and a buffered aqueous layer at pH 7.4, these being the conditions deemed most appropriate for assessing behavior in biological situations or pharmacological applications. 1-Octanol, with its small but significant water content in partition experiments, is the most appropriate organic phase, certainly a better model than chloroform or cyclohexane (*197*) for biological membranes. It has long been accepted as a rough model for basic capillary permeability (*251*).

A selection of distribution (partition) coefficients for 3-hydroxy-4-pyranones and their M^{3+} complexes is set out in Table XIX, and for hydroxypyridinones and their M^{3+} complexes in Table XX. The data in Table XIX have been taken from a range of publications on

TABLE XIX

DISTRIBUTION COEFFICIENTS FOR SELECTED 3-HYDROXY-4-PYRANONES AND THEIR COMPLEXES

		LH	FeL_3	CrL_3	AlL_3	GaL_3	InL_3
R^6	R^2						
H	Me	1.1	0.3	0.09	[a]	~0.3	1.0
H	Et	4.2	5.2		1.3	3.9	21
CH_2OH	H	0.2	0.3		0.09	0.08	0.12
CH_2OEt	H	0.4	3.4			1.9	8.9
CH_2O^nBu	H	6.0	440			350	490

[a]There is too big a disagreement between published values for an acceptable mean value to be quoted here.

TABLE XX

DISTRIBUTION COEFFICIENTS FOR SELECTED HYDROXYPYRIDINONES AND THEIR COMPLEXES

		LH	FeL_3	AlL_3[a]	GaL_3	InL_3
3-Hydroxy-4-pyridinones						
R^1	R^2					
Me	Me	0.2	[b]	0.0009	<0.01	
Et	Et	2	0.5	2.4		
^{n}Bu	Me	3	23	17	17	58
C_6H_4-4′-X	Me					
X = H		12		130	140	180
X = Me		53		450	>1000	>1000
X = OMe		19		290	350	>1000
X = NO_2		5.8		29	31	>1000
$CH_2CH_2NH_2$	H	0.22	0.05		0.01	0.03
3-Hydroxy-2-pyridinone						
R^1	$R^{4,5,6}$					
Me	H	0.52	0.10		0.07	0.39
^{n}Bu	H	15	440		430	1590
1-Hydroxy-2-pyridinone						
$R^3 = R^4 = R^5 = R^6 = H$		0.09	0.95			

[a]Values of D are available for aluminum complexes of eight 3-hydroxy-4-pyridinones, with alkyl or alkoxy groups at R^1, methyl or ethyl at R^2, and their respective parent hydroxypyridinones (*253*).

[b]There is too big a disagreement between published values for an acceptable mean value to be quoted here.

3-hydroxy-4-pyranones (*21,193*) and their iron (*21*), chromium (*252*), aluminum (*233,253*), gallium (*21,233*), and indium (*21,143*) complexes. The data in Table XX are again derived from a variety of sources, dealing with 3-hydroxy-4-pyridinones (*21,34,193,197,255*) and their iron (*21,203*), aluminum (*193,197*), gallium (*21*), and indium (*21,143,254*) complexes,[4] and with 3-hydroxy-2-pyridinone, 1-hydroxy-2-pyridinone, variously alkyl-substituted derivatives, and their iron(III) complexes (*255,256*). In all cases the values quoted for complexes refer to the tris-ligand species; in cases where more than one determination has been reported the values in Tables XIX and XX represent either a mean or a best estimate. The variation of distribution coefficient with the nature of the central metal is very similar for all the 3+ cations; perhaps the range of values for In^{3+} is somewhat larger than for the smaller 3+ cations – the values for In^{3+} are generally considerably larger than for the other M^{3+} documented here.

Both Table XIX and Table XX show that distribution coefficients cover a much larger range for complexes than for the parent ligands, and that variation of ligand substituent can have a very large effect on

[4]Data for the substituted 1-C_6H_4-4′-X ligands and complexes are from Ref. (*36*).

distribution coefficients. Overall, distribution coefficients for hydroxypyridinones cover a range of at least $-3.30 < \log_{10} P < +3.40$, thanks to substituents from hydrophilic sulfonate or carboxylate to lipophilic long-chain alkyl or alkylphenyl groups. There is an unexpectedly small effect of F-for-H substitution. Although the CF_3 group does confer significantly greater lipophilicity than the CH_3 group – P for the 3-hydroxy-4-pyridinone with $R^1 = CH_2C_6H_4$-4′-CH_3 and $R^2 = Et$ is 160, that for its 4′-CH_3 analogue is 250 (*257*) – it has been reported that partition coefficients for 2-CF_3-3-OH-4-pyranone and maltol are essentially equal, at 0.64 and 0.66 (*50*). 2-Alkyl-3-hydroxy-4-pyridinones with a ribonucleoside unit at the 1-position are less lipophilic ($D = 0.04$, 0.05 for the 2-Me, 2-Et compounds) (*60*) than 1-$CH_2OCH_2CH_2OH$ model ligands ($D = 0.10$, 0.26) (*61*).

$\log_{10} D$ values for a selection of iron(III)-tris-ligand complexes are listed alongside values for the respective ligands in Table XXI. There are large numbers of partition or distribution coefficients available for iron(III) complexes and for their parent ligands (*21,43,45,57,64,70,79,206*). Most of the data in Table XXI are taken from an extensive 1998 compilation (*203*), with one or two later additions (*65,201*). Despite the distinctly wider range of log D values for the complexes than for the ligands there is a tolerably good, though not linear, correlation with log D for the respective ligands. For the majority of tris(3-hydroxy-4-pyridinonate)iron(III) complexes the wide range of log D values (Table XXI) is accompanied by remarkably little variation in stability constants, indeed the initial report of this phenomenon shows a range of greater than 10^4 in P at constant log β_3 of 36 (*45*). The situation is not quite so clear-cut nowadays, especially

TABLE XXI

Distribution Coefficients for Selected 3-Hydroxy-4-pyridinones and their Iron(III) Complexes, Arranged in Order of Increasing Lipophilicity

R^1	R^2	Log D(LH)	Log D(FeL_3)
$CH_2CH_2CO_2H$	Me	−2.92	−4.00
$CH_2CH_2NH_2$	Me	−1.55	−3.22
$CH_2CH_2CONHEt$	Me	−0.82	−2.70
$CH_2CH_2CO_2Et$	Me	−0.77	−2.70
Me	Me	−0.77	−2.60
H	Me	−0.50	−2.38
$CH_2CH_2CONEt_2$	Me	−0.36	−1.70
Et	Et	0.23	−0.62
$CHMe_2$	Et	0.73	1.01
^{n}Bu	Et	1.22	2.36
^{n}Hx	Me	1.89	3.40
^{n}Oct[a]	Me	2.88	

[a] Insolubility precluded the determination of log D for the n-decyl analogue.

for ligands containing pH-sensitive substituents such as amino or carboxylato groups. Thus comparison of 1-($CH_2CH_2CO_2H$)- with 1-($CH_2CH_2CH_2NH_2$)-2-Me-3-hydroxy-4-pyridinone, and of their gallium complexes, reveals that whereas the lipophilicities of the two 3-hydroxy-4-pyridinones are similar, $P = 0.013$ and 0.30, respectively, the former substituent confers a much higher chelating power under model physiological conditions, with pGa = 20.1 for the carboxylato compound, 15.1 for the amino compound (*258*). There are data for several hexadentate ligands and their complexes (*259*).

C.4. Transfer chemical potentials

Transfer chemical potentials, often determined from solubility measurements, provide an alternative way of probing the solvation of complexes, in particular giving information on differences in solvation between water and organic solvents and on solvation in solvent mixtures (*260*). They provide a quantitative picture of relative and preferential solvation. Much effort has been devoted to kinetically inert coordination complexes, such as those of cobalt(III), chromium(III), and low-spin iron(II), and particularly to the special case of aqua-cations (*260,261*). For all these charged complexes the necessity to use an extra-thermodynamic assumption to derive single-ion values brings small but significant uncertainties to the values and their interpretation. Studies on uncharged species, such as these hydroxypyranone and hydroxypyridinone complexes, avoid these difficulties and thus make them interesting and valuable probes for solvation. They also provide some insights of possible pharmacological relevance.

We have investigated a number of Fe^{3+} (*114,237*), Cr^{3+} (*114*), Al^{3+}, Ga^{3+}, In^{3+} (*236*), Co^{3+} (*144*), and Gd^{3+} (*235*) complexes in several series of binary aqueous solvent mixtures. Several of these complexes show maxima in plots of solubility against solvent composition (Fig. 9), suggesting synergic solvation. That there are both hydrophilic and lipophilic areas on their peripheries, encouraging solvation by water and by an organic cosolvent in appropriate regions, leads to increased solubility in solvent mixtures. This increased stabilization due to solvation is reflected in minima in transfer chemical potential trends (Fig. 10). This trend parallels that of such compounds as tryptophan, sulfathiazole, and Fe(diimine)$_2$(CN)$_2$ complexes (*239*). Whereas in the case of arachidic acid (eicosanoic acid, Me(CH_2)$_{18}$$CO_2$H) and the Fe(diimine)$_2(CN)_2$ complexes the hydrophilic and lipophilic areas on the peripheries correspond to the cyanide ligands and the diimine ligands, respectively and separately, in the case of the hydroxypyranonate and hydroxypyridinonate complexes the hydrophilic and lipophilic areas are present on each ligand.

Solubilities in the water-1-octanol system (*262*) provide a link between solubilities and transfer chemical potentials on the one hand and distribution coefficients and possibilities of crossing membranes

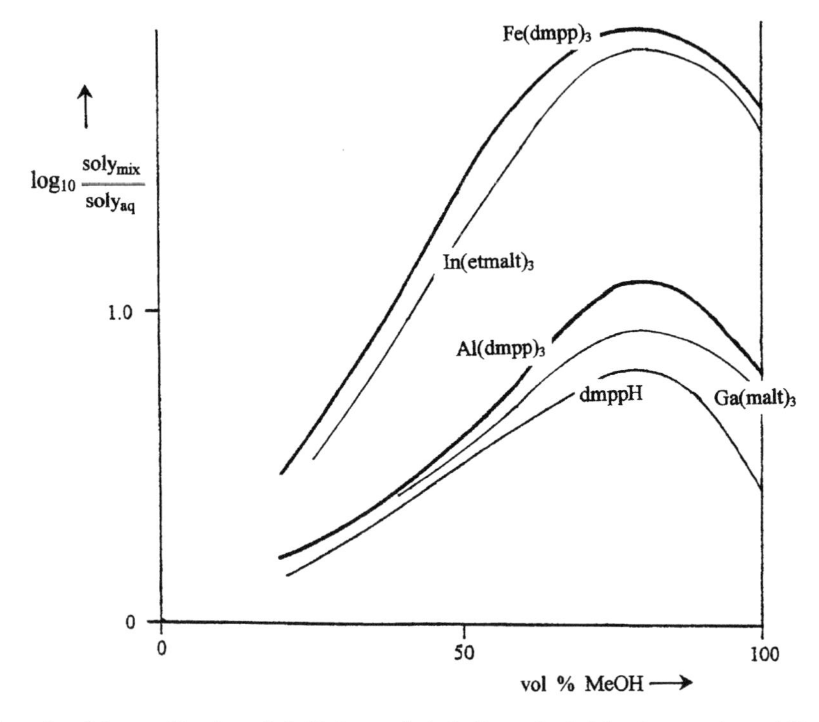

FIG. 9. Normalized solubilities of 1,2-dimethyl-3-hydroxy-4-pyridinone, dmppH, of its aluminum(III) and iron(III) complexes $Al(dmpp)_3$ and $Fe(dmpp)_3$, and of the 3-hydroxy-4-pyronate complexes $Ga(malt)_3$ and $In(etmalt)_3$ in methanol–water mixtures at 298.2 K (data from Refs. (*114*) and (*234*)).

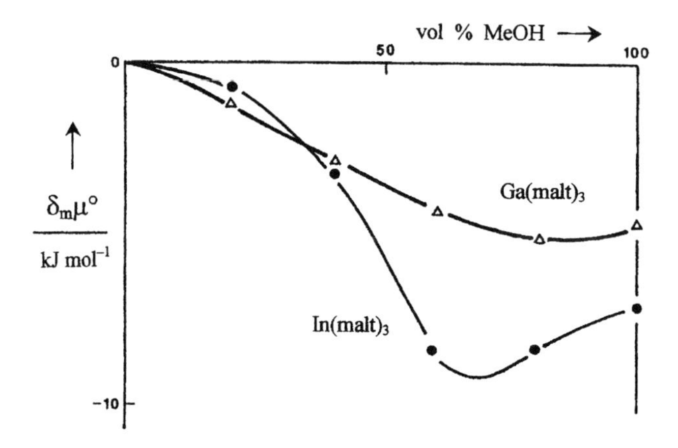

FIG. 10. Transfer chemical potentials for tris(maltolato)gallium(III) and tris(maltolato)indium(III) to methanol–water mixtures at 298.2 K (data from Ref. (*236*)).

on the other. The most striking feature of solubilities in octanol–water mixtures is the marked increase in solubility and thus decrease in transfer chemical potential attendant on transfer from 1-octanol itself into the wet octanol that is perforce the "non-aqueous" phase in partition experiments. This marked stabilization in wet octanol suggests that a hydroxypyridinone complex may have solvation properties that maximize its interaction with a membrane to such an extent that it prefers to remain within, or on the surface of, the membrane rather than being facilitated to move through the membrane. The other complication is that a complex may act as a salting-in or salting-out agent (*262*), possibly leading to a significant increase in the water content of the organic medium, thereby adding another variable in any discussion of the relevance of distribution coefficient to transmembrane transfer.

D. Kinetics and Mechanisms

D.1. General

There is a marked paucity of data on kinetic parameters for the formation and dissociation of homoleptic metal complexes ML_n. Such reactions are generally slow for linear or tripodal hexadentate ligands, but can be very fast for bidentate ligands reacting with, or dissociating from, labile metal cations. These substitution-labile centers include Mg^{2+}, Al^{3+}, and the majority of 2+ and 3+ cations of the first row of the d-block (*263*). However the expected slow dissociation has been demonstrated for dissociation of ethylmaltol from $Cr(etmalt)_3$, where the high crystal field stabilization for the d^3 metal ion results in a high crystal field activation energy and thus slow substitution (*114*). The main interest in kinetic studies of complexes of hydroxypyranones and hydroxypyridinones lies in the determination of rate constants and mechanisms for substitution involving the other ligands in ternary complexes. These investigations, detailed in the following sections, fall into two groups, replacement of halide or thiocyanate (X^-) in $M^{IV}L_2X_2$ and water exchange at ternary aqua complexes of the lanthanides. Gadolinium is of particular interest in view of the potential of such complexes as contrast agents for MRI (see Section IV.C.5 later).

D.2. Metal(IV) complexes

Kinetic and mechanistic studies of nucleophilic substitution at metal(IV) centers are fairly rare (*263*). Platinum(IV) has the substitution-inert low-spin d^6 configuration, and presumably undergoes nucleophilic substitution by an associative mechanism thanks to its high charge and large size. However there are actually very few data, probably thanks to the tendency for platinum(IV) to oxidize ligands. Substitution kinetics at metal(IV) centers may be more conveniently studied for complexes of the type ML_2X_2, where M = e.g., Sn, Ti, V, or

Zr; $L^- =$ a bidentate hydroxypyranonate or hydroxypyridinonate ligand; and X is a halide or alkoxide leaving group (*264,265*). Replacement of chloride by thiocyanate, pyrazine, or water (in acetonitrile solution) in 3-hydroxy-4-pyranonate or 3-hydroxy-4-pyridinonate tin(IV) complexes in all cases follows a two-term rate law of the type

$$-\mathrm{d}[\mathrm{SnL_2X_2}]/\mathrm{d}t = \{k_1 + k_2[\mathrm{nucleophile}]\}[\mathrm{SnL_2X_2}]$$

as illustrated in Fig. 11, indicating parallel associative (k_2) and dissociative (k_1) reaction pathways (*265*) – behavior reminiscent of platinum(II). Similar behavior was observed in several analogous metal(IV) reactions. The activation volume, $\Delta V^‡$, for thiocyanate

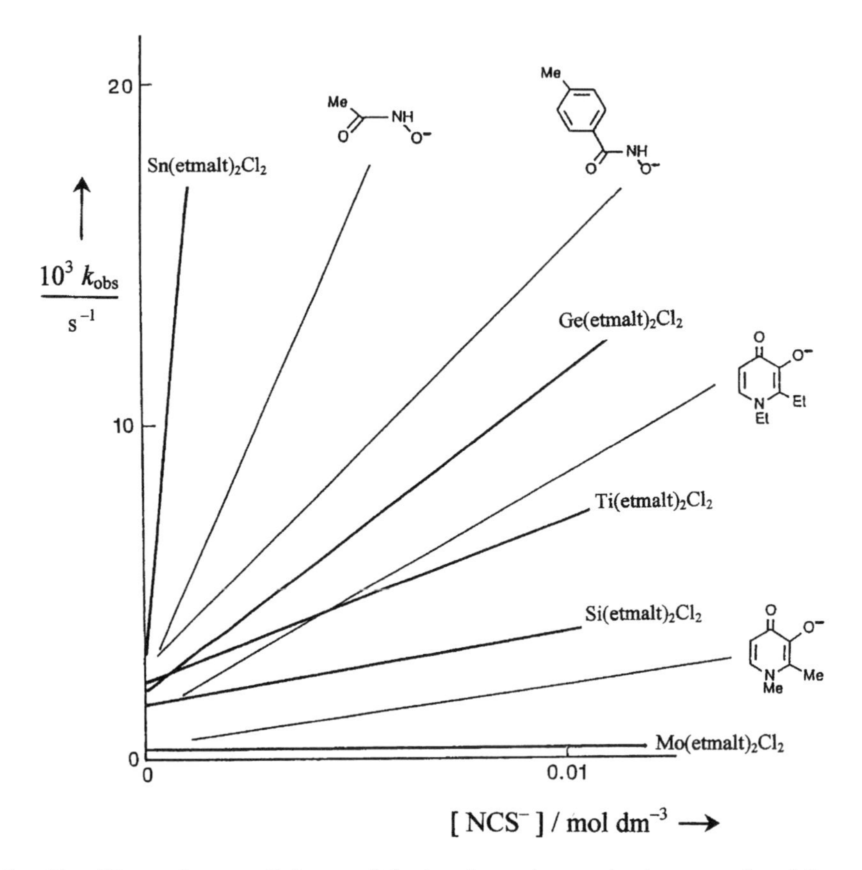

FIG. 11. Dependences of observed first-order rate constants on nucleophile concentration for thiocyanate susbtitution at dichlorobis(ethylmaltolato)metal(IV) complexes $M(etmalt)_2Cl_2$, and at a series of tin(IV) complexes SnL_2Cl_2 with L = the ligands whose formulae are shown against the thin line plots. The data refer to reactions in acetonitrile solution at 298.2 K (data from Refs. (*264*) and (*265*)).

replacing chloride in $Ti(etmalt)_2Cl_2$, in acetonitrile solution, is $-12\,cm^3\,mol^{-1}$, consistent with associative activation (*264*). This value, which is identical to that for substitution of chloride by thiocyanate at vanadium(IV) (*266*), is close to the “theoretical” value of $-10\,cm^3\,mol^{-1}$ for associative attack in organic systems unaffected by solvational complications (*267*). The non-leaving hydroxypyranonate or hydroxypyridinonate ligands have a significant effect on reactivity – k_2 for replacement of chloride by thiocyanate in TiL_2Cl_2 is nearly 500 times higher for L^- = maltolate or ethylmaltolate than for L^- = 1,2-dimethyl-3-hydroxy-4-pyridinonate (*264*).

D.3. Lanthanide complexes

Water exchange rates are one of the main factors determining the relaxivity of gadolinium complexes and their suitability as contrast agents for MRI (*268*), hence the recent activity in determining kinetics of water exchange at ternary aqua-hydroxypyridinonate complexes of gadolinium (*269*). Data on water exchange at gadolinium complexes of relevance to MRI (*270*) include kinetic parameters for two hexadentate 3-hydroxy-2-pyridinonate (hopo) derivatives, one whose ligand has three hopo units connected to a nitrogen bridgehead by serine units (*220*), a second whose tripodal ligand has hopo units in two legs, and a catechol binding moiety with a water-solubilizing triethyleneglycol group attached. The activation volume for exchange of the two water ligands in the latter, $\Delta V^{\ddagger} = -5\,cm^3\,mol^{-1}$, provides the best evidence for the operation of an associative interchange, I_a, mechanism, as earlier proposed on structural grounds. The rate constant for this water exchange is $5.3 \times 10^7\,s^{-1}$ at 298 K, with $\Delta H^{\ddagger} = 25.9\,kJ\,mol^{-1}$ (*222*); the rate constant for water exchange at a hydroxymethyl derivative of [Gd(tren-Me-3,2-hopo)] is $7 \times 10^7\,s^{-1}$ at 298 K (*271*). Water exchange at these complexes is thus considerably faster than at $[Gd(dota)(H_2O)]^-$, where $k_{298} = 4 \times 10^6\,s^{-1}$, but not as rapid as with $[Gd(trita)(H_2O)]^-$, where $k_{298} = 2.7 \times 10^8\,s^{-1}$, thanks to the steric compression around the coordinated water site caused by the insertion of an extra CH_2 group into one of the CH_2CH_2 units of the macrocyclic dota ring (*272*). All these values have been put into the general context of kinetic parameters for water exchange at metal centers (*273*).

IV. Uses and Applications

A. Introduction

A.1. General

The uses and applications of these ligands and their complexes are many and varied. We deal firstly with a variety of uses of hydroxypyranones, then with applications of hydroxypyranone and hydroxypyridinone ligands and their complexes in solvent extraction and analysis, next with actual and potential uses of these ligands and their complexes in diagnosis and therapy, and finally with a selection of miscellaneous uses and applications of various complexes of this group of ligands.

A.2. Hydroxypyranones

Maltol and ethylmaltol, especially the former, are widely used food additives (Veltol, E636, and Veltol Plus, E637, respectively) particularly in the baking industry; maltol has also been shown to stimulate plant growth. Kojic acid is used as a preservative, particularly to prevent discoloration, for instance of vegetables. Kojic acid is also extensively used cosmetically, to lighten skin color. Although there are concerns in some quarters about possible undesirable side effects, it is claimed that normal levels of consumption pose no health risk to humans (*274*), and likewise that its long-term dermatological use (in one trial for 14 years) also has no ill effects (*275*). Kojic acid, like some 3-hydroxy-4-pyridinones (*276*), inhibits the copper-containing enzyme tyrosinase, which causes melanization in humans and leads to brown coloration in plants and food (*277*). Several flavanols with a 3-hydroxy-4-pyranone nucleus (see Fig. 3), especially the quercetin (**17**) glycoside rutin (*278*), maintain a toehold in pharmacopaeias as capillary protectants – they are thought to reduce capillary leakage. Maltol, which is a degradation product of streptomycin (*14*), and kojic acid are weak antibiotics (*279*).

Flavonol (**16**) and morin (**20**) are used as colorants; morin is the dye calico yellow, used in textile dyeing and calico printing. The nature of the mordant needed for dyeing wool affects the color of the product material – mordanting with aluminum, chromium, or tin results in various shades of yellow, while an iron mordant gives an olive-brown color.

B. Analysis and Extraction

Many hydroxypyranones and hydroxypyridinones and their metal complexes have been of importance in analytical chemistry, solvent extraction, and metal separation. Here their excellent chelating properties in conjunction with the possibility of synthesizing strongly lipophilic derivatives make this class of ligands particularly useful.

Hydroxypyranones and hydroxypyridinones give a range of colors, from orange through red to violet, with iron(III)-containing solutions (*27*). Early applications of the use of kojic acid for the colorimetric determination of iron (*280*), e.g., in iron ores (*281*), followed and complemented the recommendation, in 1930, that iron(III) be used for the determination of kojic acid – iron(III) was later used for the colorimetric determination of mimosine (*282*). Other early colorimetric analyses included methods developed for the determination of molybdate (*283*) and of vanadate (*284*) by maltol. The vanadate determination is carried out in aqueous methanol, giving red chloroform-extractable $VO(OMe)(malt)_2$ which on acidification gives the blue-violet $VO(OH)(malt)_2$ (*cf.* Section II.B.5) on which the analytical method is based (*172*).

Kojic acid has been proposed as a reagent for the spectrophotometric determination of neodymium, holmium, and erbium (*285*), and has been used as an indicator in chelatometric titration of iron (*286*). The three 3-hydroxy-4-pyranones maltol, kojic acid, and meconic acid (**6**, **7**, and **8** in Fig. 2), and their flavone analogues morin and quercetin, have all been used for the spectrophotometric determination of iron (*287,288*), maltol and morin for vanadium (*289*), and morin, quercetin, quercetagetin, and flavonol (**20**, **17**, **19**, and **16,** respectively in Fig. 3) for zirconium (*225,290*). Fluorometric determination of zirconium by flavanol provides one of the earliest examples of this type of analytical application (*291*). An intense blue-green fluorescence reported for reaction of gold(III) with kojic acid was proposed as the basis of an analytical method for the determination of gold, but as the fluorophore appears to be a product of gold(III) oxidation of the kojic acid this method should perhaps be viewed with caution (*292*). A review of flavones as analytical reagents concentrates on 3-hydroxy derivatives, i.e. flavanols, and covers gravimetric, spectrophotometric, and fluorometric methods (*11*).

An early inorganic application of 3-hydroxy-4-pyranones and 3-hydroxy-4-pyridinones was to solvent extraction, e.g. of hafnium(IV) by the former (*293*). However a concurrent wide-ranging investigation of extraction by 1-(4′-tolyl)-6-carbethoxy-3-hydroxy-4-pyridinone (*294*) established the superiority of hydroxypyridinones for this application. The excellent chelating properties exhibited by the 3-hydroxy-4-pyridinone ligands in conjunction with the possibility of synthesizing strongly lipophilic derivatives such as the 1-(4′-tolyl) compound make this class of ligands particularly useful as extractants from aqueous media. Some 1-aryl-3-hydroxy-4-pyridinone derivatives have been demonstrated to be effective as extractants from aqueous media into e.g. chloroform, for a variety of transition metals. Several metal ion separations have been achieved, as e.g., hafnium(IV) from zirconium(IV) and tantalum(V) from niobium(V). The species extracted from the aqueous layer is generally a binary complex ML_n, but in a number of cases extraction is of a ternary complex containing chloride or thiocyanate as well as the 3-hydroxy-4-pyridinonate. Thus, for instance, titanium(IV) may be extracted as $TiL_2(NCS)_2$ or iron(III) as FeL_2Cl (*295*). The complexes extracted are often strongly colored (metal–ligand or ligand–metal charge-transfer) and suitable for spectrophotometric analysis. The molar extinction coefficient[5] for

[5]For a 1 cm length of a solution of concentration 1 mol dm^{-3}.

red-brown $Fe(malt)_3$ is 4610 at 412 nm (*111*), for $Fe(pmpp)_3$ is 6300 at 470 nm (*296*). Thus solvent extraction followed by spectrophotometric determination can provide analytical methods of high specificity and sensitivity (*33*).

References to solvent extraction of numerous metals by 1-phenyl- and 1-(4′-tolyl)-3-hydroxy-4-pyridinones are listed in the center column of Table XXII (*295,297–314*), while references to spectrophotometric determinations of many of these metals by these reagents are given in the right-hand column of this Table (*296,315–323*).

C. Uses in Diagnosis and Therapy

C.1. General

In recent years hydroxypyranones and hydroxypyridinones have been increasingly investigated for the control of metal ion levels in the body (*324–327*). The 3-hydroxy-4-pyranones maltol and ethylmaltol are of relatively low toxicity (*328,329*), and indeed have the pharmacological advantage of being permitted food additives. Hydroxypyridinones are particularly attractive for pharmaceutical purposes since their structure allows tailoring of many of their properties, as outlined in Section II.A.3.b earlier. They have been used in, or tested or proposed for, chelation therapy to remove excess of several toxic elements. This will be illustrated later in the sections

TABLE XXII

Solvent Extraction and Spectrophotometric Estimation of Metals using 1-Phenyl- and 1-(p-Tolyl)-3-hydroxy-4-pyridinones and their Complexes

Metal	Solvent extraction	Spectrophotometric analysis
***p*-Block**		
Gallium(III)	(*297,298*)	
***d*-Block**		
Iron(III)	(*295*)	(*296,315*)
Titanium(IV)	(*299,300*)	(*316,317*)
Hafnium(IV)	(*301*)	
Zirconium(IV)	(*302,303*)	
Vanadium(V)		(*318–321*)
Niobium(V)	(*304,305*)	(*322*)
Tantalum(V)	(*306,307*)	
Tungsten(V)	(*308*)	
Molybdenum(VI)	(*309*)	(*323*)
Tungsten(VI)	(*310*)	
***f*-Block**		
Thorium(IV)	(*311*)	
Protoactinium(V)	(*312*)	
Uranium(VI)	(*313,314*)	

on iron and on the actinides. Conversely they can serve, in the form of complexes with appropriate metals, to remedy deficiencies and to introduce specific metal ions for the purposes of diagnosis or therapy. Examples include the administration of iron to anemic patients, of insulin mimics to diabetics, and of gadolinium as a contrast agent for MRI – these applications are outlined in the following sections.

C.2. *Iron: Supplementation, regulation, and overload*

Iron is an essential element, being involved in several aspects of human metabolism (*330*), such as redox chemistry in many guises, regulatory processes (*331*), and combating infection (*332*). Perturbation of iron levels, whether decrease or increase, can be serious or fatal; iron deficiency or iron-overload conditions affect millions of people around the world. The human body does not have an efficient iron-excretion mechanism, but, as was established in the 1930s (*333*), relies on a balance between intake and excretion to maintain a healthy level of iron (*334*). The case of iron is particularly interesting in the context of the present review, since hydroxypyranones and hydroxypyridinones have been investigated both for the amelioration of iron overload and complementarily as agents to facilitate the administration of iron to patients suffering from iron deficiency.

(a) Supplementation. This comes at three levels – as a low-level dietary supplement or tonic to maintain well-being, as a prescription pharmaceutical to correct minor iron deficiency, or to treat serious clinical conditions such as iron-deficiency anemia. We shall deal with these in order in the following paragraphs.

The common practice of adding iron to breakfast cereals or to vitamin supplements exemplifies the first. Here the first requirements are cosmetic, that the iron-containing compound added should not cause discoloration or adversely affect flavor. It is also an advantage for the added iron-containing compound to be sparingly water-soluble, but for the iron to be reasonably bioavailable and not be incompatible with other constituents (*335*). There is a great deal of inorganic and physical chemistry involved in these matters, most of which is buried in the technical and patent literature.

For the second level, the correction of modest iron deficiency, various iron salts have been recommended or prescribed as iron supplements. These salts have included iron(II) succinate, fumarate, gluconate, and lactate, and iron(III) citrate – or, better (*336*), ammonium iron(III) citrate as iron(III) citrate seems to be a poorly characterized compound (*337*). There are two main problems. The first is the ease of oxidation of the iron(II) salts, particularly the lactate. The second is that absorption of iron from these sources is very inefficient (*57*), as species such as Fe^{2+}(aq), Fe^{3+}(aq), and simple hydrolysed entities such as $FeOH^{2+}$(aq) cross membranes such as those that constitute the wall of the gastrointestinal tract only with great difficulty. Admittedly these salts contain organic anions which form weak complexes which may diffuse across the wall of the gastrointestinal tract slightly less reluctantly, but absorption is poor. In cases of severe or chronic iron deficiency, a more effective means for iron supplementation is required. More complicated iron compounds have been developed to reduce these difficulties, such as ferric albuminate, ferric sodium edetate, and saccharated ferric oxide, but there has long been a need for a more efficient means of iron supplementation. That hydroxypyranones and hydroxypyridinones might be suitable vehicles was first realized in the 1970s, with many iron(III) complexes (some of which, e.g. fluoro-derivatives, may have been aspirational rather than actual) patented during the 1980s for the treatment of anemia.

The use of iron(II) sulfate as a detoxicant for mimosine-affected sheep is an unusual example of iron supplementation (*7*). It provides an interesting complement and link to the following paragraphs on the use of hydroxypyridinones as detoxicants for iron-burdened humans.

(b) Overload. There are several widespread diseases in which severe iron overload is a major and intractable problem. These include hemochromatosis, sickle cell disease, and various forms of thalassemia. The repeated blood transfusions needed in the management of various anemias lead to iron overload. Whatever the cause, long-term iron overload leads to a build-up of iron in the tissues of essential organs, leading to failure of heart, liver, or pancreas. Treatment of these conditions with an iron chelator can prolong life by many years; reduction of iron levels delays the onset of several heart dysfunctions, and also lessens the risk of death from various bacterial infections (*cf.* later). For many years desferrioxamine (DFO), isolated from *Streptomyces pilosus*, has been the only available and permitted agent for the removal of excess of iron from the body, but it is expensive and is awkward to administer – it is not effective when taken orally. It can have undesirable side effects as it is an audiovisual neurotoxin (*338*), causing ocular problems and hearing loss, and is a growth retardant (*339,340*). For these reasons, and indeed because life expectancy is still only into the thirties (*341*) even with current chelation therapy, there have been extensive efforts to find suitable synthetic ligands for the treatment of iron overload.

Hydroxypyranones and hydroxypyridinones are promising candidates as chelating agents for the treatment of iron overload, since many are readily available, they form stable complexes with Fe^{3+}, and some are permitted food additives. Hydroxypyridinones are to be preferred to hydroxypyranones as the former form the more stable complexes and are less readily metabolized. An extensive literature on synthetic iron chelators has been built up over the past 20 years – we can cite only a very small proportion here. In the 1980s it was established that hydroxypyridinones, usually in the form of

1,2-dimethyl-3-hydroxy-4-pyridinone,[6] were able to mobilize iron *in vitro* from a range of sources. Thus L1 was shown to remove iron from transferrin (*342*), lactoferrin (*65,343*), ferritin (*344*), or hemosiderin (*345*), though not from heme or cytochromes (*345*). Maltol (*198*) and 1,2-dihydroxypyridinone (*198,346*) were, like DFO, also able to remove iron from transferrin or ferritin. A biological evaluation of several 3-hydroxy-4-pyridinones indicated key roles for membrane permeability and stability constants in mobilizing iron from hepatocytes, the main cellular reservoir of excess of iron (*45*). 3-Hydroxy-4-pyridinones mobilize iron more effectively than their 3,2- or 1,2-analogues, or than 3-hydroxy-4-pyranones, from transferrins, as expected from their relative affinities for Fe^{3+} (*347–349*). Hydroxypyridinones containing a hydrophilic sulfonate or carboxylate group were found to be slightly better at mobilizing iron than some of their more lipophilic analogues, despite the charge on such ligands (*350*).

There is some qualitative information on iron mobilization. Thus it has been found that the rates of removal of iron decrease in the order

$$\text{transferrin} > \text{ferritin} > \text{hemosiderin}$$

in all cases in a timescale of several hours (*256*). The slow removal of iron from ovotransferrin is a one-step process (*349*), but from other forms of transferrin is more complex, approximating to biphasic kinetics with the two reactions having rather similar rate constants. The latter pattern applies to several iron-removal reactions by 1-methyl-3-hydroxy-2-pyridinone and maltol as well as L1 (*347*). The extent of iron removal is greatest for L1, with iron preferentially removed from the C-terminal site; mimosine shows a slight preference for iron removal from the N-terminal site (*342,351*). Significant differences in the affinities of Fe^{3+} for the C- and N-terminal sites have been established in several similar systems (*352,353*). The removal of iron from transferrin in combined chelation therapy has been discussed (*354*). An order of effectiveness has been established, and a mechanism proposed, for the removal of iron from ferritin by a series of 3-hydroxy-4-pyridinones (*355*). Removal of iron from ferritin is much slower than from transferrin (*342*), with rate constants of between 1.5×10^{-5} and $7.5 \times 10^{-5}\,s^{-1}$ reported for transfer of iron to hexadentate hydroxypyrimidinone and hydroxypyrazinone ligands (*209*).

The first step in progressing from *in vitro* mobilization experiments to clinical trials is the setting up of biological model systems (*356*). Next comes the testing of oral efficacy in animals such as mice, rats, and rabbits (*43,249*), alongside toxicity studies. We cite a few of the

[6]1,2-Dimethyl-3-hydroxy-4-pyridinone is by far the most commonly encountered ligand. It is generally referred to as Deferiprone or L1 by workers in this area; we shall therefore use the designation L1 for this compound (or for its anion) in this section, rather than using dmpp or a form of hopo as elsewhere in this review.

many toxicity studies (*249,357*), from which the general consensus appears to be that L1, and other hydroxypyridinones, are relatively non-toxic, but that large doses or long-term treatments may well cause problems. L1 does present some toxicity concerns, with reported adverse effects including increased susceptibility to infectious diseases, nausea and vomiting, joint pain, and, in high-dose or long-term regimes, ocular toxicity. However the prevalence of these has been disputed (*358*). There has also been disagreement over the relative toxicities, as well as the mobilizing efficiencies, of methyl and ethyl variants of hydroxypyridinones (*357,359–363*); the overall aim is to maximize mobilization while minimizing toxicity.

There have been extensive clinical trials of L1, whose early progress may be glimpsed through the selection of references cited here (*364*). Subsequent long-term trials have involved administering doses of L1 of up to 3 g twice a day, with a maximum dosage of 16 g administered over 24 h in one intensive test (*365*), and have involved comparisons with other iron chelators (*366*). From *in vitro* experiments to long-term trials and clinical usage the most frequent comparison is between L1 and DFO (*367*). Interestingly, combination therapy involving both L1 and DFO has been found useful in certain situations (*368*), including treatment of β-thalassemia (*369*) and management of severe cardiac siderosis (*370*). A routine of four days treatment with L1, then two or three days with DFO, each week has been recommended (*371,372*).

There have been many reviews on the use of hydroxypyridinones, particularly L1, as iron chelators. Many deal with L1 and other hydroxypyridinones in the overall context of chelation therapy for iron overload, some concentrating on the more chemical aspects, others on clinical practice. We cite a selection here, in chronological order from 1989 through 2002 so that interested readers may follow the development of this subject. The citations have been chosen to range from short articles to lengthy and fully referenced reviews, and to feature contributions from a range of the groups active in this area (*373*).

The choice of iron chelators on the basis of both molecular and cellular criteria was discussed in 2003 (*374*). One 2005 review is concerned with the design of orally active iron chelators (*375*), another considers the prospects for effective clinical use of several hydroxypyridinones, dealing with novel species such as the 1-allyl compound as well as with the established deferiprone (L1) and desferrioxamine (Desferal, DFO) (*376*). A review dated 2006 deals with relevance of iron mobilization from both transferrin and other iron-containing proteins by L1 to the treatment of various anemias and other iron-overload conditions (*377*). Two 2007 reviews concentrate on L1, as the only hydroxypyridinone in general clinical use. One author concludes that, on balance, L1 is to be preferred to DFO. This conclusion is on the grounds that, despite the not infrequent occurrence of minor side effects, the incidence of serious side effects is lower for L1 (*370*). The other author, whose detailed review cites 129 references, concludes that the evidence is still not unequivocal, but thinks it likely that DFO is to be preferred for the treatment of thalassemia and sickle cell disorders, though L1 is an acceptable alternative for DFO-intolerant patients (*368*).

Thus the only hydroxypyridinone which is at the time of writing established in clinical practice as an iron chelator is L1. Other hydroxypyridinones are being actively evaluated in a search for analogues which will avoid the disadvantages of L1, or which are effective in situations where L1 is ineffective. Thus some comparative tests of ethyl vs. methyl, e.g., on certain rat and monkey models, indicate that the diethyl compound is, like DFO, effective, but that the dimethyl compound is not (*361*); the diethyl compound (depp; CP94) can be more than 50% orally bioavailable (*378*) and may be less toxic (see earlier). 1-Allyl-2-methyl-3-hydroxy-4-pyridinone (L1NAll) is also showing promise (*376,379,380*).

The improved chelating properties of 2-(1′-hydroxyalkyl) derivatives (*70*) have led to efforts to improve performance by administering ester prodrugs whose hydrolysis produces hydroxyethyl groups. The lipophilic prodrugs reach the required site, there hydrolyzing to more hydrophilic and more effective iron mobilizers (*62*); the optimal prodrug in the group tested was 1-[2′-(pivaloyloxy)ethyl]-2-methyl-3-hydroxy-4-pyridinone (*63*). The superior chelating properties of hexadentate ligands – pFe from 27 upward, *cf.* pFe ~20 for L1 (Section II.A.3.d) – are also being assessed. Tren-Me-3,2-hopo (**25** in Fig. 4) showed potential as an iron chelator (*381*), as subsequently have other tripodal (*138*) and linear (*81*) hexadentate hydroxypyridinones. Multidentate hydroxypyridinones show promise for removing potentially toxic iron which is not bound by transferrin in plasma (*382*).

(c) Iron in other disorders. Iron is an essential element for very nearly all bacteria, so its availability is critical for bacterial virulence and limiting its supply will discourage or prevent their growth (*383*). Iron-chelating drugs such as hydroxypyridinones may therefore prove effective antibacterial agents. Indeed antibacterial properties have been reported for some synthetic cephalosporins containing 3-hydroxy-4-pyridinone units (*384*), and subsequently were established for five simple 2-methyl-3-hydroxy-4-pyridinones (*68*). These strongly inhibited both *Escherichia coli* and *Listeria inocua*; they were demonstrated to mobilize iron from species such as lactoferrin. These antibacterial properties can have a beneficial side effect in treating iron overload, for thalassemia often proves fatal not in itself but through associated severe bacterial infection. Iron chelators used in treating thalassemia thus also reduce the risk of death by this secondary cause. 3-Hydroxy-4-pyridinones have been assessed as antimalarial drugs for some time (*360,385,386*). In this role the more lipophilic chelators are most effective – 1,2-diethyl-3-hydroxy-4-pyridinone is effective, but the 1,2-dimethyl compound (L1) is not (*360*).

L1 could be useful in those cancer chemotherapy treatments which cause an elevation of iron levels (*382*). Thus it may help combat undesirable side effects of antitumor treatment using the anthracycline doxorubicin (Adriamycin) by sequestering free iron and iron bound to doxorubicin and thus preventing the generation of harmful hydroxyl radicals by reduction of hydrogen peroxide (from superoxide generated by the reduced free radical semiquinone form of anthracycline reacting with oxygen) by iron(II)-anthracycline complexes (*387*). L1, like DFO, has a cytoprotective effect, eliminating lipid-derived free radicals (*388*). 3-Hydroxy-4-pyridinones, particularly those of a hydrophilic nature, have marked antiinflammatory properties. They have been suggested for the treatment, by injection, of arthritis. Again this seems to involve the chelation of iron deposited in the joints, thus reducing the formation of free radicals and consequent tissue damage (*389*).

C.3. Aluminum

Until the early 1970s aluminum was generally felt to be fairly harmless (*390*). Since then it has been shown or suggested to cause problems in many clinical and environmental circumstances. The metabolism of aluminum, and disorders and diseases which may be caused or exacerbated by aluminum, were exhaustively reviewed (58 pages) and documented (959 references) in 1986 (*391*). It has been implicated in several types of neurotoxicity (*392*), in adverse side effects of dialysis (*393*) or of repeated albumin infusion (*394*), in Alzheimer's disease (*395*), and in bone disorders (*396*). Transferrin is the main, probably the sole, carrier of aluminum in the blood (*397,398*), so mobilization of aluminum from transferrin, demonstrated for DFO, is the chief criterion for an aluminum chelator. DFO has for many years been the only well-established chelator for reducing aluminum levels in the body (*202,399*), being beneficial in the treatment of patients with aluminum overload from, e.g., dialysis problems or bone disorders (*400,401*). Therefore members of the hydroxypyranone and hydroxypyridinone group of ligands have been assessed alongside DFO (and a number of other potential chelators) for the removal of aluminum from the body (*253,402*). They have been reviewed and assessed in relation to general aspects of health, metabolism, toxicology, and the environmental impact of aluminum (*193*). Maltol, kojic acid, and 1,2-dimethyl-3-hydroxy-4-pyridinone were among 28 ligands examined as chelating agents for Al^{3+}, and to a lesser extent Ga^{3+} and In^{3+}, in a simple blood plasma model medium; speciation patterns permitted direct comparisons between their binding efficiencies in such a medium (*205*). Quantitative speciation studies in model biological systems have been carried out for 1,2-dimethyl- and 1,2-diethyl-3-hydroxy-4-pyridinone, as well as for DFO for comparison (*403*). Pharmacokinetic studies of five

3-hydroxy-4-pyridinonate complexes of aluminum in rats indicated that although the 1,2-dimethyl complex appeared to be non-toxic the same could not be said for three of the other complexes (*404*). Adverse effects of the most promising chelator, 1,2-dimethyl-3-hydroxy-4-pyridinone, have generally been studied in relation to its use to treat iron overload (*cf.* earlier), but they have also been considered in relation to its possible use in aluminum chelation therapy (*402,405*). The possibilities of using chelators such as hydroxypyridinones to remove excess of both aluminum and iron from the brain have been considered (*406*). It is likely to be much more difficult to remove iron, as it is firmly bound in ferritin and hemosiderin in the brain, whereas the aluminum is more accessible. As the problem seems to be that excessive aluminum may interfere with the normal functions of the iron this seems to make the situation more amenable to treatment than is usually the case.

The opening years of the present century have seen considerable activity, especially in Lisbon, in the development of new hydroxypyridinone ligands for the chelation of aluminum in pharmaceutical, clinical, or environmental applications. New bidentate 3-hydroxy-4-pyridinones have included a variety of groups on the ring-nitrogen. Derivatives have included 1-$(CH_2)_nCO_2H$ with $n = 2$–4 (*407*), 1-$CH_2CH_2NEt_2$ (*408*), peptide-mimetic groups such as 1-$(CH_2)_3CONHCH_2CH_2NEt_2$ (*408*) and 1-$(CH_2)_3CONHCH(NH_2)CH_2C_6H_5$ (*409*), groups containing a piperazine spacer, e.g., 1-$(CH_2)_3C_4H_8N_2$-C_6H_4-4′-X with X = H (*409,410*) or Cl (*410*), and simpler arylalkylamino groups such as 1-$(CH_2)_3NHCH_2C_6H_4$-4′-X with X = H or NO_2 (*410*). A number of hydroxypyrimidinone and hydroxypyrazinone ligands have also been synthesized (*411*), along with multifunctional ligands containing 3-hydroxy-4-pyridinone groups attached to the aminocarboxylates ida (*412*) and edta (*413*). Stability constants, pM values, and distribution coefficients have been determined for many of these ligands and their complexes, sometimes with iron and gallium as well as with aluminum. The gallium complexes are relevant to biodistribution studies, for which ^{67}Ga was used. In contrast to all this activity relating to hydroxypyridinones as potential chelators, a recent review of the current status and potential of aluminum chelation therapy for the treatment of Alzheimer's disease contains rather few references to hydroxypyridinones (*414*).

C.4. Diabetes

It was reported in 1980 that vanadate showed insulin activity (*415*); at about the same time it was demonstrated that vanadate is reduced to vanadium(IV) *in vivo* (*416,417*). Subsequently it was found that orally administered vanadate restored blood glucose levels and cardiac function to normal in rats (*418*). Vanadyl sulfate, less toxic than vanadate, also lowers blood sugar levels and ameliorates cardiac problems associated with diabetes (*419*). It also enhances the effects of insulin (*420*). However both vanadate and vanadyl salts are very poorly absorbed, so an uncharged and less hydrophilic complex of vanadium might well prove more easily absorbed and thus more efficacious. This reasoning led to the successful testing and patenting of BMOV, as an oral hypoglycemic agent (*161*). BMOV lowers blood sugar levels (*421*) and consequently has a favorable secondary effect on heart problems which often provide complications in diabetic patients (*422*); its effectiveness has been compared with that of vanadyl sulfate (*423*).

Several other oxovanadium(IV) complexes have been synthesized for assessment as potential insulin-mimetic agents (*163,424*). These include bis-oxalato-oxovanadate(IV) (*425*) as well as bis-kojate and several bis-hydroxypyridinonate analogues (*426*) of BMOV. Assessment has ranged from *in vitro* modeling (*427*), including probing possible interactions with blood serum (*153*), to biochemical and morphological studies on rat livers. The modeling studies have been carried out on both hydroxypyridinonate complexes and on BMOV; the liver studies on both BMOV (*428*) and its kojate analogue (*429*).

Vanadium(III) (*118*) and vanadium(V) (*430*) complexes, like vanadium(IV), enhance insulin action. The vanadium(III) complexes are more resistant to aerial oxidation than expected, so need not be ruled out on that count. However the vanadium(III) and vanadium(V) complexes tested so far have not proved as effective as BMOV (*431*).

Maltolate complexes of cobalt(II), copper(II), zinc(II), chromium(III), and molybdenum(VI) have also been tested for antihypoglycemic activity. Complexes of MoO_2^{2+}, e.g. *cis*-$MoO_2(malt)_2$ (*173*), appear to rank next to vanadium complexes in effectiveness in lowering blood sugar levels; they may also have a role in preventing heart defects developing as a result of diabetes (*174,432*). Co^{2+} complexes are moderately effective, while those of Zn^{2+}(*433*), and Cr^{3+} are relatively ineffective (*431*) – though $Cr(malt)_3$ was patented for the treatment of diabetes as long ago as 1988.

C.5. Magnetic resonance imaging: Gadolinium complexes

NMR imaging techniques rely on the use of paramagnetic species, of which the Gd^{3+} cation offers particularly suitable magnetic properties, to enhance signals (*434,435*). Multidentate hydroxypyridinonate complexes of gadolinium have been much investigated as contrast agents for MRI (*269*), as they appear to offer considerable potential advantages. However it is necessary to maximize stability, selectivity, relaxivity, and solubility in the search for Gd^{3+} complexes suitable for use as MRI contrast agents (*268*). Selectivity, and the maximization of stability to avoid the release of toxic Gd^{3+} (aq), have been documented in Section II.A.3.e earlier, where the incorporation of hydrophilic entities such as –CH_2OH (*271*), –OH, –NH_2, –CO_2H (*224*), or $-CH_2OCH_2-$ (*82,222*) to improve water-solubility has already been mentioned. During the course of the investigations of the synthesis, characterization, and stability of hexadentate hydroxypyridinone ligands and gadolinium complexes cited in Section II.A.3.e the relaxivities of the various complexes have also been monitored (*82,217,220,269,271,436*). The factors which determine relaxivity (*268,269*) are the number of water molecules coordinated to the Gd^{3+}, the exchange rates for such molecules (*270,273*), electronic relaxation, and molecular tumbling. Hexadentate hydroxypyridinonate ligands offer a marked advantage over currently used contrast agents such as the dota and dtpa complexes, for these latter have one water coordinated to the Gd^{3+}, whereas the hexadentate hydroxypyridinonates leave two coordination sites available for binding water to the Gd^{3+} to bring it to its preferred coordination number of eight. That this does indeed happen has been demonstrated for the tren-Me-3,2-hopo complex, for which the crystal structure determination by X-ray diffraction showed two waters coordinated to the Gd^{3+} (*217,436*). The biodistribution and efficacy of a number of the mixed (hopo/tam) hexadentate ligand complexes of Gd^{3+} have been monitored, and the effects of varying the nature of ligand substituents assessed (*437*).

C.6. Radioisotopes

The use of these types of ligands for the introduction of appropriate radioisotopes for diagnosis or therapy has also been proposed (*325,438,439*), e.g. in oral administration of Ga^{3+} in the treatment of bone cancers (*440*), or in administration of ^{67}Ga and ^{111}In for radiodiagnosis or radiotherapy (*441*). ^{67}Ga-labeled complexes of 1-alkyl-2-methyl-3-hydroxy-4-pyridinones were proposed as potential imaging agents, but biodistribution studies showed too rapid renal elimination (*199*). The more lipophilic 1-aryl-2-methyl-3-hydroxy-4-pyridinones were then suggested (*36*) and evaluated for their potential for myocardial imaging, on the basis of biodistribution studies in a range of animals, from mice to a dog (*442*). Technetium(IV) complexes $[TcL_3]^+$ with $L^- =$ 1-alkyl-, 1-phenyl-, 1-benzyl-, or 1-cyclohexyl-2-methyl-3-hydroxy-4-pyridinonate have been proposed as potential renal imaging agents (*443*).

C.7. Actinides

The requirements for a successful agent for the removal of radioactive actinides from the body are even more stringent than for the use of lanthanides in diagnosis. For gadolinium it is necessary that the hydroxypyridinonate complex is very stable, to avoid significant release of toxic Gd^{3+} (aq), is sufficiently soluble, and has an appropriate HLB. For actinide elimination it is also necessary for the chelator to be sufficiently soluble and to have suitable targeting properties. It is also desirable that the chelator does not have such a high affinity for, e.g., Ca^{2+}, Zn^{2+}, or Fe^{3+} that these ions are seriously depleted during treatment. Nonetheless despite these strict requirements considerable success has been achieved in the design and synthesis of hydroxypyridinones with potential to act as actinide chelators. The much-studied 1,2-dimethyl-3-hydroxy-4-pyridinone (dmpp; L1) has been suggested as a decorporation agent for uranium and other actinides (*231*), but hexa- and octadentate hydroxypyridinones seem more likely candidates, and indeed several such ligands show promise for the decorporation of uranium, plutonium, and americium, and other actinides – in a range of oxidation states from +3 for americium up to +6 (*177*). The demonstration of the successful removal of americium from mice by tren-Me-3,2-hopo, where acute toxicity and tissue damage were minimal, provided an early example (*444*). Both 3,2-hopo and 1,2-hopo ligands look promising, including linear as well tripodal (tren-based) hexadentates, and an octadentate ligand derived from ethane-1,2-diamine and containing four hopo units (Fig. 12(a)) (*177*). However the decorporation of plutonium has been the main focus of attention. Although hexadentate and octadentates ligands are to be preferred in the light of their greater binding power, even bidentate ligands such as 4-carbamoyl-1-methyl-3-hydroxy-2-pyridinone have been assessed as sequestrants for plutonium(IV). Many of these ligands proved more effective than $CaNa_3$dtpa or $ZnNa_3$dtpa, both by injection and by oral administration. The dtpa salts can remove plutonium from body fluids, but are ineffective at removing plutonium which has reached the liver or skeleton (*445*). With sufficiently strongly binding ligands plutonium can be mobilized from its complexes with transferrin and ferritin (*346*), a hybrid desferrioxamine-hydroxypyridinonate ligand (Fig. 12(b)) (*446*) forming a particularly stable complex with plutonium(IV) (*447*). The synthesis of several hydroxypyridinones of thorium and of cerium has been carried out to provide less hazardous analogues for use in the search for effective chelators for

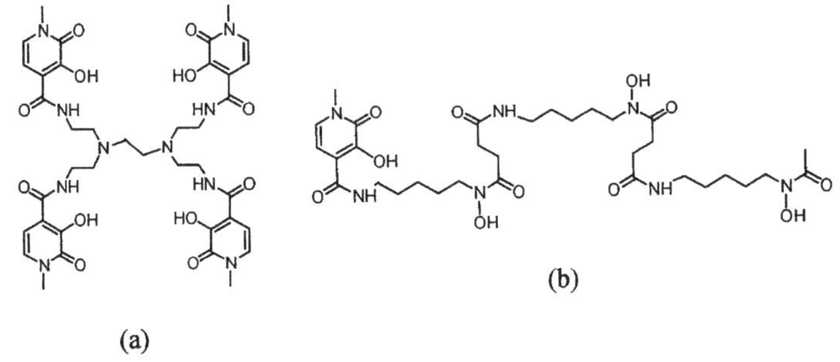

(a)

(b)

FIG. 12. (a) An octadentate tetra(3-hydroxy-2-pyridinone) chelator on an ethane-1,2-diamine template and (b) a hybrid desferrioxamine-3-hydroxy-2-pyridinone chelator.

plutonium(IV) (*147,148*) – though there also seems to be a need for an effective chelator for thorium, as the only currently recommended agent, dtpa, is only partially effective (*177*).

D. Miscellaneous

Early synthetic work on the conversion of kojic acid into 3-hydroxy-4-pyridinones was instigated in the hope of finding potential anesthetics (*25*), but the results of this search proved disappointing. However, various investigators over the subsequent seven and a half decades have found a range of specialized applications and potential uses for several hydroxypyridinones, a few of which are mentioned later.

1,2-Dimethyl-3-hydroxy-4-pyridinone, dmpp, has been suggested for treatment of manganese overload (*109*), which can occur amongst workers in mining, battery manufacturing, and welding (*448*). Manganese is neurotoxic (*449,450*), probably acting by modulating iron redox chemistry (*450*). As $Mn(dmpp)_2(H_2O)_2$ is readily oxidizable to Mn(III), dmpp may facilitate reduction of manganese levels.

Trispyrazolylborates are models for tris-histidine active sites in zinc enzymes, e.g., the matrix metalloproteinases involved in breakdown of extracellular matrices. Inhibition of these metalloproteinases may prove valuable in the treatment of, *inter alios*, cancer and arthritis, so efforts are being made to find appropriate ligands to block the zinc active site. The search has recently moved on from hydroxamates to hydroxypyridinones – 1-hydroxy-2-pyridinone is a cyclic analogue of hydroxamic acid. As reported in Section II.B.2 earlier, hydroxypyridinones form stable five-coordinate complexes on reaction with hydrotris(3,5-phenylmethylpyrazolyl)borate zinc hydroxide. Modeling studies suggest that hydroxypyridinonate ligands should be able to access the active site in the enzyme with ease (*110*).

Hydroxypyranonate and hydroxypyridinonate complexes have been patented for the treatment of zinc deficiency (*451*) and are used in zinc supplements. Maltolate and ethylmaltolate complexes of zinc, copper, and tin have been included as essential constituents in patents for formulations for oral care (*96*). The arylmercury(II) complexes RHgL, where R = $4\text{-}XC_6H_4$, X = Me, OMe, NO_2, and L = malt, koj, have been tested for antibacterial activity (*101*).

$Cr(malt)_3$ has been suggested as an uncharged substitute for the commonly used tris(oxalato)chromate(III) for the broadening of accessible nitroxyl spin labels or probes in experiments involving biological substrates where the charge on $[Cr(ox)_3]^{3-}$ makes it unsuitable. The maltolate complex is effective, but less so than $[Cr(ox)_3]^{3-}$ (*452*). Interest has been shown recently in the non-linear optical properties of maltolate complexes, specifically of calcium, cadmium, and lead (*453*).

There have been a few attempts to utilize hydroxypyranonate and hydroxypyridinonate complexes in homogenous catalysis, where the modest water-solubility of some of their organometallic derivatives could prove attractive. $Ru(malt)_2(PPh_3)_2$ catalyzes the dimerization of phenylacetylene to a 1:1 mixture of the E and Z forms of PhC≡CCH = CHPh (*101*). Ruthenium 3-hydroxy-2-pyridinone and 3-hydroxy-4-pyridinone complexes act as catalysts for oxidation of alcohols (*103*). The complex $RuL_2(AsPh_3)_2$, where L^- = 4-hydroxy-3-phenylazo-benzo-2-pyranonate, acts as a catalyst for oxidation of benzyl alcohol, cyclohexanol, or cinnamyl alcohol by hydrogen peroxide (*454*). This final reference deals with a non-chelating hydroxypyridinone, and serves to remind us of the restriction of the present review to those hydroxypyranones and hydroxypyridinones which can chelate metal ions.

V. Nomenclature and Abbreviations

Abbreviations for 3-hydroxy-4(1*H*)-pyranones, such as maltol and kojic acid, are generally based on their common names (see Section II.A.2). The situation for hydroxypyridinones is more complicated, with abbreviations depending on the research group involved. Thus the most commonly encountered compound, 1,2-dimethyl-3-hydroxy-4(1*H*)-pyridinone, from which the (1*H*) is often omitted, and pyridinone often shortened to pyridone, is variously referred to as L1, CP20, dmpp, or a slight variation of the last. The isomers of the hydroxypyridinones are often referred to as 3,4-, 3,2-, and 1,2-hopo or -HOPO, or 3,4-, 3,2-, and 1,2-HP. Nomenclature and abbreviation complications increase on going from bidentate to tetra-, hexa-, and octadentate hydroxypyridinones. For the most fully studied series, the tripodal hexadentates, the ligand abbreviation generally specifies the cap or scaffold unit followed by the hydroxypyridinone, e.g., tren-3,2-HOPO. Such usage implies three identical arms to the tripod; when in recent developments it has been found advantageous to have dissimilar arms, then the format is of the type tren-$(3,2\text{-HOPO})_2$ (TAM), with ligands with three identical arms now shown as e.g., tren-$(3,2\text{-HOPO})_3$. The reader is advised to consult Gordon *et al.* (*177*), where formulae and abbreviations for over a hundred actinide chelators, many of them hydroxypyridinones, are systematically laid out.

Abbreviations for ligands which appear only once are generally defined at the relevant point in the text; abbreviations which appear in more than one place are listed and defined.

acac	acetylacetonate (pentane-2,4-dionate)
depp	1,2-diethyl-3-hydroxy-4-pyridinonate

dmpp	1,2-dimethyl-3-hydroxy-4-pyridinonate (also known as L1)
dota	1,4,7,10-tetraazadodecane-*N,N′,N″,N‴*-tetraacetate
dtpa	diethylenetriaminepentaacetate
edta	ethylenediamine-*N,N,N′,N′*-tetraacetate
empp	1-ethyl-2-methyl-3-hydroxy-4-pyridinonate
etmalt	ethyl maltolate; 2-ethyl-3-hydroxy-4-pyranonate
hopo	hydroxypyridinone or hydroxypyridinonate (see earlier)
ida	iminodiacetate
koj	kojate; 6-hydroxymethyl-3-hydroxy-4-pyranonate
malt	maltolate; 2-methyl-3-hydroxy-4-pyranonate
mec	meconate
tam	2,3-dihydroxyterephthalamide
tren	tris(aminoethyl)amine
trpn	tris(amino-*n*-propyl)amine

ACKNOWLEDGEMENTS

The authors are most grateful to the organizers of the XVII Spanish–Italian Congress on the Thermodynamics of Metal Complexes for their invitation to present a Plenary Lecture on hydroxypyranone and hydroxypyridinone complexes, from which this present review has been developed and elaborated; JB is particularly grateful to Dr. Pilar López and Professor Francisco Sánchez for their hospitality. The authors also wish to thank Colin Hubbard for his encouragement and assistance in polishing and submitting the chapter.

REFERENCES

1. (a) Brand, J. *Ber. Deutschen Chem. Ges.* **1894**, *27*, 806–810. (b) Kiliani, H.; Bazlen, M. *Ber. Deutschen Chem. Ges.* **1894**, *27*, 3115–3120.
2. Nickell, L. G.; Gordon, P. N. *Agr. Biol. Chem.* **1963**, *27*, p. 65.
3. (a) Arnarp, J.; Bielawski, J.; Dahlin, B.-M.; Dahlman, O.; Enzell, C. R.; Pettersson, T. *Acta Chem. Scand.* **1990**, *44*, 916–926. (b) Arnarp, J.; Bielawski, J.; Dahlin, B.-M.; Dahlman, O.; Enzell, C. R.; Pettersson, T. *Acta Chem. Scand.* **1990**, *44*, 963–967.
4. Saito, K. *Jpn. J. Bot.* **1907**, *21*, 7–11.
5. Haitinger, L.; Lieben, A. *Monatsh.* **1884**, *5*, 339–366.
6. (a) Bickel, A. F. *J. Am. Chem. Soc.* **1947**, *69*, 1801–1803. (b) Adams, R.; Jones, V. V. *J. Am. Chem. Soc.* **1947**, *69*, 1803–1805. (c) Bickel, A. F. *J. Am. Chem. Soc.* **1947**, *69*, 1805–1806. (d) Adams, R.; Govindachari, T. R. *J. Am. Chem. Soc.* **1947**, *69*, 1806–1808. (e) Berson, J. A.; Jones, W. M.; O'Callaghan, L. F. *J. Am. Chem. Soc.* **1956**, *78*, 622–623.
7. Matsumoto, H.; Smith, E. G.; Sherman, G. D. *Arch. Biochem. Biophys.* **1951**, *33*, 201–211.
8. Stünzi, H.; Perrin, D. D.; Teitei, T.; Harris, R. L. N. *Aust. J. Chem.* **1979**, *32*, 21–30.
9. Yoshida, R. R. Ph.D. Thesis. University of Minnesota. 1944.
10. Stünzi, H.; Harris, R. L. N.; Perrin, D. D.; Teitei, T. *Aust. J. Chem.* **1980**, *33*, 2207–2220.

11. Katyal, M. *Talanta* **1968**, *15*, 95–106.
12. Stacey, M.; Turton, L. M. *J. Chem. Soc.* **1946**, 661–665.
13. Spielman, M. A.; Freifelder, M. *J. Am. Chem. Soc.* **1947**, *69*, 2908–2909.
14. Schenk, J. R.; Spielman, M. A. *J. Am. Chem. Soc.* **1945**, *67*, 2276–2277.
15. Patton, S. *J. Dairy Sci.* **1950**, *33*, 102–106.
16. Patton, S. *J. Biol. Chem.* **1950**, *184*, 131–134.
17. Brennan, T. M.; Weeks, P. D.; Brannegan, D. P.; Kuhla, D. E.; Elliott, M. L.; Watson, H. A.; Wlodecki, B. *Tetrahedron Lett.* **1978**, 331–334.
18. Ohura, R.; Katayama, A.; Takagishi, T. *Textile Res. J.* **1991**, *61*, 242–246.
19. Yabuta, T. *J. Chem. Soc.* **1924**, *125*, 575–587.
20. Becker, H.-D. *Acta Chem. Scand.* **1962**, *16*, 78–82.
21. Ellis, B. L.; Duhme, A. K.; Hider, R. C.; Hossain, M. B.; Rizvi, S.; van der Helm, D. *J. Med. Chem.* **1996**, *39*, 3659–3670.
22. See, e.g., Gilchrist, T. L. *"Heterocyclic Chemistry"*; 2nd edn. Longmans: London, **1992**; Chapter 5.5,
23. See p. 178 of Ref. (*22*).
24. Mullock, E. B.; Suschitzky, H. *J. Chem. Soc. (C)* **1967**, 828–830.
25. Armit, J. W.; Nolan, T. J. *J. Chem. Soc.* **1932**, 3023–3031.
26. Adams, R.; Johnson, J. L. *J. Am. Chem. Soc.* **1949**, *71*, 705–708.
27. Heynes, K.; Vogelsang, G. *Chem. Ber.* **1954**, *87*, 1377–1384.
28. Hahn, V.; Kukolja, S. *Croat. Chem. Acta* **1961**, *33*, 137–144.
29. Severin, T.; Loidl, A. *Z. Lebensm. Unters.-Forsch.* **1976**, *161*, 119–124.
30. Imafuku, K.; Ishizaka, M.; Matsumura, H. *Bull. Chem. Soc. Jpn.* **1979**, *52*, 107–110.
31. Spenser, I. D.; Notation, A. D. *Can. J. Chem.* **1962**, *40*, 1374–1379.
32. Tamhina, B.; Jakopčić, K.; Zorko, F.; Herak, M. J. *J. Inorg. Nucl. Chem.* **1974**, *36*, 1855–1857.
33. Jakopčić, K.; Tamhina, B.; Zorko, F.; Herak, M. J. *J. Inorg. Nucl. Chem.* **1977**, *39*, 1201–1203.
34. Kontoghiorghes, G. J.; Sheppard, L. *Inorg. Chim. Acta* **1987**, *136*, L11–L12.
35. Molenda, J. J.; Jones, M. M.; Basinger, M. A. *J. Med. Chem.* **1994**, *37*, 93–98.
36. Zhang, Z.; Rettig, S. J.; Orvig, C. *Inorg. Chem.* **1991**, *30*, 509–515.
37. Burgess, J.; de Castro, B.; Oliveira, C.; Rangel, M. *J. Chem. Res.* **1996**, (*S*) 234–235, (*M*) 1338–1351,
38. Schlindwein, W.; Waltham, E.; Burgess, J.; Binsted, N.; Nunes, A.; Leite, A.; Rangel, M. *Dalton Trans.* **2006**, 1313–1321.
39. Harris, R. N. L. *Aust. J. Chem.* **1976**, *29*, 1329–1334.
40. Färber, M.; Osiander, H.; Severin, T. *J. Heterocycl. Chem.* **1994**, *31*, 947–956.
41. Fischer, B. E.; Hodge, J. E. *J. Org. Chem.* **1964**, *29*, 776–781.
42. Nelson, W. O.; Karpishin, T. B.; Rettig, S. J.; Orvig, C. *Can. J. Chem.* **1988**, *66*, 123–131.
43. Dobbin, P. S.; Hider, R. C.; Hall, A. D.; Taylor, P. D.; Sarpong, P.; Porter, J. B.; Xiao, G.; van der Helm, D. *J. Med. Chem.* **1993**, *36*, 2448–2458.
44. (a) Griffin, W. C. *J. Soc. Cosmet. Chem.* **1954**, *5*, 249–256. (b) McGowan, J. C. *Tenside Surf. Det.* **1990**, *27*, p. 229. (c) Van de Waterbeemd, H.; Testa, B. *Adv. Drug Res.* **1987**, *16*, 85–225. (d) Rekker, R. F.; Mannhold, R. (Eds.) *"Calculation of Drug Lipophilicity"*; VCH: Weinheim, **1992**.
45. Porter, J. B.; Gyparaki, B. L. C.; Huehns, E. R.; Sarpong, P.; Saez, V.; Hider, R. C. *Blood* **1988**, *72*, 1497–1503.
46. Hider, R. C.; Hall, A. D. *Perspect. Bioinorg. Chem.* **1991**, *1*, 209–253.
47. Kontoghiorghes, G. J.; May, A. *Biol. Met.* **1990**, *3*, 183–187.
48. Porter, J. B.; Hider, R. C.; Huehns, E. R. *Sem. Hematol.* **1990**, *27*, 95–100.
49. Alshehri, S.; Burgess, J.; Shaker, A. *Transition Met. Chem.* **1998**, *23*, 689–691.
50. Hider, R. C.; Huehns, E. R.; Porter, J. B. UK Patent Application, GB 2 242 191A, 1991.
51. Price, C. J. M. Phil. Thesis. Leicester. 1994.

52. Burgess, J.; Cambon, A.; Waltham, E., *unpublished observations.*
53. Burgess, J.; Coe, P.; Patel, M., *unpublished observations.*
54. Molenda, J. J.; Basinger, M. A.; Hanusa, T. P.; Jones, M. M. *J. Inorg. Biochem.* **1994**, *55*, 131–146.
55. Zhang, Z.; Rettig, S. J.; Orvig, C. *Can. J. Chem.* **1992**, *70*, 763–770.
56. Burgess, J.; Fawcett, J.; Parsons, S. A. *Acta Cryst.* **2001,** *E57*, o1016–o1018.
57. Liu, Z. D.; Khodr, H. H.; Lu, S. L.; Hider, R. C. *J. Pharm. Pharmacol.* **2000**, *52*, 263–272.
58. Burgess, J.; Fawcett, J.; Russell, D. R.; Zaisheng, L. *Acta Cryst. C* **1998**, *54*, 430–433.
59. Burgess, J.; Fawcett, J.; Russell, D. R.; Waltham, E. *Acta Cryst. C* **1998**, *54*, 2011–2015.
60. Liu, G.; Bruenger, F. W.; Barrios, A. M.; Miller, S. C. *Nucleosides Nucleotides* **1995**, *14*, 1901–1904.
61. Liu, G.; Men, P.; Kenner, G. H.; Miller, S. C.; Bruenger, F. W. *Nucleosides Nucleotides Nucleic Acids* **2004**, *23*, 599–611.
62. Liu, Z. D.; Liu, D. Y.; Lu, S. L.; Hider, R. C. *J. Pharm. Pharmacol.* **1999**, *51*, 555–564.
63. Rai, B. L.; Liu, D. Y.; Lu, S. L.; Hider, R. C. *Eur. J. Med. Chem.* **1999**, *34*, 475–485.
64. Abeysinghe, R. D.; Roberts, P. J.; Cooper, C. E.; MacLean, K. H.; Hider, R. C.; Porter, J. B. *J. Biol. Chem.* **1996**, *271*, 7965–7972.
65. Liu, Z. D.; Kayyali, R.; Hider, R. C.; Porter, J. B.; Theobald, A. E. *J. Med. Chem.* **2002**, *45*, 631–639.
66. Feng, M.-H.; van der Does, L.; Bantjes, A. *J. Biomater. Sci., Polym. Ed.* **1992**, *4*, 145–154.
67. Feng, M.-H.; van der Does, L.; Bantjes, A. *J. Biomater. Sci., Polym. Ed.* **1992**, *4*, 99–105.
68. Gagnon, M. K. J.; St. Germain, T. R.; McNamara, R. A.; Vogels, C. M.; Westcott, S. A. *Aust. J. Chem.* **2000**, *53*, 693–697.
69. Feng, M.-H.; van der Does, L.; Bantjes, A. *J. Med. Chem.* **1993**, *36*, 2822–2827.
70. Piyamongkol, S.; Liu, Z. D.; Hider, R. C. *Tetrahedron* **2001**, *57*, 3479–3486.
71. Li, M.-S.; Gu, L.-Q.; Huang, Z.-S.; Xiao, S.-H.; Ma, L. *Tetrahedron* **1999**, *55*, 2237–2244.
72. Patel, M. K.; Fox, R.; Taylor, P. D. *Tetrahedron* **1996**, *52*, 1835–1840.
73. Ledl, F.; Osiander, H.; Pachmayr, O.; Severin, T. *Z. Lebensm. Unters.-Forsch.* **1989**, *188*, 207–211.
74. Fox, R. C.; Taylor, P. D. *Synth. Commun.* **1999**, *29*, 989–1001.
75. Hodge, J. E.; Nelson, E. C. *Cereal Chem.* **1961**, *38*, 207–221.
76. Lutz, T. G.; Clevette, D. J.; Rettig, S. J.; Orvig, C. *Inorg. Chem.* **1989**, *28*, 715–719.
77. Lutz, T. G.; Clevette, T. J.; Hoveyda, H. R.; Karunaratne, V.; Nordin, A.; Sjöberg, S.; Winter, M.; Orvig, C. *Can. J. Chem.* **1994**, *72*, 1362–1369.
78. Scarrow, R. C.; Raymond, K. N. *Inorg. Chem.* **1988**, *27*, 4140–4149.
79. Rai, B. L.; Khodr, H.; Hider, R. C. *Tetrahedron* **1999**, *55*, 1129–1142.
80. Meyer, M.; Telford, J. R.; Cohen, S. M.; White, D. J.; Xu, J.; Raymond, K. N. *J. Am. Chem. Soc.* **1997**, *119*, 10093–10103.
81. Clarke Jurchen, K. M.; Raymond, K. N. *J. Coord. Chem.* **2005**, *58*, 55–80.
82. Johnson, A. R.; O'Sullivan, B.; Raymond, K. N. *Inorg. Chem.* **2000**, *39*, 2652–2660.
83. Hider, R. C.; Taylor, P. D.; Walkinshaw, M.; Wang, J. L.; van der Helm, D. *J. Chem. Res.* **1990**, (*S*) 316–317, (*M*) 2520–2555,
84. Chan, H.-K.; Ghosh, S.; Venkataram, S.; Rahman, Y. E.; Grant, D. J. W. *J. Pharm. Sci.* **1992**, *81*, 353–358.
85. Xiao, G. Y.; van der Helm, D.; Hider, R. C.; Dobbin, P. S. *J. Chem. Soc., Dalton Trans.* **1992**, 3265–3271.

86. Burgess, J.; Fawcett, J.; Patel, M. S.; Russell, D. R. *J. Chem. Res.* **1993**, (S) 50–51, (M) 0214–0246.
87. Burgess, J.; Fawcett, J.; Russell, D. R.; Hider, R. C.; Hossain, M. B.; Stoner, C. R.; van der Helm, D. *Acta Cryst.* **1996**, *C52*, 2917–2920.
88. Brown, S.; Burgess, J.; Fawcett, J.; Llewellyn, M. A.; Parsons, S. A.; Russell, D. R.; Waltham, E. *Acta Cryst.* **1995**, *C51*, 1335–1338.
89. Orvig, C.; Rettig, S. J.; Zhang, Z. *Acta Cryst.* **1994**, *C50*, 1511–1514.
90. Zhang, Z.; Hui, T. L. T.; Orvig, C. *Can. J. Chem.* **1989**, *67*, 1708–1710.
91. Simpson, L.; Rettig, S. J.; Trotter, J.; Orvig, C. *Can. J. Chem.* **1991**, *69*, 893–900.
92. (a) Annan, T. A.; Peppe, C.; Tuck, D. G. *Can. J. Chem.* **1990**, *68*, 1598–1605. (b) Habeeb, J. J.; Tuck, D. G.; Walters, F. H. *J. Coord. Chem.* **1978**, *8*, 27–33.
93. Kiliani, H.; Bazlen, M. *Ber. Deutschen Chem. Ges.* **1894**, *27*, 3115–3120.
94. Morita, H.; Shimomura, S.; Kawaguchi, S. *Bull. Chem. Soc. Jpn.* **1976**, *49*, 2461–2464.
95. Gérard, C. *Bull. Soc. Chim. Fr.* **1979**, I-451–I-456.
96. Barrett, M. C.; Mahon, M. F.; Molloy, K. C.; Steed, J. W.; Wright, P. *Inorg. Chem.* **2001**, *40*, 4384–4388.
97. Eljammal, A.; Howell, P. L.; Turner, M. A.; Li, N.; Templeton, D. M. *J. Med. Chem.* **1994**, *37*, 461–466.
98. Greaves, S. J.; Griffith, W. P. *Polyhedron* **1988**, *7*, 1973–1979.
99. Griffith, W. P.; Mostafa, S. I. *Polyhedron* **1992**, *11*, 2997–3005.
100. Bhatia, S.; Kauschik, N. K.; Sodhi, G. S. *Bull. Chem. Soc. Jpn.* **1989**, *62*, 2693–2696.
101. Marwaha, S. S.; Kaur, J.; Sodhi, G. S. *J. Inorg. Biochem.* **1994**, *54*, 67–74.
102. Fryzuk, M. D.; Jonker, M. J.; Rettig, S. J. *Chem. Commun.* **1997**, 377–378.
103. El-Hendawy, A. M. *Transition Met. Chem.* **1992**, *17*, 250–255.
104. Carter, L.; Davies, D. L.; Fawcett, J.; Russell, D. R. *Polyhedron* **1993**, *12*, 1599–1602.
105. Capper, G.; Carter, L. C.; Davies, D. L.; Fawcett, J.; Russell, D. R. *J. Chem. Soc., Dalton Trans.* **1996**, 1399–1403.
106. Lang, R.; Polborn, K.; Severin, T.; Severin, K. *Inorg. Chim. Acta* **1999**, *294*, 62–67.
107. Habereder, T.; Warchhold, M.; Nöth, H.; Severin, K. *Angew. Chem. Int. Ed.* **1999**, *38*, 3225–3228.
108. Ahmed, S. I.; Burgess, J.; Fawcett, J.; Parsons, S. A.; Russell, D. R.; Laurie, S. H. *Polyhedron* **2000**, *19*, 129–135.
109. Hsieh, W.-Y.; Zaleski, C. M.; Pecoraro, V. L.; Fanwick, P. E.; Liu, S. *Inorg. Chim. Acta* **2006**, *359*, 228–236.
110. Puerta, D. T.; Cohen, S. M. *Inorg. Chem.* **2003**, *42*, 3423–3430.
111. Morita, H.; Hayashi, Y.; Shimomura, S.; Kawaguchi, S. *Chem. Lett.* **1975**, 339–342.
112. Dutt, N. K.; Sarma, U. U. M. *J. Inorg. Nucl. Chem.* **1975**, *37*, 1801–1802.
113. Agarwal, R. C.; Gupta, S. P.; Rastogi, D. K. *J. Inorg. Nucl. Chem.* **1974**, *36*, 208–211.
114. Burgess, J.; Hubbard, C. D.; Patel, M. S.; Radulović, S.; Thuresson, K.; Parsons, S. A.; Guardado, P. *Transition Met. Chem.* **2002**, *27*, 134–144.
115. Schlindwein, W.; Rangel, M.; Burgess, J. *J. Synchrotron. Radiat.* **1999**, *6*, 579–581.
116. Ahmet, M. T.; Frampton, C. S.; Silver, J. *J. Chem. Soc., Dalton Trans.* **1988**, 1159–1163.
117. Finnegan, M. M.; Rettig, S. J.; Orvig, C. *J. Am. Chem. Soc.* **1986**, *108*, 5033–5035.
118. Melchior, M.; Rettig, S. J.; Liboiron, B. D.; Thompson, K. H.; Yuen, V. G.; McNeill, J. H.; Orvig, C. *Inorg. Chem.* **2001**, *40*, 4686–4690.
119. Clarke, E. T.; Martell, A. E.; Reibenspies, J. *Inorg. Chim. Acta* **1992**, *196*, 177–183.

120. Odoko, M.; Yamamoto, K.; Hosen, M.; Okabe, N. *Acta Cryst. C* **2003**, *59*, 121–123.
121. Burgess, J.; Fawcett, J.; Parsons, S. A.; Russell, D. R. *Acta Cryst. C* **1994**, *50*, 1911–1913.
122. Yin, H.; Li, F.; Wang, D. *J. Coord. Chem.* **2007**, *60*, 1133–1141.
123. Hsieh, W. Y.; Liu, S. *Inorg. Chem.* **2005**, *44*, 2031–2038.
124. Abbott, A. P.; Capper, G.; Davies, D. L.; Fawcett, J.; Russell, D. R. *J. Chem. Soc., Dalton Trans.* **1995**, 3709–3713.
125. Burgess, J.; Parsons, S. A.; Singh, K.; Waltham, E.; López, P.; Sánchez, F.; Rangel, M.; Schlindwein, W. *Transition Met. Chem.*, (in press).
126. El-Hendawy, A. M.; El-Shahawi, M. S. *Polyhedron* **1989**, *8*, 2813–2816.
127. (a) Marcus, R. A. *J. Chem. Phys.* **1956**, *24*, 966–978. (b) Marcus, R. A. *Ann. Rev. Phys. Chem.* **1964**, *15*, 155–196. (c) Cannon, R. D. *"Electron Transfer Reactions"*; Butterworth: London, **1980**.(d) Sutin, N. *Progr. Inorg. Chem.* **1983**, *30*, 441–498. (e) Marcus, R. A.; Sutin, N. *Biochem. Biophys. Acta* **1985**, *811*, 265–322.
128. Clarke, E. T.; Martell, A. E. *Inorg. Chim. Acta* **1992**, *191*, 57–63.
129. Clarke, E. T.; Martell, A. E. *Inorg. Chim. Acta* **1992**, *196*, 185–194.
130. Clevette, D. J.; Nelson, W. O.; Nordin, A.; Orvig, C.; Sjöberg, S. *Inorg. Chem.* **1989**, *28*, 2079–2081.
131. Orvig, C.; Rettig, S. J.; Trotter, J.; Zhang, Z. *Can. J. Chem.* **1990**, *68*, 1803–1807.
132. Nelson, W. O.; Karpishin, T. B.; Rettig, S. J.; Orvig, C. *Inorg. Chem.* **1988**, *27*, 1045–1051.
133. Nelson, W. O.; Rettig, S. J.; Orvig, C. *Inorg. Chem.* **1989**, *28*, 3153–3157.
134. Burgess, J.; Fawcett, J.; Patel, M. S.; Russell, D. R., *unpublished structure determination*.
135. Scarrow, R. E.; Riley, P. E.; Abu-Dari, K.; White, D. L.; Raymond, K. N. *Inorg. Chem.* **1985**, *24*, 954–967.
136. Xiao, G.; van der Helm, D.; Hider, R. C.; Dobbin, P. S. *Inorg. Chem.* **1995**, *34*, 1268–1270.
137. Moore, E. G.; Xu, J.; Jocher, C. J.; Werner, E. J.; Raymond, K. N. *J. Am. Chem. Soc.* **2006**, *128*, 10067–10068.
138. Jurchen, K. M. C.; Raymond, K. N. *Inorg. Chem.* **2006**, *45*, 1078–1090.
139. Anderson, B. F.; Buckingham, D. A.; Robertson, G. B.; Webb, J. *Acta Cryst.* **1983**, *C39*, 723–725.
140. Burgess, J.; Fawcett, J.; Parsons, S. A.; Russell, D. R. *Acta Cryst.* **1994**, *C50*, 1911–1913.
141. Zaremba, K.; Lasocha, W.; Adamski, A.; Stanek, J.; Pattek-Janczyk, A. *J. Coord. Chem.* **2007**, *60*, 1537–1546.
142. Nelson, W. O.; Rettig, S. J.; Orvig, C. *J. Am. Chem. Soc.* **1987**, *109*, 4121–4123.
143. Matsuba, C. A.; Nelson, W. O.; Rettig, S. J.; Orvig, C. *Inorg. Chem.* **1988**, *27*, 3935–3939.
144. Burgess, J.; Fawcett, J.; Llewellyn, M. A.; Parsons, S. A.; Russell, D. R. *Transition Met. Chem.* **2000**, *25*, 541–546.
145. Aspinall, H. C. *"Chemistry of the f-Block Elements"*; Gordon and Breach: Amsterdam, **2001**.
146. Kappel, M. J.; Nitsche, H.; Raymond, K. N. *Inorg. Chem.* **1984**, *24*, 605–611.
147. Xu, J.; Radkov, E.; Ziegler, M.; Raymond, K. N. *Inorg. Chem.* **2000**, *39*, 4156–4164.
148. Xu, J.; Whisenhunt, D. W.; Veeck, A. C.; Uhlir, L. C.; Raymond, K. N. *Inorg. Chem.* **2003**, *42*, 2665–2674.
149. Beauchamp, A.; Benoit, R. L. *Can. J. Chem.* **1966**, *44*, 1615–1624.
150. Denekamp, C. I. F.; Evans, D. F.; Parr, J.; Woollins, J. D. *J. Chem. Soc., Dalton Trans.* **1993**, 1489–1493.
151. Evans, D. F.; Wong, C. Y. *Polyhedron* **1991**, *10*, 1131–1138.
152. Taylor, P. D. *Chem. Commun.* **1996**, 405–406.

153. Buglyó, P.; Kiss, T.; Kiss, E.; Sanna, D.; Garribba, E.; Micera, G. *J. Chem. Soc., Dalton Trans.* **2002**, 2275–2282.
154. Edwards, D. S.; Liu, S. N.; Poirier, M. J.; Zhang, Z. H.; Webb, G. A.; Orvig, C. *Inorg. Chem.* **1994**, *33*, 5607–5609.
155. Denekamp, C. I. F.; Evans, D. F.; Slawin, A. M. Z.; Williams, D. J.; Wong, C. Y.; Woollins, J. D. *J. Chem. Soc., Dalton Trans.* **1993**, 2375–2382.
156. Barrett, M. C.; Mahon, M. F.; Molloy, K. C.; Wright, P. *Main Group Met. Chem.* **2000**, *23*, 663–667.
157. Alshehri, S.; Burgess, J.; Fawcett, J.; Parsons, S. A.; Russell, D. R. *Polyhedron* **2000**, *19*, 399–405.
158. Otera, J.; Kawasaki, Y.; Tanaka, T. *Inorg. Chim. Acta.* **1967**, *1*, 294–296.
159. Stewart, C. P.; Porte, A. L. *J. Chem. Soc., Dalton Trans.* **1972**, 1661–1666.
160. Bechmann, W.; Uhlemann, E.; Kirmse, R.; Köhler, K. *Z. Anorg. Allg. Chem.* **1987**, *544*, 215–224.
161. McNeill, J. H.; Yuen, V. G.; Hoveyda, H. R.; Orvig, C. *J. Med. Chem.* **1992**, *35*, 1489–1491.
162. Burgess, J.; de Castro, B.; Oliviera, C.; Rangel, M.; Schlindwein, W. *Polyhedron* **1997**, *16*, 789–794.
163. Rangel, M. *Transition Met. Chem.* **2001**, *26*, 219–223.
164. Rangel, M.; Leite, A.; Amorim, M. J.; Garribba, E.; Micera, G.; Lodyga-Chruscinska, E. *Inorg. Chem.* **2006**, *45*, 8086–8097.
165. Sakurai, H.; Tamura, A.; Fugono, J.; Yasui, H.; Kiss, T. *Coord. Chem. Rev.* **2003**, *245*, 31–37.
166. Caravan, P.; Gelmini, L.; Glover, N.; Herring, F. G.; Li, H.; McNeill, J. H.; Rettig, S. J.; Setyawati, I. A.; Shuter, E.; Sun, Y.; Tracey, A. S.; Yuen, V. G.; Orvig, C. *J. Am. Chem. Soc.* **1995**, *117*, 12759–12770.
167. Buglyó, P.; Kiss, E.; Fábián, I.; Kiss, T.; Sanna, D.; Garribba, E.; Micera, G. *Inorg. Chim. Acta* **2000**, *306*, 174–183.
168. Sun, Y.; James, B. R.; Rettig, S. J.; Orvig, C. *Inorg. Chem.* **1996**, *35*, 1667–1673.
169. Archer, C. M.; Dilworth, J. R.; Jobanputra, P.; Harman, M. E.; Hursthouse, M. B.; Karaulov, A. *Polyhedron* **1991**, *10*, 1539–1543.
170. Luo, H. Y.; Rettig, S. J.; Orvig, C. *Inorg. Chem.* **1993**, *32*, 4491–4497.
171. Jungnickel, J. E.; Klinger, W. *Fresenius' Z. Anal. Chem.* **1964**, *206*, 275–281.
172. Weidemann, C.; Priebsch, W.; Rehder, D. *Chem. Ber.* **1989**, *122*, 235–243.
173. Lord, S. J.; Epstein, M. A.; Paddock, R. L.; Vogels, C. M.; Hennigar, T. L.; Zaworotko, M. J.; Taylor, N. J.; Driedzic, W. R.; Broderick, T. L.; Westcott, S. A. *Can. J. Chem.* **1999**, *77*, 1249–1261.
174. Epstein, N. A.; Horton, J. L.; Vogels, C. M.; Taylor, N. J. *Aust. J. Chem.* **2000**, *53*, 687–691.
175. Santos, M. A.; Gama, S.; Pessoa, J. C.; Oliveira, M. C.; Tóth, I.; Farkas, E. *Eur. J. Inorg. Chem.* **2007**, 1728–1737.
176. Brown, E. J.; Whitwood, A. C.; Walton, P. H.; Duhme-Klair, A.-K. *Dalton Trans.* **2004** 2458–2462.
177. Gordon, A. E. V.; Xu, J.; Raymond, K. N.; Durbin, P. *Chem. Rev.* **2003**, *103*, 4207–4282.
178. (a) Beck, M. T. *"Chemistry of Complex Equilibria"*; Van Nostrand Reinhold: London, **1970**.(b) Beck, M. T.; Nagypál, I. *"Chemistry of Complex Equilibria"*; Ellis Horwood: Chichester, **1990**.
179. Makni, C.; Regaya, B.; Aplincourt, M.; Kappenstein, C. *C. R. Hebd. Seances Acad. Sci., Ser. C* **1975**, *280*, 117–120.
180. Gérard, C.; Hugel, R. P. *J. Chem. Res.* **1978**, (*S*) 404–405, (*M*) 4875–4889,
181. Choux, G.; Benoit, R. L. *Bull. Soc. Chim. Fr.* **1967**, 2920–2923.
182. Ma, R.; Riebenspies, J. J.; Martell, A. E. *Inorg. Chim. Acta* **1994**, *223*, 21–29.
183. Li, Y. J.; Martell, A. E. *Inorg. Chim. Acta* **1993**, *214*, 103–111.

184. Jakusch, T.; Gajda-Schrantz, K.; Adachi, Y.; Sakurai, H.; Kiss, T.; Horváth, L. *J. Inorg. Biochem.* **2006**, *100*, 1521–1526.
185. Masoud, M. S.; El-Thana, S. A.; El-Enein, A. *Transition Met. Chem.* **1989**, *14*, 155–156.
186. Murakami, Y. *Bull. Chem. Soc. Jpn.* **1962**, *35*, 52–56.
187. Clark, N. J.; Willeford, B. R. *J. Am. Chem. Soc.* **1957**, *79*, 1296–1297.
188. Bryant, B. E.; Fernelius, W. C. *J. Am. Chem. Soc.* **1954**, *76*, 5351–5352.
189. Kido, H.; Fernelius, W. C.; Haas, C. G. *Anal. Chim. Acta* **1960**, *23*, 116–123.
190. Gérard, C.; Ducauze, C. *Bull. Soc. Chim. Fr.* **1975**, 1955–1958.
191. Gérard, C.; Hugel, R. P. *Bull. Soc. Chim. Fr.* **1975**, 2404–2408.
192. Hedlund, T.; Öhman, L.-O. *Acta Chem. Scand. A* **1988**, *42*, 702–709.
193. Yokel, R. A. *J. Toxicol. Environ. Health* **1994**, *41*, 131–174.
194. Murakami, Y. *J. Inorg. Nucl. Chem.* **1962**, *24*, 679–688.
195. Yoneda, H.; Choppin, G. R.; Bear, J. L.; Quagliano, J. V. *Inorg. Chem.* **1964**, *3*, 1642–1644.
196. Gérard, C.; Hugel, R. P. *J. Chem. Res.* **1980**, (*S*) 314, (*M*) 3927–3957,
197. Santos, M. A. *Coord. Chem. Rev.* **2002**, *228*, 187–203.
198. Kontoghiorghes, G. J. *Inorg. Chim. Acta* **1987**, *135*, 145–150.
199. Clevette, D. J.; Lyster, D. M.; Nelson, W. O.; Rihela, T.; Webb, G. A.; Orvig, C. *Inorg. Chem.* **1990**, *29*, 667–672.
200. Rai, B. L.; Dekhordi, L. S.; Khodr, H.; Jin, Y.; Liu, Z.; Hider, R. C. *J. Med. Chem.* **1998**, *41*, 3347–3359.
201. Liu, Z. D.; Khodr, H. H.; Liu, D. Y.; Lu, S. L.; Hider, R. C. *J. Med. Chem.* **1999**, *42*, 4814–4823.
202. Sommer, L.; Losmanova, A. *Coll. Czech. Chem. Commun.* **1961**, *26*, 2781–2805.
203. Martell, A. E.; Motekaitis, R. J.; Smith, R. M. *Polyhedron* **1990**, *9*, 171–187.
204. Orvig, C. In: "*Coordination Chemistry of Aluminum*"; Ed. Robinson, G.H.; VCH: Weinheim, **1993**.
205. Clevette, D. J.; Orvig, C. *Polyhedron* **1990**, *9*, 151–161.
206. Streater, M.; Taylor, P. D.; Hider, R. C.; Porter, J. *J. Med. Chem.* **1990**, *33*, 1749–1755.
207. Xiao, G.; van der Helm, D.; Hider, R. C.; Rai, B. L. *J. Phys. Chem.* **1996**, *100*, 2345–2352.
208. Sun, Y.; Motekaitis, R. J.; Martell, A. E. *Inorg. Chim. Acta* **1998**, *281*, 60–63.
209. Ohkanda, J.; Katoh, A. In: "*Reviews on Heteroatom Chemistry*"; Ed. Oae, S.; MYU: Tokyo, **1998**, pp. 87–118.
210. Martin, R. B. *Clin. Chem.* **1986**, *32*, 1797–1806.
211. Motekaitis, R. J.; Martell, A. E. *Inorg. Chim. Acta* **1991**, *183*, 71–80.
212. Liu, Z. D.; Piyamongkol, S.; Liu, D. Y.; Khodr, H. H.; Lu, S. L.; Hider, R. C. *Bioorg. Med. Chem.* **2001**, *9*, 563–573.
213. Abergel, R. J.; Raymond, K. N. *Inorg. Chem.* **2006**, *45*, 3622–3631.
214. Zhou, T.; Neubert, H.; Liu, D. Y.; Liu, Z. D.; Ma, Y. M.; Kong, X. L.; Luo, W.; Mark, S.; Hider, R. C. *J. Med. Chem.* **2006**, *49*, 4171–4182.
215. Martin, R. B. In: "*Ciba Foundation Symposium 169: Aluminium in Biology and Medicine*"; Eds. Chadwick, D.J.; Whelan, J.; Wiley: Chichester, **1992**, pp. 5–25.
216. Cacheris, W. P.; Quay, S. C.; Rocklage, S. M. *Magn. Reson. Imaging* **1990**, *8*, 467–481.
217. Xu, J.; Franklin, S. J.; Whisenhunt, D. W.; Raymond, K. N. *J. Am. Chem. Soc.* **1995**, *117*, 7245–7246.
218. Jocher, C. A.; Moore, E. G.; Xu, J.; Avedano, S.; Botta, M.; Aime, S.; Raymond, K. N. *Inorg. Chem.* **2007**, *46*, 9182–9191.
219. O'Sullivan, B.; Doble, D. M. J.; Thompson, M. K.; Siering, C.; Xu, J.; Botta, M.; Aime, S.; Raymond, K. N. *Inorg. Chem.* **2003**, *42*, 2577–2583.
220. Cohen, S. M.; Xu, J.; Radkov, E.; Raymond, K. N.; Botta, M.; Barge, A.; Aime, S. *Inorg. Chem.* **2000**, *39*, 5747–5756.

221. Doble, D. M. J.; Melchior, M.; O'Sullivan, B.; Siering, C.; Xu, J.; Pierre, V. C.; Raymond, K. N. *Inorg. Chem.* **2003**, *42*, 4930–4937.
222. Thompson, M. K.; Botta, M.; Nicolle, G.; Helm, L.; Aime, S.; Merbach, A. E.; Raymond, K. N. *J. Am. Chem. Soc.* **2003**, *125*, 14274–14275.
223. Pierre, V. C.; Melchior, M.; Doble, D. M. J.; Raymond, K. N. *Inorg. Chem.* **2004**, *43*, 8520–8525.
224. Pierre, V. C.; Botta, M.; Aime, S.; Raymond, K. N. *Inorg. Chem.* **2006**, *45*, 8355–8364.
225. Grimaldi, F. S.; White, C. E. *Anal. Chem.* **1953**, *25*, 1886–1890.
226. Chruscinska, E.; Garribba, E.; Micera, G.; Panzanelli, A. *J. Inorg. Biochem.* **1999**, *75*, 225–232.
227. Chiacchierini, E.; Havel, J.; Sommer, L. *Coll. Czech. Chem. Commun.* **1968**, *33*, 4240–4247.
228. Chiacchierini, E.; Bartusek, M. *Coll. Czech. Chem. Commun.* **1969**, *34*, 530–536.
229. Song, B.; Saatchi, K.; Rawji, G. H.; Orvig, C. *Inorg. Chim. Acta* **2002**, *339*, 393–399.
230. Sommer, L.; Kuřilová-Navrátilová, L.; Šepel, T. *Coll. Czech. Chem. Commun.* **1966**, *31*, 1288–1314.
231. Pashalidis, I.; Kontoghiorghes, G. J. *J. Radioanal. Nucl. Chem.* **1999**, *242*, 181–184.
232. Burgess, J. *"Ions in Solution"*; Ellis Horwood: Chichester, **1988**; pp. 104–105,
233. Finnegan, M. M.; Lutz, T. G.; Nelson, W. O.; Smith, A.; Orvig, C. *Inorg. Chem.* **1987**, *26*, 2171–2176.
234. Burgess, J.; Patel, M. S. *Inorg. Chim. Acta* **1990**, *170*, 241–243.
235. Ahmed, S.; Burgess, J.; Parsons, S. A. *Polyhedron* **1994**, *13*, 23–25.
236. Beatty, E.; Burgess, J.; Patel, M. S. *Can. J. Chem.* **1994**, *72*, 1370–1375.
237. Alshehri, S.; Burgess, J.; Darcey, K. A.; Patel, M. S. *Transition Met. Chem.* **1994**, *19*, 119–122.
238. Jones, M. M.; Molenda, J. J.; Hanusa, T. P.; Voehler, M. W. *J. Inorg. Biochem.* **1996**, *62*, 127–136.
239. Burgess, J.; Drasdo, D. N.; Singh, K. *Transition Met. Chem.* **1994**, *19*, 113–114.
240. (a) Narbutt, J. *J. Phys. Chem.* **1991**, *95*, 3432–3435. (b) Narbutt, J.; Moore, P. *J. Solution Chem.* **1991**, *20*, 1227–1235. (c) Narbutt, J. *J. Radioanal. Nucl. Chem.* **1992**, *163*, 59–67. (d) Narbutt, J. *Polish J. Chem.* **1993**, *67*, 293–303.
241. McGowan, J. C.; Mellors, A. *"Molecular Volumes in Chemistry and Biology"*; Ellis Horwood: Chichester, **1986**; Chapter 5.
242. Burgess, J.; Patel, M. S. *J. Chem. Soc., Dalton Trans.* **1992**, 2647–2648.
243. Zai-Sheng, L.; Burgess, J.; Lane, R. *Transition Met. Chem.* **2002**, *27*, 239–243.
244. (a) Setschenow, J. *Z. Phys. Chem.* **1889**, *4*, 117–128. (b) Long, F. A.; McDevit, W. F. *Chem. Rev.* **1952**, *51*, 119–169. (c) Sergeeva, V. F. *Russ. Chem. Rev.* **1965**, *34*, 309–318.
245. (a) Brønsted, J. N. *Z. Phys. Chem.* **1932**, *162*, p. 128. (b) Blandamer, M. J.; Burgess, J.; McGowan, J. C. *J. Chem. Soc., Dalton Trans.* **1980**, 616–619.
246. Perez-Tejeda, P.; Maestre, A.; Delgado-Cobos, P.; Burgess, J. *Can. J. Chem.* **1990**, *68*, 243–246.
247. Fujita, T.; Iwasa, J.; Hansch, C. *J. Am. Chem. Soc.* **1964**, *86*, 5175–5180.
248. Cantwell, F. F.; Mohammed, H. Y. *Anal. Chem.* **1979**, *51*, 218–223.
249. Porter, J. B.; Morgan, J.; Hoyes, K. P.; Burke, L. C.; Huehns, E. R.; Hider, R. C. *Blood* **1990**, *76*, 2389–2396.
250. Tomlinson, E. *J. Pharm. Sci.* **1982**, *71*, 602–604.
251. Levin, V. A. *J. Med. Chem.* **1980**, *23*, 682–684.
252. Tapparo, A.; Perazzolo, M. *Int. J. Env. Anal. Chem.* **1989**, *36*, 13–16.
253. Yokel, R. A.; Datta, A. K.; Jackson, E. G. *J. Pharmacol. Exp. Ther.* **1991**, *257*, 100–106.
254. Ishii, H.; Numao, S.; Odashima, T. *Bull. Chem. Soc. Jpn.* **1989**, *62*, 1817–1821.
255. Kontoghiorghes, G. J. *Inorg. Chim. Acta* **1988**, *151*, 101–106.

256. Kontoghiorghes, G. J.; Sheppard, L.; Barr, J. *Inorg. Chim. Acta* **1988**, *152*, 195–199.
257. Burgess, J.; Melquiond, S.; Waltham, E., *unpublished results.*
258. Santos, M. A.; Grazina, R.; Neto, A. Q.; Cantino, G.; Gano, L.; Patricio, L. *J. Inorg. Biochem.* **2000**, *78*, 303–311.
259. See, e.g., Refs. (*79,197,206*).
260. Blandamer, M. J.; Burgess, J. *Transition Met. Chem.* **1988**, *13*, 1–18.
261. (a) Marcus, Y. *Pure Appl. Chem.* **1983**, *55*, 977–1021. (b) Marcus, Y. *"Ion Solvation"*; Wiley: New York, **1986**. (c) Kalidas, C.; Hefter, G.; Marcus, Y. *Chem. Rev.* **2000**, *100*, 819–852.
262. Ahmed, S.; Burgess, J.; Capper, G.; Fellowes, N. C.; Patel, M. S. *Polyhedron* **1993**, *12*, 1145–1148.
263. Tobe, M. L.; Burgess, J. *"Inorganic Reaction Mechanisms"*; Addison-Wesley-Longman: Harlow, **1999**.
264. Burgess, J.; Parsons, S. *Appl. Organomet. Chem.* **1993**, *7*, 343–351.
265. Burgess, J.; Parsons, S. *Polyhedron* **1993**, *12*, 1959–1966.
266. Alshehri, S.; Burgess, J.; Parsons, S. A.; Casey, A. T. *Internat. J. Chem. Kinet.* **1997**, *29*, 835–838.
267. Asano, T.; le Noble, W. J. *Chem. Rev.* **1978**, *78*, 407–489.
268. Caravan, P.; Ellison, J. J.; McMurry, T. J.; Lauffer, R. B. *Chem. Rev.* **1999**, *99*, 2293–2352.
269. Aime, S.; Botta, M.; Terreno, E. *Adv. Inorg. Chem.* **2005**, *57*, 173–237.
270. Dunand, F. A.; Helm, L.; Merbach, A. E. *Adv. Inorg. Chem.* **2003**, *54*, 1–69.
271. Hajela, S.; Botta, M.; Giraudo, S.; Xu, J.; Raymond, K. N.; Aime, S. *J. Am. Chem. Soc.* **2000**, *122*, 11228–11229.
272. Ruloff, R.; Tóth, E.; Scopelliti, R.; Tripier, R.; Handel, H.; Merbach, A. E. *Chem. Commun.* **2002**, *1*, 2630–2631.
273. Helm, L.; Nicolle, G. M.; Merbach, A. E. *Adv. Inorg. Chem.* **2005**, *57*, 327–379.
274. Burdock, G. A.; Soni, M. G.; Carabin, I. G. *Reg. Toxicol. Pharmacol.* **2001**, *33*, 80–101.
275. Yamamoto, K.; Ebihara, T.; Nakayama, H.; Okubo, A.; Higa, Y. *Nishinihon J. Dermatol.* **1998**, *60*, 6–14.
276. Hider, R. C.; Lerch, K. *Biochem. J* **1989**, *257*, 289–290.
277. Cabanes, J.; Chazarra, S.; Garcia-Carmona, F. *J. Pharm. Pharmacol.* **1994**, *46*, 982–985.
278. (a) Krewson, C. F.; Couch, J. F. *J. Am. Pharm. Assoc.* **1952**, *41*, 582–587. (b) Naghski, J.; Krewson, C. F. *J. Am. Pharm. Assoc.* **1953**, *42*, 66–68.
279. (a) Cook, A. H.; Lacey, M. S. *Nature* **1945**, *155*, 790–791. (b) Jennings, M. A.; Williams, T. I. *Nature* **1945**, *155*, p. 302.
280. Moss, M. L.; Mellon, M. G. *Ind. Eng. Chem., Anal. Ed.* **1941**, *13*, 612–614.
281. Mehlig, J. P.; Shepherd, M. J. *Analyt. Chem.* **1949**, *21*, 642–643.
282. Matsumoto, H.; Sherman, G. D. *Arch. Biochem. Biophys.* **1951**, *33*, 195–200.
283. Jungnickel, J. E.; Klinger, W. *Fresenius' Z. Anal. Chem.* **1964**, *202*, 107–112.
284. Jungnickel, J. E.; Klinger, W. *Fresenius' Z. Anal. Chem.* **1964**, *203*, 257–260.
285. Zhou, S.-F.; Li, Z. *Talanta* **1990**, *37*, 341–345.
286. Sommer, L.; Kolarik, Z. *Chem. Listy* **1956**, *50*, p. 1323. Coll. Czech. Chem. Commun. **1956**, *21*, p. 1645.
287. Marczenko, Z. *"Spectrophotometric Determination of Elements"*; Ellis Horwood: Chichester, **1976**; p. 316.
288. McBryde, W. A.; Atkinson, G. F. *Can. J. Chem.* **1961**, *39*, 510–525.
289. See p. 596 of Ref. (*287*).
290. See p. 617 of Ref. (*287*).
291. Alford, W. C.; Shapiro, L.; White, C. E. *Anal. Chem.* **1951**, *23*, 1149–1152.
292. Naik, D. V. *Anal. Chim. Acta* **1979**, *106*, 147–150.
293. Hála, J.; Smola, J. *J. Inorg. Nucl. Chem.* **1972**, *34*, 1039–1042.
294. Janko, M.; Herak, M. J. *Croat. Chem. Acta* **1971**, *43*, 179–182.

295. Herak, M. J.; Tamhina, B. *Croat. Chem. Acta* **1973**, *45*, 237–241.
296. Tamhina, B.; Herak, M. J. *Croat. Chem. Acta* **1973**, *45*, 603–610.
297. Tamhina, B.; Herak, M. J.; Jakopčić, K. *J. Less-Common Metals* **1973**, *33*, 289–294.
298. Herak, M. J.; Tamhina, B.; Jakopčić, K. *J. Inorg. Nucl. Chem.* **1973**, *35*, 1665–1669.
299. Tamhina, B.; Vojković, V. *Mikrochim. Acta* **1986**, *4*, 137–147.
300. Tamhina, B.; Vojković, V.; Herak, M. J. *Croat. Chem. Acta* **1977**, *49*, 533–537.
301. Vojković, V.; Tamhina, B. *Solvent Extr. Ion Exch.* **1986**, *4*, 27–39.
302. Vojković, V.; Tamhina, B. *Solvent Extr. Ion Exch.* **1987**, *5*, 245–253.
303. Janko, M.; Herak, M. J. *Mikrochim. Acta* **1972**, (2), 198–207.
304. Tamhina, B.; Ivsić, A.; Gojmerac, A. *Solvent Extr. Ion Exch.* **1987**, *5*, 909–922.
305. Tamhina, B.; Ivsić, A.; Gojmerac, A. *Vestn. Slov. Kem. Drus.* **1986**, *33*(Suppl.), 201–202.
306. Vojković, V.; Ivsić, A. G.; Tamhina, B. *Solvent Extr. Ion Exch.* **1996**, *14*, 479–490.
307. Vojković, V.; Gojmerac, A.; Tamhina, B. *Croat. Chem. Acta* **1997**, *70*, 667–677.
308. Tamhina, B.; Brajenović, N. *Process Metall.* **1992**, *7A*, 1027–1032.
309. Tamhina, B.; Herak, M. J. *Mikrochim. Acta* **1997**, *1*, 47–56.
310. Tamhina, B.; Herak, M. J. *J. Inorg. Nucl. Chem.* **1977**, *39*, 391–393.
311. Tamhina, B.; Gojmerac, A.; Herak, M. J. *J. Inorg. Nucl. Chem.* **1978**, *40*, 335–338.
312. Herak, M. J.; Janko, M. *J. Inorg. Nucl. Chem.* **1972**, *34*, 2627–2632.
313. Tamhina, B.; Herak, M. J. *J. Inorg. Nucl. Chem.* **1976**, *38*, 1505–1507.
314. Tamhina, B.; Gojmerac, A.; Herak, M. J. *Mikrochim. Acta* **1976**, 569–578.
315. Herak, M. J.; Janko, M.; Tamhina, B. *Mikrochim. Acta* **1973**, 783–795.
316. Vojković, V.; Tamhina, B.; Herak, M. J. *Fresenius' Z. Anal. Chem.* **1977**, *285*, 266–267.
317. Tamhina, B.; Vojković, V. *Mikrochim. Acta* **1986**, *4*, 137–147.
318. Tamhina, B.; Herak, M. J. *Mikrochim. Acta* **1975**, *2*, 45–54.
319. Tamhina, B.; Vojković, V.; Herak, M. J. *Croat. Chem. Acta* **1976**, *48*, 183–188.
320. Tamhina, B.; Vojković, V.; Herak, M. J. *Microchem. J.* **1980**, *25*, 8–13.
321. Vojković, V.; Tamhina, B.; Herak, M. J. *Fresenius' Z. Anal. Chem.* **1975**, *276*, 377–378.
322. Ivsic, A. G.; Tamhina, B. *Talanta* **1991**, *38*, 1403–1407.
323. Tamhina, B.; Herak, M. J. *Microchem. J.* **1977**, *22*, 144–148.
324. Hider, R. C.; Hall, A. D. *Progr. Med. Chem.* **1991**, *28*, 41–173.
325. Burgess, J. *Transition Met. Chem.* **1993**, *18*, 439–448.
326. Burgess, J. *Atti Accad. Peloritana dei Pericolanti 1993* **1995**, *71*, 171–185.
327. Thompson, C. H.; Barta, C. A.; Orvig, C. *Chem. Soc. Rev.* **2006**, *35*, 545–556.
328. Rennhard, H. H. *J. Agric. Food Chem.* **1971**, *19*, 152–154.
329. Singh, S.; Epemolu, R. O.; Dobbin, P. S.; Tilbrook, G. S.; Ellis, B. L.; Damani, L. A.; Hider, R. C. *Drug Metab. Dispos.* **1992**, *20*, 256–261.
330. Crichton, R. R. "*Inorganic Biochemistry of Iron Metabolism*"; 2nd edn. Wiley: New York, **2001**.
331. Silver, S.; Walden, W. (Eds.) "*Metal Ions in Gene Regulation*"; Chapman & Hall: London, **1998**.
332. (a) 2nd edn.; Bullen, J. J.; Griffiths, E. (Eds.) "*Iron and Infection* Wiley: Chichester, **1999**.(b) Bullen, J. J.; Rogers, H. J.; Spalding, P. B.; Ward, C. G. *FEBS Immunol. Med. Microbiol.* **2005**, *43*, 325–330.
333. (a) Widdowson, E. M.; McCance, R. A. *J. Hyg.* **1936**, *36*, 13–23. (b) McCance, R. A.; Widdowson, E. M. *Lancet* **1937**, *230*, 680–684.
334. McCance, R. A.; Widdowson, E. M. In: "*The Composition of Foods*"; 5th edn.; Eds. Holl, B.; Welch, A.A.; Unwin, I.D.; Buss, D.H.; Paul, A.A.; Southgate, D.A.T.; Royal Society of Chemistry: Cambridge, **1995**.
335. See, e.g., Hallberg, L. *Baillières Clin. Haematol.* **1994**, *7*, 805–814.
336. Stearns, G.; Stringer, D. *J. Nutr.* **1937**, *13*, 127–141.

337. Martin, R. B. *J. Inorg. Biochem.* **1986**, *28*, 181–187.
338. Rubinstein, M.; Dupont, P.; Doppee, J.-P.; Dehon, C.; Ducobu, J.; Hainaut, J. *Lancet* **1985**, *325*, 817–818.
339. Porter, J. B.; Huehns, E. R. *Baillières Clin. Haematol.* **1989**, *2*, 459–474.
340. Kruck, T. P. A.; Fisher, E. A.; McLachlan, D. R. C. *Clin. Pharmacol. Ther.* **1990**, *48*, 439–446.
341. Olivieri, N. *Baillières Clin. Haematol.* **1998**, *11*, 147–162.
342. Kontoghiorghes, G. J. *Biochim. Biophys. Acta* **1986**, *869*, 141–146.
343. Kontoghiorghes, G. J. *Biochim. Biophys. Acta* **1986**, *882*, 267–270.
344. Kontoghiorghes, G. J. *Biochem. J* **1986**, *233*, 299–302.
345. Kontoghiorghes, G. J.; Chambers, S.; Hoffbrand, A. V. *Biochem. J.* **1987**, *241*, 87–92.
346. Taylor, D. M.; Kontoghiorghes, G. J. *Inorg. Chim. Acta* **1986**, *125*, L35–L38.
347. Kontoghiorghes, G. J. *Biochim. Biophys. Acta* **1987**, *924*, 13–18.
348. (a) Kontoghiorghes, G. J. *Clin. Chim. Acta* **1987**, *163*, 137–141. (b) Kontoghiorghes, G. J. *Mol. Pharmacol.* **1986**, *30*, 670–673.
349. Stefanini, S.; Chiancone, E.; Cavallo, S.; Saez, V.; Hall, A. D.; Hider, R. C. *J. Inorg. Biochem.* **1991**, *44*, 27–37.
350. Molenda, J. J.; Jones, M. M.; Cecil, K. M.; Basinger, M. A. *Chem. Res. Toxicol.* **1994**, *7*, 815–822.
351. Kontoghiorghes, G. J.; Evans, R. W. *FEBS Lett.* **1985**, *189*, 141–144.
352. Evans, R. W.; Williams, J. *Biochem. J.* **1978**, *173*, 543–552.
353. Aisen, P.; Listowsky, I. *Ann. Rev. Biochem.* **1980**, *49*, 357–393.
354. Devanur, L. D.; Evans, R. W.; Evans, P. J.; Hider, R. C. *Biochem. J.* **2008**, *409*, 439–447.
355. Brady, M. C.; Lilley, K. S.; Treffry, A.; Harrison, P. M.; Hider, R. C.; Taylor, P. D. *J. Inorg. Biochem.* **1989**, *35*, 9–22.
356. See, e. g., Hershko, C. *Baillières Clin. Haematol.* **1989**, *2*, 293–321.
357. (a) See, e.g., Porter, J. B.; Hoyes, K. P.; Abeysinghe, R.; Huehns, E. R.; Hider, R. C. *Lancet 2* **1989**, p. 156. (b) Berdoukas, V. *Lancet* **1991**, *337*, p. 672. (c) Berdoukas, V.; Bentley, P.; Frost, H.; Schnebli, H. P. *Lancet* **1993**, *341*, p. 1088. (d) Mehta, J.; Singhal, S.; Mehta, B. C. *Blood* **1993**, *81*, 1970–1971. (e) Cohen, A. R.; Galanello, R.; Piga, A.; Dipalma, A.; Vullo, C.; Tricta, F. *Br. J. Haematol.* **2000**, *108*, 305–312.
358. Hoffbrand, A. V.; Bartlett, A. N.; Veys, P. A.; O'Connor, N. T. J.; Kontoghiorghes, G. J. *Lancet* **1989**, *ii*, 457–458.
359. Hider, R. C.; Singh, S.; Porter, J. B.; Huehns, E. R. *Ann. NY Acad. Sci.* **1990**, *612*, 327–338.
360. Hershko, C.; Theanacho, E. N.; Spira, D. T.; Peter, H. H.; Dobbin, P.; Hider, R. C. *Blood* **1991**, *77*, 637–643.
361. Bergeron, R. J.; Streiff, R. R.; Wiegand, J.; Luchetta, G.; Creary, E. A.; Peter, H. H. *Blood* **1992**, *79*, 1882–1890.
362. Kontoghiorghes, G. J.; Barr, J.; Nortey, P.; Sheppard, L. *Am. J. Hematol.* **1993**, *42*, 340–349.
363. Huehns, E. R.; Porter, J. B.; Hider, R. C. *Hemoglobin* **1988**, *12*, 593–600.
364. (a) Kontoghiorghes, G. J.; Aldouri, M.; Sheppard, L.; Hoffbrand, A. V. *Lancet 1* **1987** 1294–1295. (b) Kontoghiorghes, G. J.; Aldouri, M. A.; Hoffbrand, A. V.; Barr, J.; Wonke, B.; Kourouclaris, T.; Sheppard, L. *Brit. Med. J.* **1987**, *295*, 1509–1512. (c) Kontoghiorghes, G. J.; Hoffbrand, A. V. *Lancet* **1989**, *334*, 1516–1517. (d) Kontoghiorghes, G. J. *Ann. NY Acad. Sci* **1990**, *612*, 339–349. (e) Olivieri, N. F.; Templeton, D. M.; Koren, G.; Chung, D.; Hermann, C.; Freedman, M. H.; McClelland, R. A. *Ann. NY Acad. Sci* **1990**, *612*, 369–377.
365. (a) Kontoghiorghes, G. J.; Bartlett, A. N.; Hoffbrand, A. V.; Goddard, J. G.; Sheppard, L.; Barr, J.; Nortey, P. *Br. J. Haematol.* **1990**, *76*, 295–300. (b) Kontoghiorghes, G. J.; Goddard, J. G.; Bartlett, A. N.; Sheppard, L. *Clin. Pharmacol. Ther.* **1990**, *48*, 255–261.

366. Al-Refaie, F. N.; Hershko, C.; Hoffbrand, A. V.; Kosaryan, M.; Olivieri, N. F.; Tondury, P.; Wonke, B. *Br. J. Haematol.* **1995**, *91*, 224–229.
367. (a) See, e.g., Olivieri, N. F.; Koren, G.; Hermann, C.; Bentur, Y.; Chung, D.; Klein, J.; St. Louis, P.; Freedman, M. H.; McClelland, R. A.; Templeton, D. M. *Lancet* **1990**, *336*, 1275–1279. (b) Sheppard, L. N.; Kontoghiorghes, G. J. *Inorg. Chim. Acta* **1991**, *188*, 177–183. (c) Florence, A.; Ward, R. J.; Peters, T. J.; Crichton, R. R. *Biochem. Pharmacol.* **1992**, *44*, 1023–1027. (d) Anderson, L. J.; Wonke, B.; Prescott, E.; Holden, S.; Walker, J. M.; Pennell, D. J. *Lancet* **2002**, *360*, 516–520.
368. Maggio, A. *Br. J. Haematol.* **2007**, *138*, 407–421.
369. Athanassiou-Metaxa, M.; Kousi, A.; Hatzipantelis, E. S.; Tsatra, I.; Ikonomou, M.; Perifanis, V.; Tsantali, H. *Haematologica* **2004**, *89*, p. ELT07.
370. Galanello, R. *Ther. Clin. Risk Manag.* **2007**, *3*, 795–805.
371. Aydinok, Y.; Nisli, G.; Kavakli, K.; Coker, C.; Kantar, M.; Çetingül, N. *Acta Haematol.* **1999**, *102*, 17–21.
372. Abdelrazik, N. *Hematology* **2007**, *12*, 577–585.
373. (a) Porter, J. B.; Huehns, E. R.; Hider, R. C. *Baillières Clin. Haematol.* **1989**, *2*, 257–292. (b) Dobbin, P. S.; Hider, R. C. *Chem. Brit.* **1990**, 565–568. (c) Porter, J. B.; Hider, R. C.; Huehns, E. R. *Sem. Hematol.* **1990**, *27*, 95–100. (d) Olivieri, N. F.; Koren, G.; St. Louis, P.; Freedman, M. H.; McClelland, R. A.; Templeton, D. M. *Sem. Hematol.* **1990**, *27*, 101–104. (e) Herschko, C.; Pinson, A.; Link, G. *Blood Rev.* **1990**, *4*, 1–8. (f) Brittenham, G. M. *Blood* **1992**, *80*, 569–574. (g) Kontoghiorghes, G. J. *Analyst* **1995**, *120*, 845–851. (h) Hoffbrand, A. V. *Sem. Hematol.* **1996**, *33*, 1–8. (i) Tilbrook, G. S.; Hider, R. C. *Met. Ions Biol. Syst.* **1998**, *35*, 691–730. (j) Anderson, C. J.; Welch, M. J. *Chem. Rev.* **1999**, *99*, 2219–2234. (k) Faa, G.; Crisponi, G. *Coord. Chem. Rev.* **1999**, *184*, 291–310. (l) Martell, A. E.; Motekaitis, R. J.; Sun, Y.; Ma, R.; Welch, M. J.; Pajeau, T. *Inorg. Chim. Acta* **1999**, *291*, 238–246. (m) Aouad, F.; Florence, A.; Zhang, Y.; Collins, F.; Henry, C.; Ward, R. J.; Crichton, R. R. *Inorg. Chim. Acta* **2002**, *339*, 470–480. (n) Liu, Z. D.; Hider, R. C. *Med. Res. Rev.* **2002**, *22*, 26–64.
374. Crichton, R. R.; Ward, R. J. *Curr. Med. Chem.* **2003**, *10*, 997–1004.
375. Hider, R. C.; Zhou, T. *Ann. NY Acad. Sci* **2005**, *1054*, 141–154.
376. Kontoghiorghes, G. J.; Eracleous, E.; Economides, C.; Kolganou, A. *Curr. Med. Chem.* **2005**, *12*, 2663–2681.
377. Kontoghiorghes, G. J. *Hemoglobin* **2006**, *30*, 183–200.
378. Epemolu, O. R.; Singh, S.; Hider, R. C.; Damani, L. A. *Drug Metab. Dispos.* **1992**, *20*, 736–741.
379. Kontoghiorghes, G. J. *Drugs Future* **1990**, *15*, 230–232.
380. Kontoghiorghes, G. J. *Drugs Future* **2005**, *30*, p. 1241.
381. Yokel, R. A.; Fredenburg, A. M.; Durbin, P. W.; Xu, J.; Rayens, M. K.; Raymond, K. N. *J. Pharm. Sci.* **2000**, *89*, 545–555.
382. Chua, A. C. G.; Ingram, H. A.; Raymond, K. N.; Baker, E. *Eur. J. Biochem.* **2003**, *270*, 1689–1698.
383. Peto, T. E. A.; Hershko, C. *Baillières Clin. Haematol.* **1989**, *2*, 435–458.
384. Sakagami, K.; Iwamatsu, K.; Atsumi, K.; Hatanaka, M. *Chem. Pharm. Bull.* **1990**, *38*,2271–2273, 3476–3479,
385. Hershko, C.; Gordeuk, V. R.; Thuma, P. E.; Theanacho, E. N.; Spira, D. T.; Hider, R. C.; Peto, T. E.; Brittenham, G. M. *J. Inorg. Biochem.* **1992**, *47*, 267–277.
386. Dehkordi, L. S.; Liu, Z. D.; Hider, R. C. *Eur. J. Med. Chem.* **2007**, August 2, e.pub.
387. Barnabé, N.; Zastre, J. A.; Venkataram, S.; Hasinoff, B. B. *Free Radic. Bio. Med* **2002**, *33*, 266–275.
388. Morel, I.; Sargent, O.; Cogrel, P.; Lescoat, G.; Pasdeloup, N.; Brissot, P.; Cillard, P.; Cillard, J. *Free Radic. Bio. Med* **1995**, *18*, 303–310.
389. Hewitt, S. D.; Hider, R. C.; Sarpong, P.; Morris, C. J.; Blake, D. R. *Ann. Rheum. Dis.* **1989**, *48*, 382–388.

390. Sorenson, J. R. J.; Campbell, I. R.; Tepper, L. B.; Lingg, L. B. *Environ. Health Perspec.* **1974**, *8*, 3–95.
391. Ganrot, P. O. *Environ. Health Perspec.* **1986**, *65*, 363–441.
392. (a) McLachlan, D. R. *Neurobiol. Aging* **1986**, *7*, 525–532. (b) Kruck, T. P. A.; McLachlan, D. R. C. *Met. Ions Biol. Syst.* **1988**, *24*, 285–314. (c) Kruck, T. P. A.; McLachlan, D. R. *Prog. Clin. Biol. Res.* **1989**, *317*, 1155–1167.
393. Alfrey, A. C.; Legendre, G. R.; Kaehny, W. D. *N. Engl. J. Med.* **1976**, *294*, 184–188.
394. Köppel, C.; Baudisch, H.; Ibe, K. *Clin. Toxicol.* **1988**, *26*, 337–356.
395. (a) Crapper, D. R.; Krishnan, S. S.; Dalton, A. J. *Science (Washington, DC)* **1973**, *180*, 511–513. (b) Ciba Foundation Symposium 169. *"Aluminum in Biology and Medicine"*, Eds. Chadwick, D. J.; Whelan, J.; Wiley-Interscience: New York, **1992**.
396. Spencer, H.; Kramer, L. *J. Am. Coll. Nutr.* **1985**, *4*, 121–128.
397. Cochran, M.; Patterson, D.; Neoh, S.; Stevens, B.; Mazzachi, R. *Clin. Chem.* **1985**, *31*, 1314–1316.
398. Rahman, H.; Skillen, A. W.; Channon, S. M.; Ward, M. K.; Kerr, D. N. S. *Clin. Chem.* **1985**, *31*, 1969–1973.
399. Kruck, T. P. A.; Fisher, E. A.; McLachlan, D. R. C. *Clin. Pharmacol. Ther.* **1993**, *53*, 30–37.
400. Malluche, H. H.; Smith, A. J.; Abreo, K.; Faugere, C. *N. Engl. J. Med.* **1984**, *311*, 140–144.
401. See, e.g., Leung, F. Y.; Hodsman, A. B.; Muirhead, N.; Henderson, A. R. *Clin. Chem.* **1985**, *31*, 20–23.
402. Yokel, R. A. *Coord. Chem. Rev.* **2002**, *228*, 97–113.
403. Desroches, S.; Biron, F.; Berthon, G. *J. Inorg. Biochem.* **1999**, *75*, 27–35.
404. Allen, D. D.; Orvig, C.; Yokel, R. A. *Toxicology* **1994**, *92*, 193–202.
405. Domingo, J. L. *Adverse Drug React. Toxicol. Rev.* **1996**, *15*, 145–165.
406. Crichton, R. R.; Florence, A.; Ward, R. J. *Coord. Chem. Rev.* **2002**, *228*, 365–371.
407. Santos, M. A.; Gil, M.; Marques, S.; Gano, L.; Cantinho, L.; Chaves, S. *J. Inorg. Biochem.* **2002**, *92*, 43–54.
408. Santos, M. A.; Gil, M.; Gano, L.; Chaves, S. *J. Biol. Inorg. Chem.* **2005**, *10*, 564–580.
409. Santos, M. A.; Gil, M.; Marques, S.; Gano, L.; Chaves, S. *Anal. Bioanal. Chem.* **2005**, *381*, 413–419.
410. Chaves, S.; Gil, M.; Marques, S.; Gano, L.; Santos, M. A. *J. Inorg. Biochem.* **2003**, *97*, 161–172.
411. Esteves, M. A.; Cachudo, A.; Chaves, S.; Santos, M. A. *Eur. J. Inorg. Chem.* **2005**, 597–605.
412. Santos, M. A.; Gama, S.; Gano, L.; Cantinho, L.; Farkas, G. *Dalton Trans.* **2004**, 3372–3781.
413. Santos, M. A.; Gama, S.; Gano, L.; Farkas, G. *J. Inorg. Biochem.* **2005**, *99*, 1845–1852.
414. Domingo, J. L. *J. Alzheimer's Dis.* **2006**, *10*, 331–341.
415. Schechter, Y.; Karlish, S. J. D. *Nature* **1980**, *284*, 556–558.
416. Cantley, L. C.; Aisen, P. *J. Biol. Chem.* **1979**, *254*, 1781–1784.
417. Sakurai, H.; Shimomura, S.; Fukuzawa, K.; Ishizu, K. *Biochem. Biophys. Res. Commun.* **1980**, *96*, 293–298.
418. Heyliger, C. E.; Tahiliani, A. G.; McNeill, J. H. *Science* **1985**, *227*, 1474–1477.
419. Pedersen, R. A.; Ramanadham, S.; Buchan, A. M. J.; McNeill, J. H. *Diabetes* **1989**, *38*, 1390–1395.
420. Ramanadham, S.; Cros, G. H.; Mongold, J. J.; Serrano, J. J.; McNeill, J. H. *Can. J. Physiol. Pharmacol.* **1990**, *68*, 486–491.
421. Yuen, V. G.; Orvig, C.; McNeill, J. H. *Can. J. Physiol. Pharmacol.* **1993**, *71*, 263–269.

422. Yuen, V. G.; Orvig, C.; Thompson, K. H.; McNeill, J. H. *Can. J. Physiol. Pharmacol.* **1993**, *71*, 270–276.
423. Yuen, V. G.; Orvig, C.; McNeill, J. H. *Can. J. Physiol. Pharmacol.* **1995**, *73*, 55–64.
424. Thompson, K. H.; McNeill, J. H.; Orvig, C. *Chem. Rev.* **1999**, *99*, 2561–2571.
425. Kordowiak, A. M.; Trzos, R.; Gryboś, R. *Hormone Metab. Res.* **1997**, *29*, 101–105.
426. Katoh, A.; Taguchi, K.; Saito, R.; Fujisawa, Y.; Takino, T.; Sakurai, H. *Heterocycles* **2003**, *60*, 1147–1159.
427. Rangel, M.; Tamura, A.; Fukushima, C.; Sakurai, H. *J. Biol. Inorg. Chem.* **2001**, *6*, 128–132.
428. Dąbroś, W.; Kordowiak, A. M.; Dziga, D.; Gryboś, R. *Pol. J. Pathol.* **1998**, *49*, 67–76.
429. Dąbroś, W.; Gryboś, R.; Miarka, A.; Kordowiak, A. M. *Pol. J. Pathol.* **2000**, *51*, 17–24.
430. Crans, D. C.; Yang, L.; Alfano, J. A.; Chi, L.-H.; Jin, W.; Mahroof-Tahir, M.; Robbins, K.; Toloue, M. M.; Chan, L. K.; Plante, A. J.; Grayson, R. Z.; Willsky, G. R. *Coord. Chem. Rev.* **2003**, *237*, 13–22.
431. Thompson, K. H.; Chiles, J.; Yuen, V. G.; Tse, J.; McNeill, J. H.; Orvig, C. *J. Inorg. Biochem.* **2004**, *98*, 683–690.
432. Tam, N. C.; Fletcher, S. P.; Vogels, C. M.; Westcott, S. A. *Transition Met. Chem.* **2003**, *28*, 103–109.
433. Yoshihawa, Y.; Ueda, E.; Kawabe, K.; Miyake, H.; Sakurai, H.; Kojima, Y. *Chem. Lett.* **2000**, 874–875.
434. Lauffer, R. B. *Chem. Rev.* **1987**, *87*, 901–927.
435. Aime, S.; Botta, M.; Fasano, M.; Terreno, E. *Chem. Soc. Rev.* **1998**, *27*, 19–29.
436. Xu, J.; Churchill, D. G.; Botta, M.; Raymond, K. N. *Inorg. Chem.* **2004**, *43*, 5492–5494.
437. Thompson, M. K.; Misselwitz, B.; Tso, L. S.; Doble, D. M. J.; Schmitt-Willich, H.; Raymond, K. N. *J. Med. Chem.* **2005**, *48*, 3874–3877.
438. Martell, A. E. "*Inorganic Chemistry in Biology and Medicine*"; ACS Meeting No. 178; Washington, DC, **1980**.
439. Zhang, Z. H.; Lyster, P. M.; Webb, G. A.; Orvig, C. *Nucl. Med. Biol.* **1992**, *19*, 327–335.
440. Clarke, M. J.; Zu, F.; Frasca, D. R. *Chem. Rev.* **1999**, *99*, 2511–2533.
441. Anderson, C. J.; Welch, M. J. *Chem. Rev.* **1999**, *99*, 2219–2234.
442. Zhang, Z. H.; Lyster, D. M.; Webb, G. A.; Orvig, C. *Nucl. Med. Biol.* **1992**, *19*, 327–331.
443. Edwards, D. S.; Liu, S. N.; Lyster, D. M.; Poirier, M. J.; Vo, C.; Webb, G. A.; Zhang, Z. H.; Orvig, C. *Nucl. Med. Biol.* **1993**, *20*, 857–863.
444. Durbin, P. W.; Kullgren, B.; Xu, J.; Raymond, K. N. *Radiat. Prot. Dosimetry* **1994**, *53*, 305–309.
445. Xu, J.; Kullgren, B.; Durbin, P. W.; Raymond, K. N. *J. Med. Chem.* **1995**, *38*, 2606–2614.
446. Raymond, K. N.; Scarrow, R. C.; White, D. L. U. S. Pat. Appl., US 796 815, **1987**.
447. White, D. L.; Durbin, P. W.; Jeung, N.; Raymond, K. N. *J. Med. Chem.* **1988**, *31*, 11–18.
448. Li, G. J.; Zhang, L. L.; Lu, L.; Wu, P.; Zheng, W. *J. Occup. Environ. Med.* **2004**, *46*, 241–248.
449. Dobson, A. W.; Erikson, K. M.; Aschner, M. *Ann. NY Acad. Sci.* **2004**, *1012*, 115–128.
450. (a) HaMai, D.; Bondi, S. C. *Ann. NY Acad. Sci.* **2004**, *1012*, 129–141. (b) HaMai, D.; Bondi, S. C. *Neurochem. Int.* **2004**, *44*, 223–229.
451. Hider, R. C.; Kontoghiorghes, G.; Silver, J.; Stockham, M. A. GB Patent Application 2 148 896, **1985**.

452. Burchfield, J.; Telehowski, P.; Rosenberg, R. C.; Eaton, S. S.; Eaton, G. R. *J. Magn. Reson.* **1994**, *104*, 69–72.
453. Sukhov, B. G.; Mukha, S. A.; Antipova, I. A.; Medvedeva, S. A.; Larina, L. I.; Grigorieva, Y. A.; Illarionov, A. I.; Starchenko, A. A.; Chuvashev, F. Yu.; Trofimov, B. A. *Mendeleev Commun.* **2007**, *17*, 154–155.
454. Shoair, A.-G. F. *J. Coord. Chem.* **2007**, *60*, 1101–1109.

LATE TRANSITION METAL-OXO COMPOUNDS AND OPEN-FRAMEWORK MATERIALS THAT CATALYZE AEROBIC OXIDATIONS

RUI CAO[a], JONG WOO HAN[a], TRAVIS M. ANDERSON[a], DANIEL A. HILLESHEIM[a], KENNETH I. HARDCASTLE[a], ELENA SLONKINA[b], BRITT HEDMAN[b], KEITH O. HODGSON[b], MARTIN L. KIRK[c], DJAMALADDIN G. MUSAEV[d], KEIJI MOROKUMA[d], YURII V. GELETII[a] and CRAIG L. HILL[a]

[a]Department of Chemistry, Emory University, Atlanta, GA 30322, USA
[b]Stanford Synchrotron Radiation Laboratory, SLAC, Stanford University, Stanford, CA 94309, USA
[c]Department of Chemistry and Chemical Biology, The University of New Mexico, Albuquerque, NM 87131, USA
[d]Cherry L. Emerson Center, Emory University, Atlanta, GA 30322, USA

I. Introduction

Transition metal oxides have uses in myriad industries from heterogeneous catalysis and transport (e.g., automobile catalytic converters) to sensors, integrated circuits and biomedical technologies, yet research progress in these areas often has been slow in part because it is hard to know and thus to tune or tailor the precise electronic and geometrical structures of metal oxide surfaces and their associated acid/base and redox chemistries. The attraction of some early transition metal oxygen-anion clusters (polyoxometalates or POMs) is that they mimic the structural and reactivity properties of transition metal oxides, yet they are discrete molecular units (polyanions with well defined counterions) that are soluble and amenable to thorough characterization by nearly all the methods (diffraction, spectroscopic, etc.) common to inorganic chemists (*1–6*). In the last few years thousands of articles describing hundreds of distinct POM structures have appeared. This has ushered in a more defensible comprehension of POM physicochemical properties and a

ADVANCES IN INORGANIC CHEMISTRY
VOLUME 60 ISSN 0898-8838 / DOI: 10.1016/S0898-8838(08)00006-8

commensurately greater sophistication for preparing, purifying and characterizing these mineral-like inorganic cluster anions.

It is now possible to prepare POMs with targeted structures, reduction potentials, surface charge densities and other properties. It is the diversity and tunability of these properties that have enabled particular POMs to realize two classes of new materials described here that have been sought and speculated about for some time: terminal-oxo compounds of the late transition metal elements and porous materials that catalyze organic oxidations using only the ambient air. This work was presented at the International Symposium on Redoxactive Metal Complexes in Erlangen, Germany (October 4–7, 2006).

II. Late Transition Metal-Oxo (LTMO) Complexes

Terminal multiply bonded oxygen ligands to the d block elements become increasingly unstable as one moves across the series from left to right and electrons are added to the metal center (*7–14*). To illustrate: the terminal-oxo complexes of d^0, d^1 and d^2 metal centers such as W(VI, V, IV), Mo(VI, V, IV), V(V or IV), etc. are quite stable and display modest reactivity. As a consequence, W-oxo, Mo-oxo, V-oxo and electronically related units are ubiquitous in nature (soils, minerals, etc.) and also occur in various synthetic materials including many classes of inorganic compounds. In contrast, the terminal-oxo complexes of d^3 and d^4 metal centers are less stable thermodynamically and correspondingly more reactive. Indeed the oxoferryl (Fe^{IV}-oxo) group is arguably the most important and prevalent potent oxidant in biology; both the low-spin ($S = 1$) and high-spin ($S = 2$) forms are powerful oxidants found in metalloenzymes (*15–18*). By the same token, high-valent Mn-oxo species are highly reactive and may be key intermediates in metalloenzymes and the oxygen evolving center (OEC) in green plant photosynthesis (*19–25*). Continuing this stability/reactivity trend, terminal metal-oxo species of elements to the right of column 8 (Fe, Ru, Os), with one exception, were unknown prior to 2004. This exception was the d^4 four-coordinate C_{3v} (mesityl)$_3$Ir^V-oxo complex reported by Hay-Motherwell *et al.* (*26,27*). No such complexes of the column 10 (Ni, Pd, Pt) or column 11 (Cu, Ag, Au) elements were known.

These findings led to the concept of the "Metal-oxo Wall" or "Ru-oxo Wall", namely terminal metal-oxo units are well known for nearly all early and mid-transition metal elements but simply unknown for the late transition metal elements (Fig. 1). The generic explanation for this phenomenon is that as one moves to the right in the d block, the metal center necessarily has more d electrons. This in turn requires an increasing population of orbitals that are antibonding with respect to the terminal metal-oxo unit. A simplified molecular orbital diagram for a six-coordinate C_{4v} transition metal-oxo unit shown in Fig. 2 explains

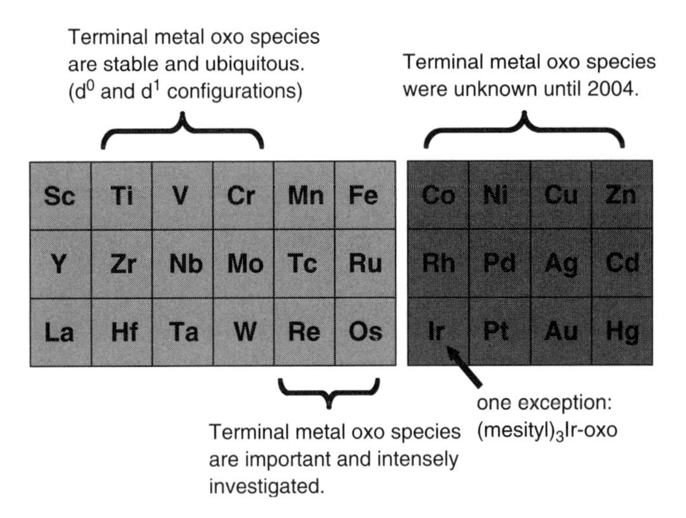

FIG. 1. The "Metal-oxo Wall" illustrates the decreasing thermodynamic stability and increasing reactivity of the terminal metal-oxo unit as one traverses these elements from left to right in the periodic table.

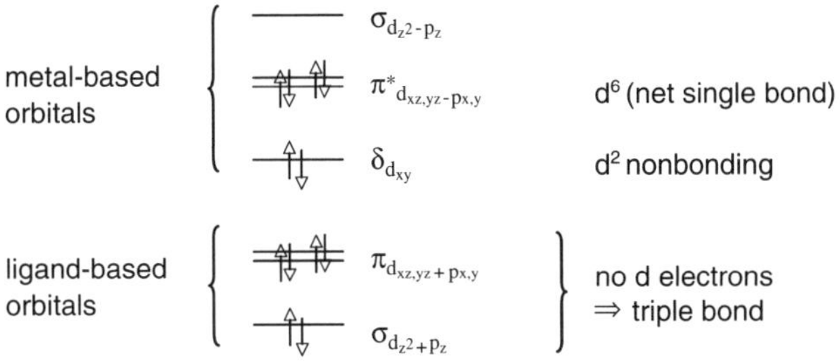

metal-based orbitals

$\sigma_{d_{z^2}-p_z}$

$\pi^*_{d_{xz,yz}-p_{x,y}}$ d^6 (net single bond)

$\delta_{d_{xy}}$ d^2 nonbonding

ligand-based orbitals

$\pi_{d_{xz,yz}+p_{x,y}}$

$\sigma_{d_{z^2}+p_z}$

no d electrons $\Rightarrow$ triple bond

FIG. 2. Significant molecular orbitals of terminal transition metal-oxo units in a six-coordinate C_{4v} ligand environment. The d^0 configuration is a formal triple bond. The highest occupied molecular orbital in the d^2 configuration is formally nonbonding (δ symmetry) so the metal-oxo bond order remains 3.0. However, d-electron counts above d^2 populate orbitals that are antibonding between the metal and the terminal multiply bonded ligand (oxo in this case, but alternatively, alkyl-imido, nitrido, sulfido, etc.). Note that all the equatorial ligand orbitals and the metal $d_{x^2-y^2}$ orbital (b_1 in C_{4v} symmetry) are ignored for simplicity.

this and the generic basis of multiple bonding in these species. Six-coordinate geometries are the most common in coordination compounds, metalloenzyme active sites and metal oxide surfaces. As shown, any d-electron count above two renders common six-coordinate transition metal-oxo complexes reactive by virtue of populating antibonding orbitals with an associated decrease in metal-oxo bond strength and increase in metal-oxo bond length. Although terminal metal-oxo complexes of the noble metal elements, palladium, silver, gold and in particular, platinum, the single most important element in abiological catalysis, were unknown prior to the work presented in this chapter, this was not for lack of effort. Organometallic, inorganic and heterogeneous catalysis research groups had endeavored to identify them as intermediates using spectroscopic methods and indeed attempted to prepare and isolate them, but without reported success.

The same general ligand electronic features that stabilize conventional low-valent organotransition metal complexes would also stabilize terminal metal-oxo complexes of the late transition metal elements (LTMO compounds), namely strong σ donation and π accepting character. Removal of antibonding (π^*) electron density on the metal-oxo unit would have the effect of increasing the net π bonding and thus stabilizing the terminal metal-oxo unit. POMs of many kinds (polytungstates, polymolybdates, polyvanadates, polyniobates, etc. and mixed metal polyanions) such as the corresponding metal oxides (WO_3, MoO_3, V_2O_5, Nb_2O_5, etc.) are electron poor and thus capable electron acceptors. In addition, POMs can function as strong multi-dentate ligands for nearly all metals (s, p, d and f block elements). However, knowing the energies of the empty metal-based delocalized orbitals in POMs, those that might be orchestrated to accept antibonding electron density from a central metal-oxo moiety, is not straightforward.

Reaction of $[A\text{-}\alpha\text{-}PW_9O_{34}]^{9-}$ and Pt(II) salts in aqueous solution under argon produces the expected sandwich complex, $[Pt_3^{II}(PW_9O_{34})_2]^{12-}$, based on NMR and FT-IR data, one containing conventional d^8 Pt(II) square planar centers (Fig. 3). Significantly, however, the same reaction when conducted in the presence of ambient air, exhibits different colors indicating different processes are operable. By judicious choice of counterion (addition of excess KCl), the first noble metal-oxo complex, a Pt(IV)-oxo compound, K_7Na_9 $[Pt(O)(OH_2)(A\text{-}\alpha\text{-}PW_9O_{34})_2]$ (**1**) was eventually isolated in pure form in 25% yield (Fig. 3) (*28*). Although variable-temperature magnetic susceptibility and NMR spectra indicated that **1** is diamagnetic, it decomposes in water even in the absence of an organic substrate (or other reducing agents).

Crystallographic studies of **1** reveal a very short Pt–O bond (1.720(18) Å) is *trans* to the longer Pt–OH_2 bond (2.29(4) Å). The central Pt atom is bound to two symmetrically equivalent polytungstate clusters each through two oxygens, defining a square equatorial plane in an approximate C_{2v} molecular geometry. The Pt–O_{oxo} bond is the shortest Pt–O bond in the literature and clearly in the range of terminal multiply bonded transition metal-oxo species. The distance to the ligand *trans* to the terminal-oxo is 2.29(4) Å strongly suggesting a dative (coordination) Pt–OH_2 bond. Known Pt–OH bond lengths range

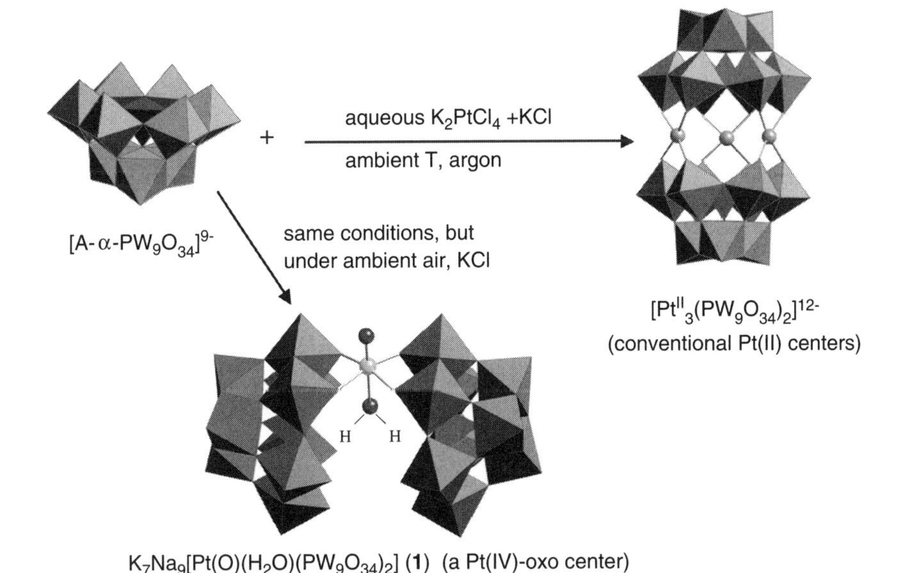

FIG. 3. Synthesis of a conventional d^8 Pt(II) complex under argon and an unconventional terminal Pt(IV)-oxo complex under air.

from 1.943 Å to 2.079 Å and average 1.998 Å. Two potassium cations located in the cavity between the two polytungstate units doubtless stabilize the entire sandwich structure by electrostatically linking the two $[A\text{-}\alpha\text{-}PW_9O_{34}]^{9-}$ frameworks.

Although we felt the distances in the Pt-oxo unit were unequivocal based on the disorder-free X-ray structures, we endeavored to use an additional structural method to assess this unprecedented structure, neutron diffraction. As an independent structural technique, neutron diffraction can determine not only the absolute structure of molecules but also the location of hydrogens. The latter are almost never located in even the best X-ray crystallographic structure determinations of polytungstates.

The neutron diffraction data of a crystal of **1** with approximate dimensions of $1 \times 2 \times 3\,\text{mm}^3$ were obtained at the Intense Pulsed Neutron Source (IPNS) at Argonne National Laboratory using the time-of-flight Laue single crystal diffractometer (SCD). Examination of the Fourier maps around O35 (terminal-oxo ligand on the Pt center) do not show the significant negative scattering density which is indicative of the existence of hydrogen atoms. Significantly, a large area of negative scattering density is seen around water oxygens in the structure (see later), showing neutron diffraction is a powerful structural method to locate hydrogen atoms in POM environments and thus confirming that no hydrogen atom is bound to the terminal-oxo ligand. Some negative density can be observed in the Fourier maps in the area surrounding O6W, the aqua ligand *trans* to the terminal-oxo oxygen, and may be safely attributed to the presence of hydrogen atoms. However, due to the positional disorder of this *trans* aqua ligand (slight displacement of this oxygen from the C_2 axis by potassium ions and phosphate oxygens that are within the van der Waals distance of O6W), the hydrogen atoms on this aqua ligand are not included in the final refinement.

An important internal check is that two hydrogens, H1a and H1b, are clearly localized and bound to the lattice water molecule O1W, and that H1a apparently is hydrogen bonded to the POM oxygen O4 at a distance of 2.52 Å. An additional region of negative scattering is seen close to O4 and may denote a hydrogen atom associated with any of a number of neighboring water oxygens. Besides O1W, there are four water oxygen atoms sufficiently close to O4 to facilitate hydrogen bonding (O8W, 2.73 Å; O21W, 2.56 Å; O10W, 2.85 Å; O17W, 3.03 Å). This map shows the kind of negative scattering density that would lead to a hydrogen atom assignment, and it is clear that no such feature is seen close to O35.

By using a ligand in which two $[A\text{-}\alpha\text{-}PW_9O_{34}]^{9-}$ units are fused together by a $[O{=}W^{VI}(OH_2)]^{4+}$ moiety, a second terminal LTMO complex, of formula, $K_{10}Na_3[Pd^{IV}(O)(OH)WO(OH_2)(PW_9O_{34})_2]$ (**2**), was isolated and purified by crystallization (*29*). Like the Pt-oxo complex, this Pd-oxo complex was thoroughly characterized by multiple X-ray structure determinations, variable-temperature magnetic measurements, several spectroscopic techniques, and multiple elemental analyses. A crystallographic disorder problem involving the atoms on the opposite side of the polyanion unit from the Pd-oxo group was not clarified until recently when a better single crystal of the terminal Pd-oxo complex, **2**, was obtained. In conjunction with this X-ray crystallographic determination, an even lower temperature of data collection was used and face indexing was applied for better absorption correction, since typically used correction methods for organic or organometallic complexes are not sufficient in POM systems with many heavy atoms. This addition/correction (JACS, 2008, 130, 2877–2877) only indicated that the original assignment of the Pd-oxo unit is correct. Importantly, a single peak at −11.70 ppm is shown in the ^{31}P NMR spectrum of **2**, a result consistent to the crystallographic studies of this complex showing the presence of two symmetry-equivalent phosphorus atoms per polyanion. Extended X-ray absorption fine structure (Pd EXAFS) studies of **2** confirmed that the very short terminal Pd-oxo bond distance (1.68 ± 0.05 Å) was the same within experimental error as that obtained in the diffraction studies (1.63 ± 0.03 Å) (*29*). The Pd XAS data for **2** were acquired at the Stanford Synchrotron Radiation Laboratory (SSRL) on focused 30-pole wiggler beam line 10–2 with the Si(220) monochromator for energy selection at the Pd *K*-edge. The Curve fitting results for the Pd *K*-edge EXAFS of **2** are shown in Table I.

TABLE I

CURVE FITTING RESULTS FOR THE PD *K*-EDGE EXAFS OF **2**

Fit number	Coordination model for the first shell	N^a	Scatterer	R (Å)^b	σ^2 (Å^2)^c	Fit error^d
1	5 O	5	O	1.97	0.0043	0.217
		4	W	3.60	0.0021	
		4	O–W	3.73	0.0021	
		2	K–O	3.98	0.0023	
		8	O–O	4.67	0.0043	
2	6 O	6	O	1.97	0.0055	0.263
		4	W	3.60	0.0021	
		4	O–W	3.73	0.0020	
		2	K–O	3.98	0.0023	
		8	O–O	4.67	0.0044	
3	1 O	1	O	1.71	0.0154	0.181
	4 O	4	O	1.96	0.0028	
		4	W	3.60	0.0027	
		4	O–W	3.73	0.0023	
		2	K–O	3.96	0.0026	
		8	O–O	4.66	0.0045	
4	1 O	1	O	1.68	0.0067	0.151
	5 O	5	O	1.96	0.0043	
		4	W	3.60	0.0023	
		4	O–W	3.73	0.0023	
		2	K–O	3.97	0.0026	
		8	O–O	4.67	0.0046	
5	1 O	1	O	1.71	0.0146	0.160
	4 O	4	O	1.96	0.0027	
	1 O	1	O	2.41	0.0047	
		4	W	3.59	0.0024	
		4	O–W	3.73	0.0023	
		2	K–O	3.95	0.0027	
		8	O–O	4.62	0.0044	

$^a N$ is the coordination number.
$^b R$ is the mean distance.
$^c \sigma^2$ is the bond length variance.
^{d}The fit error is defined as $[\Sigma(\chi_{exp}-\chi_{calc})^2 k^6 / \Sigma \chi_{exp}{}^2 k^6]^{1/2}$, where χ is the EXAFS data point. The accuracies estimated for bond lengths are approximately ± 0.03 Å; the resolution is 0.14 Å.

In contrast to the terminal-oxo unit in the platinum complex, **1**, this unit in palladium complex, **2**, is sterically protected because it points inside a central metal-oxide-like d^0 oxidatively resistant cavity (Fig. 4). Thus it is not surprising that **2** is far more stable in unbuffered aqueous solution (a period of at least days) than **1**. Nonetheless, while steric encapsulation is a defensible rationale for the relative hydrolytic stabilities of **1** and **2**, it is clearly not the entire story. More recent data suggest that the stability of LTMO units is not purely steric in nature because more robust polytungstate ligands, ones that have stabilizing $[O{=}W^{VI}(OH_2)]^{4+}$ groups linking the same chelating {A-PW_9} units

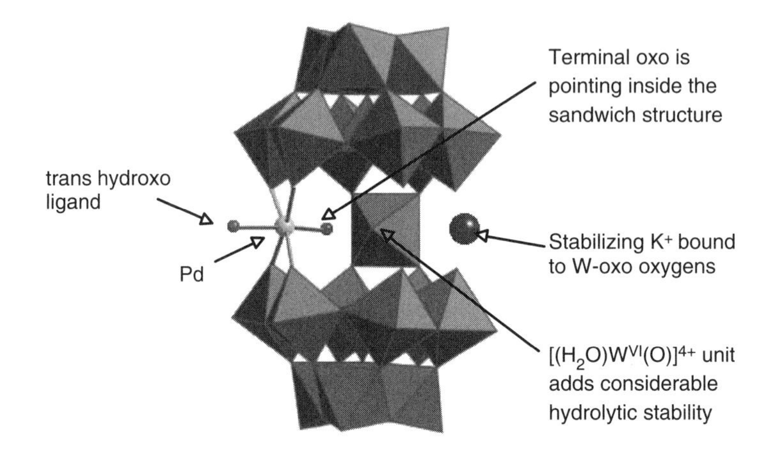

FIG. 4. X-ray crystal structure of the Pd-oxo polyanion of **2** showing the central K^+ and $[O{=}W^{VI}(OH_2)]^{4+}$ units that stabilize the complex.

present in **2** yet have the terminal noble metal-oxo group pointing away from and not into a cavity of the polytungstate are also significantly more stable than **1**. At the same time recent experimental results clearly indicate that steric protection around the terminal-oxo oxygen significantly retards reactivity toward organic substrates. The structures and reactivities of such LTMO complexes will be published shortly.

All crystallographic studies unambiguously rule out cationic Pd counterions to the polyanion of **2**, and chemical and physiochemical studies (e.g., ^{31}P NMR and electronic absorption spectroscopy, cyclic voltammetry method) suggest that the only one Pd atom established by repeated elemental analyses is located at the bridging site. There are three lines of evidence for OH^- as the ligand *trans* to the terminal oxo unit. First, the bond length of 1.99(2) Å is more consistent with a Pd–OH than Pd=O or Pd–OH_2. Second, the total number of countercations (10 K^+ and 3 Na^+), found by elemental analyses and established by X-ray crystallography, are consistent with the overall 13− negative charge on the polyanion unit. Notably, anions other than **2** are clearly absent in the lattice (and the absence of Cl^- was confirmed by elemental analyses), consistent with the +4 oxidation state of Pd and an overall charge of 13− on the molecule. Third, the protonation and deprotonation process followed by pH-dependent UV-Vis and ^{31}P NMR titration studies strongly suggest the *trans* OH^- ligand (Fig. 5).

The existence of a multiply bonded terminal Pd=O unit is further confirmed by ^{17}O NMR spectroscopy. Previous studies revealed that

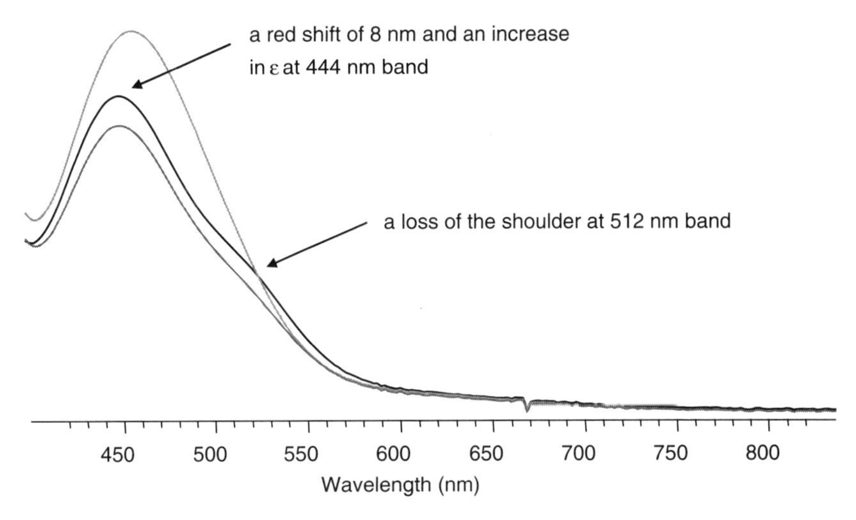

FIG. 5. Titrations of **2**, the terminal Pd-oxo complex, in aqueous solution, monitored by UV-visible spectroscopy. Black Line, fresh **2** in H_2O (pH = 6.30); red line, pH = 3.5; blue line, final solution (pH = 6.25). These spectroscopic changes are completely reversible. Similar reversible changes are also noted in the ^{31}P NMR spectra of this complex as a function of pH.

the oxygens in many POMs can exchange with the oxygen in water (*30–33*). As expected, we found that ^{17}O enrichment in **2** was readily accomplished by one of two methods both involving ^{17}O-enriched water. However, it is worth noting that nonstatistical ^{17}O enrichment of all the symmetry-distinct oxygen sites could render accurate quantification (integration of the NMR spectra) problematical. The direct use of ^{17}O-enriched water as the solvent results in rapid oxygen exchange at mild temperature in the case of $K_{10}[P_2W_{20}O_{70}(OH_2)_2]$ $\{P_2W_{20}\}$, an isostructural all-tungsten analogue of **2**. On the other hand, ^{17}O enrichment of **2** is less efficient than in $\{P_2W_{20}\}$ due to the significantly lower solubility of the LTMO complex in water. An alternative isotopic enrichment procedure entailing the use of nonaqueous solvents, was used. The stability of these POMs in water and in nonaqueous solvents was investigated and confirmed using ^{31}P NMR.

The ^{17}O chemical shifts for **2** and $\{P_2W_{20}\}$ agree well with the general correlation between downfield chemical shift and oxygen π-bond order (*30,34*). In the structure of **2**, there are 11 classes of symmetry-equivalent terminal-oxo oxygens: a Pd=O (multiple bond) and 10 W=O (triple bond) units. Differences in the π-bond order and in the metal center itself (Pd versus W) result in *ca.* 160 ppm upfield shift for the terminal-oxo oxygen on Pd relative to those on the W atoms. The relatively similar coordination environments of the several symmetry-distinct bridging oxygens of the polyoxoanion framework in **2** result in several similar chemical shift values from 370 to 450 ppm. A very strong peak at 326 ppm is assigned to the oxygen of Pd–OH_2 unit (the original –OH ligand on Pd protonates during extraction from water to CH_2Cl_2 because the pH is lowered to 3.0 with the addition of HCl to facilitate this extraction process). The polytungstate isostructural to **2**, namely $K_{10}[P_2W_{20}(OH_2)_2O_{70}]$, was also prepared and examined by ^{17}O NMR spectroscopy. As expected, the chemical shift ranges of the terminal-oxo oxygens on W and the bridging oxygens of the framework are similar to those of **2**. Importantly, however, there are no peaks around 560 and 326 ppm, which are assigned to the oxygens of Pd=O and Pd–OH_2, respectively. The ^{17}O NMR experiments run on a ^{17}O-enriched sample of **2** also suggest the solid-state structure is maintained in solution. All these NMR data and the experimentally and theoretically documented correlation between ^{17}O NMR chemical shift and oxygen π-bond order as well as the fact that no peak shown up around 560 ppm in $K_{10}[P_2W_{20}(OH_2)_2O_{70}]$ indicate that there is π-bonding and multiple bonding character between the Pd and its terminal-oxo oxygen.

Five lines of evidence suggest that the oxidation state of the Pd of **2** is +4. First, 13 countercations (10 K^+ and 3 Na^+) can be located in all X-ray data sets and are found in repeated elemental analyses (no Cl^- is detectable). This is consistent with the negative charge of the polyanion of **2**, $[Pd^{IV}O(OH)P_2W_{19}O_{69}(OH_2)]^{13-}$. Second, structural analysis of the Pd atom, including X-ray single crystal diffraction and Pd EXAFS, confirms the multiply bonded terminal-oxo ligand and indicates the existence of a d^6 Pd(IV) configuration. Third, four electrons are needed to obtain Pd(0) in controlled potential coulometry which is consistent with the reduction of Pd(IV) to metallic Pd(0). Fourth, the diamagnetism of **2** (based on room temperature magnetic susceptibility measurements and ^{31}P NMR chemical shift and peak width) argues for a d^6 Pd(IV) and against a d^8 Pd(II) or d^4 Pd(VI) (for a local C_{4v} metal center). Fifth, the electronic absorption spectrum of **2** is also consistent with a six-coordinated local C_{4v} Pd(IV) with d^6 electronic configuration.

Interestingly, Kortz and co-workers reported that reaction of Pd(II) precursors with various lacunary (defect) polytungstates under oxidizing conditions (exposed to air) led only to conventional Pd POM complexes, i.e., those containing only the expected d^8 Pd(II) square planar centers (*35–37*). Comparison of the Kortz and Hill group synthetic conditions leads to three inferences. First, the use of B-type trivacant Keggin-type POM units, $[B\text{-}XW_9O_{33}]^{9-}$, X = As(III) and Sb(III), (versus the A-type used to prepare the LTMO complexes described earlier) does not facilitate formation of isolable LTMO complexes. The likely explanation for this is that the lone pair of electrons on the central heteroatom, X, in the $[B\text{-}XW_9O_{34}]^{9-}$ polyanions crowds and thus destabilizes the ligand on the central late transition metal that protrudes into the cavity defined by the two $\{XW_9\}$ units. This ligand can be either the terminal-oxo or the ligand *trans* to it; in either case the LTMO unit is destabilized. The ligands *trans* to the terminal-oxo in **1** and **2** are aqua and hydroxo, respectively (see Figs. 3 and 4). Second, Kortz and co-workers prepared $[Cs_2K(H_2O)_7Pd_2WO(H_2O)(A\text{-}\alpha\text{-}SiW_9O_{34})_2]^{9-}$ under conditions very similar to those used in our laboratory and no Pd-oxo compound was noted (*36*). This likely reflects electronic differences between $[A\text{-}\alpha\text{-}PW_9O_{34}]^{9-}$ and the isostructural Si analogue that Kortz used, $[A\text{-}\alpha\text{-}SiW_9O_{34}]^{10-}$, and indicates that lone pairs on the POM heteroatom disrupting the linearity of LTMO unit, $[(H_2O)M(O)]^{n+}$ is not the only issue precluding LTMO formation. Third, the counterion used in the LTMO compound synthesis is important. For example, Na, K and Cs can and usually do lead to different complexes in solution and in the crystalline solid state. Our recent work, showing that the kind and concentration of counteractions largely control speciation of Pd-substituted POMs, will be published soon.

Dioxygen- (and air-) based oxidation catalyzed by supported Au(0) nanoclusters and films has been one of the most intensely investigated research areas in the past 5–7 years (*38–43*). O_2-based oxidations of CO to CO_2 and several organic substrates have been examined in depth, and considerable emphasis has been placed on characterization of the Au catalysts themselves (*38–43*). Success at generating terminal Pt- and Pd-oxo complexes under aerobic conditions suggested that the terminal metal-oxo compounds of the Group 11 elements (Cu, Ag, Au), and in particular Au, might also be accessible, although with higher d-electron counts, the synthesis and isolation of Au-oxo compounds could well be even more problematical than for the Pt and Pd analogues.

Au-oxo complexes can indeed be prepared, isolated and characterized using procedures similar to those employed for the Pt- and Pd-oxo complexes, but appropriate pH values and ionic strengths are required to keep all species in solution and to maintain the stability of the initial and final polytungstate units (*44*). To date, two terminal Au-oxo complexes have been prepared and fully characterized: $K_{15}H_2[Au(O)(OH_2)P_2W_{18}O_{68}]$ (**3**), a complex that is isostructural to the Pt-oxo complex, **1**, and $K_7H_2[Au(O)(OH_2)P_2W_{20}O_{70}(OH_2)_2]$ (**4**). The Au-oxo complexes generally appear to be less stable than the Pt-oxo and Pd-oxo complexes; they are certainly less stable to reduction. Complexes **1** and **3**, **2**, and **4** have, respectively, zero, one or two $[O{=}W^{VI}(OH_2)]^{4+}$ bridges between the $[A\text{-}\alpha\text{-}PW_9O_{34}]^{9-}$ units, and stability in solution increases significantly with the presence of each $[O{=}W^{VI}(OH_2)]^{4+}$ bridge. This is not surprising given that each bridge reduces the number of active sites (including the LTMO units) in the compounds that are sterically accessible to other reactants in solution and simultaneously reduces the negative charge density on all the POM oxygens, a factor that has always correlated with POM reactivity (*6*). For example, the most negatively charged polytungstates, effectively nucleophiles, are well documented to react most rapidly and completely with a range of organic and organometallic electrophiles (*6,45*).

The difference between **3** and **4** with their isostructural all-tungsten complexes, $K_{14}[P_2W_{19}O_{69}(OH_2)]\cdot24H_2O$ and $K_4H_2[P_2W_{21}O_{71}(OH_2)_3]\cdot 28H_2O$, respectively, in the thermogravimetric analysis (TGA) is worthy of note (Fig. 6): both **3** and **4** lose an additional *ca.* 0.6% weight at high temperature; whereas the weights of $\{P_2W_{19}\}$ and $\{P_2W_{21}\}$ remain constant (*44*). These results suggest that both **3** and **4** likely decompose by evolving dioxygen at high temperature (the contents of the terminal-oxo oxygen on Au and its *trans* aqua ligand are 0.59% and 0.58% for **3** and **4**, respectively). The thermal evolution of O_2 is under investigation.

The purity of the terminal Au-oxo complexes, **3** and **4**, was established by several methods including ^{31}P NMR (**3** and **4** have only one phosphorus peak at −8.55 and −13.15 ppm, respectively), cyclic voltammetry, electronic absorption spectroscopy, vibrational spectroscopy, detailed magnetic measurements and elemental analysis on all elements (triplicate analyses for Au) (*44*). The single peak in the ^{31}P NMR spectra is consistent with the C_{2v} symmetry of **3** and **4** established by multiple X-ray crystallographic structure determinations and a neutron diffraction study on **3** at liquid He

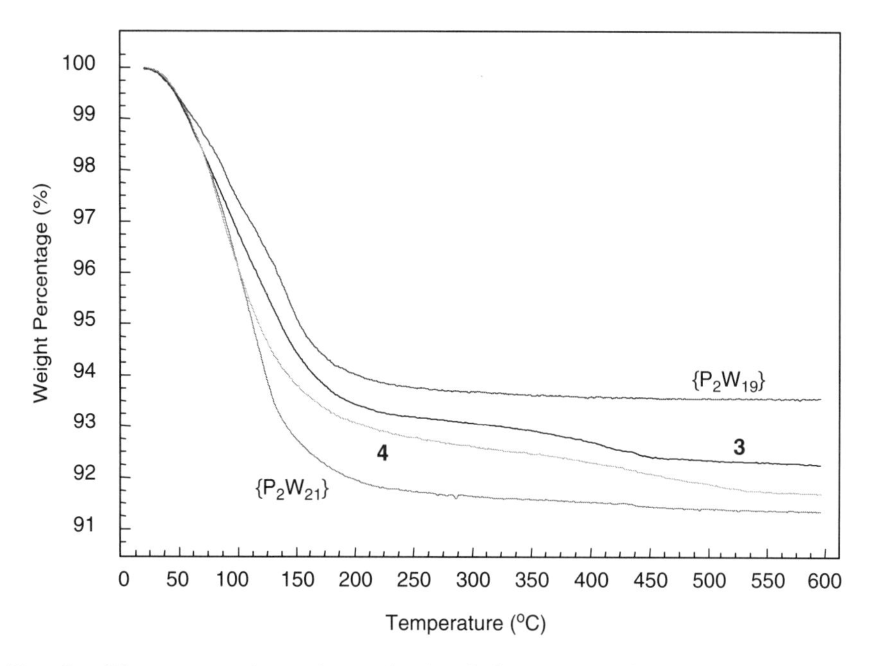

FIG. 6. Thermogravimetric analysis of the terminal Au-oxo complexes, **3** (black), **4** and their isostructural all-tungsten analogues (noted as $\{P_2W_{19}\}$ and $\{P_2W_{21}\}$, respectively).

temperature. Differential scanning calorimetry (DSC) and TGA show that **3** and **4** are both fairly stable as pure single crystals.

The structure of Au-oxo complex **3** is unambiguously determined by both X-ray single crystal diffraction and neutron diffraction, and the latter confirms the terminal-oxo ligand on the Au center (*44*). Similar to **1**, the very short Au–O_{oxo} bond (1.763(17) Å) is *trans* to the longer Au–OH_2 bond (2.29(4) Å) (Fig. 7), and the central Au atom is flanked by two symmetrically equivalent {A-PW_9} polytungstate cluster ligands each through two oxygens, defining an O_4 square equatorial plane around the Au atom. Covalent single Au–O bonds with bridging oxo groups are typically in the range of 1.90–2.10 Å, exemplified by many structurally characterized gold oxides (*46*), and recently reported molecular gold complexes (*47–49*).

In neutron diffraction studies, the Fourier map does not show significant negative scattering density around the terminal Au-oxo oxygen O35, a result similar to the previous neutron diffraction on the terminal Pt-oxo complex **1**, thus ruling out the possibility that a

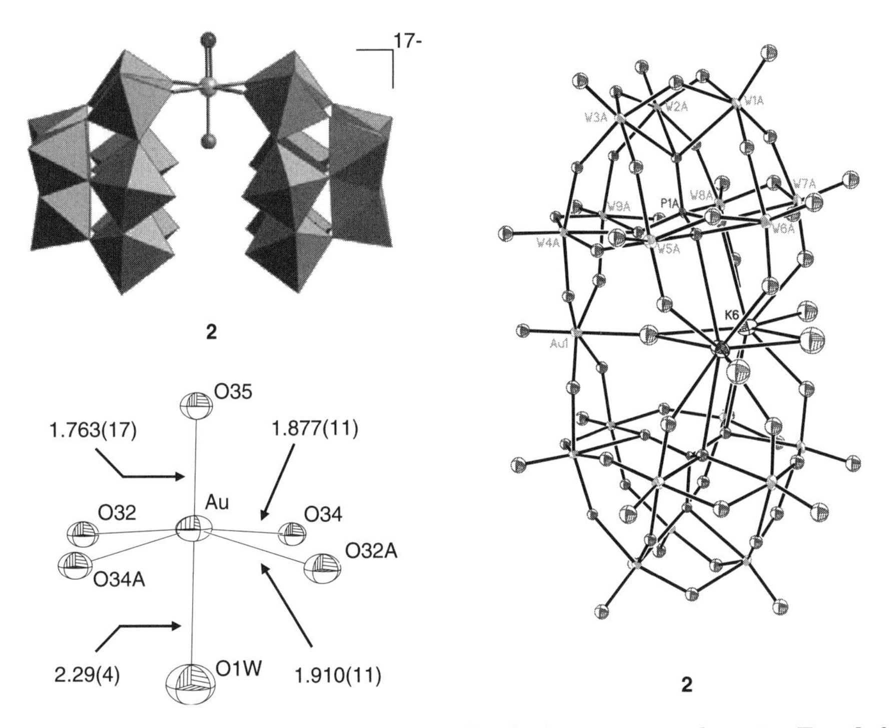

FIG. 7. Structure of the singly-bridged Au-oxo complex, **3**. Top left, combination polyhedral/ball-and-stick representation; right, thermal ellipsoid plot. The PO_4 and WO_6 polyhedra (or W atom) are shown in lighter and darker gray. Bottom left, X-ray structure of the coordination polyhedron around the Au atom in **3**. The central Au atom is domed (displaced out of the equatorial O_4 plane toward the terminal oxo) by 0.31(1) Å.

hydrogen atom is bound to the oxo ligand. The lack of a hydrogen on the oxo oxygen, Au–O_{oxo} distance, and displacement of the Au atom out of the O_4 equatorial plane (henceforth referred to as the "doming" distance) are three arguments from diffraction studies alone suggesting that these complexes are terminal Au-oxo complexes and not Au-hydroxo complexes (*44*). The *trans* aqua ligand is confirmed by the presence of hydrogen atoms in the area surrounding O1W. The crescent-shaped scattering density proximal to O1W indicates disorder of this aqua ligand about the twofold axis. This is also consistent with the observation in the X-ray structure of **3** that oxygen O1W has large thermal parameters resulting from slight displacement of this oxygen from the C_2 axis by the repulsion of potassium ions and phosphate oxygens O27. The latter are within van der Waals distance of O1W.

Unlike **3**, the polyanion unit of $[AuO(OH_2)P_2W_{20}O_{70}(OH_2)_2]^{9-}$ (**4**) is disordered. This type of disorder (resulting from crystallographically imposed D_{3h} symmetry on the C_{2v} polyanion), is also present in $[P_2W_{21}O_{71}(OH_2)_3]^{6-}$, an all-tungsten isostructural complex in which an O=W(OH_2) group replaces the O=Au(OH_2) group (*44*). It can be well modeled: the two W atoms on the equatorial plane, as well as their inward-oriented oxo groups and outward-oriented aqua ligands are evenly divided with respect to the three interior *mm* sites. The latter have an ideal site-occupancy factor of 2/3. The Au atom, together with its inward-directed aqua and outward-directed oxo ligands, is disordered over the three exterior *mm* sites. These have an ideal site-occupancy factor of 1/3. All the resulting key distances in **4** including Au–O_{oxo} (1.77(4) Å), Au–OH_2 (2.32(6) Å, aqua ligand *trans* to oxo ligand), Au–OW (1.963(11) Å, Au–O bonds in the equatorial plane) and Au–C_t (0.32(1) Å, the doming of Au) are almost identical within experimental error to those in the three X-ray crystal structures of **3**. In other words, the Au(O)(OH_2)(O–W)$_4$ coordination spheres are nearly identical in these two structures. The purity of the samples from which the single crystal of **4** for diffraction were obtained were confirmed by ^{31}P NMR (*44*). No polytungstophosphates, including $[P_2W_{21}O_{71}(OH_2)_3]^{6-}$ {P_2W_{21}}, were extant in either sample. Importantly, addition of authentic samples of {P_2W_{20}} (the starting monovacant polytungstate ligand) or/and {P_2W_{21}} (the isostructural all-tungsten polytungstate) to the aqueous solution of **4** has no effect on the Au-oxo complex peak and gives the new characteristic chemical shifts of {P_2W_{20}} or/and {P_2W_{21}} at −12.30 or/and −13.25 ppm. This finding, in turn, shows, first, that the Au atom is not a countercation to the polyanion of **4**, but rather incorporated into the polytungstate framework, and second, the polyanion is stable and does not undergo metal exchange or any other kind of isomerization in solution. Table II summarizes and compares key distances in the LTMO compounds prepared and characterized to date.

TABLE II

BOND DISTANCES (Å) IN THE $[M–O_6]$ CORES OF THE LATE TRANSITION METAL-OXO (LTMO) COMPLEXES TO DATE (M = PT, PD AND AU) FROM X-RAY DIFFRACTION STUDIES

Complex	Atoms	Distances (Å)
	Pt1 – O35	1.720(18)
Pt-oxo (**1**)	Pt1 – O6W	2.29(4)
	Pt1 – O29/O29A	1.911(10)
	Pt1 – O30/O30A	1.905(11)
	Pd1 – O70	1.60(2)
	Pd1 – O71	1.99(2)
Pd-oxo (**2**)	Pd1 – O32	1.937(10)
	Pd1 – O33	2.007(11)
	Pd1 – O37	1.958(11)
	Pd1 – O38	1.993(11)
	Au1 – O35	1.763(17)
Au-oxo (**3**)	Au1 – O1W	2.29(4)
	Au1 – O32/O32A	1.910(11)
	Au1 – O34/O34A	1.877(11)
	Au1 – O10	1.77(4)
Au-oxo (**4**)	Au1 – O11	2.32(6)
	Au1 – O5	1.963(11)

Variable-temperature measurements of the magnetic susceptibility at 0.1 and 1.0 T between 2 and 290 K indicated temperature-independent diamagnetism (*44*). That is, flat χ versus T curves, for both **3** ($\chi_{dia/TIP}$(**3**) $= -10.2 \times 10^{-4}$ emu mol^{-1}) and **4** ($\chi_{dia/TIP}$(**4**) $= -9.8 \times 10^{-4}$ emu mol^{-1}) were obtained. By comparison these numbers for the corresponding non-functionalized isopolytungstate cluster compounds are as follows: $\{P_2W_{19}\}$ ($\chi_{dia/TIP}\{P_2W_{19}\} = -10.0 \times 10^{-4}$ emu mol^{-1}) and $\{P_2W_{21}\}$ ($\chi_{dia/TIP}\{P_2W_{21}\} = -9.5 \times 10^{-4}$ emu mol^{-1}) provided that differences in the numbers of crystal water molecules and potassium countercations are appropriately noted. The diamagnetic susceptibilities for $\{P_2W_{19}\}$ and $\{P_2W_{21}\}$ on a molar basis have been determined under the same conditions. We note that substitution of a $W^{VI}(O)(OH_2)$ group in $\{P_2W_{19}\}$ and $\{P_2W_{21}\}$ by an $Au^{III}(O)(OH_2)$ group in **3** and **4** results in a slight increase in the absolute values. This is explained to some extent by an increase in the number of countercations. The temperature-independent paramagnetic (TIP) contributions are sizable for all polytungstates. Using diamagnetic values (calculated from tabulated Pascal constants) for **3**, **4**, $\{P_2W_{19}\}$ and $\{P_2W_{21}\}$ significantly overestimates the actual observed susceptibilities.

The oxidation state of Au in both Au-oxo complexes **3** and **4** was thoroughly investigated by several chemical and physicochemical methods (*44*). First, bulk electrolysis (coulometry at controlled potential) confirms the Au(III) oxidation state assignment in both **3**

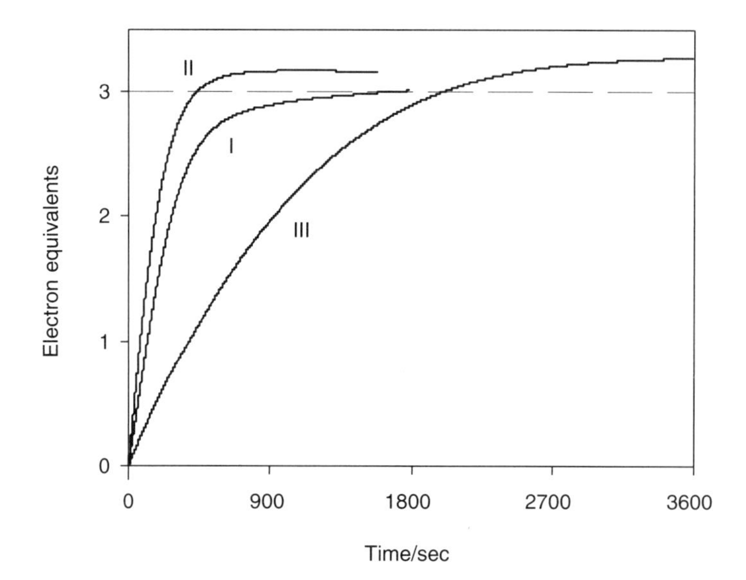

FIG. 8. Plots of the corrected electron equivalents, $\mathbf{n} = (\mathbf{C}-\mathbf{C_b})/FN$, passed during bulk electrolysis at controlled potential versus time. (I) 3.37 mM $(NH_4)AuCl_4$, 50 mM Na_2SO_4 buffer, pH *ca.* 2, 0.4 M NaCl, $E = 320$ mV; (II) 2.0 mM **3**, 50 mM Na_2SO_4 buffer, pH *ca.* 2, 0.4 M NaCl, $E = 300$ mV; (III) 4.2 mM **4**, 0.2 M $Na(CH_2ClCO_2)$ buffer, pH *ca.* 3, $E = 350$ mV.

and **4**: three electrons (**n** was found to be *ca.* 3.2) are needed for the complete conversion of gold to Au(0) (Fig. 8, bright Au mirrors are produced in the bulk electrolysis experiments). Second, chemical titration of the more stable Au-oxo complex **4** in solution by well-characterized one-electron reducing agents, including 2,2′-azino-bis(3-ethyl benzothiazoline-6-sulfonic acid) ($ABTS^{2-}$), $[Fe^{II}(CN)_6]^{4-}$ and $[SiW_{12}O_{40}]^{5-}$, confirms the existence of a d^8 Au(III) center in each molecule. The reaction of **4** with $ABTS^{2-}$ or $[Fe^{II}(CN)_6]^{4-}$ is fast and can be followed by UV-Vis spectroscopy since the oxidized products, intensively green colored $ABTS^{\bullet -}$ or the yellow $[Fe^{III}(CN)_6]^{3-}$, are stable under argon and their yields can be quantified by their characteristic electronic absorption features ($ABTS^{\bullet -}$ has maxima at 417, 645 and 728 nm, and the $[Fe^{III}(CN)_6]^{3-}$ has a maximum at 420 nm). Moreover, the reduction of **4** by $[SiW_{12}O_{40}]^{5-}$ is monitored by using solution reduction potentials (Fig. 9) (*44*). The titration end-point determined from intersection of two lines shown on Fig. 9 proves that 3.05 ± 0.15 equivalents of $[SiW_{12}O_{40}]^{5-}$ are required to reduce 1 equivalent of **4**. This finding is an average of two independent experiments and consistent with the reduction of Au(III) to Au(0). Third, X-ray absorption spectroscopy (XAS) was used to study the gold oxidation state in Au-oxo complexes. The similar near-edge region (position and intensity of the white line) at the Au L_2 edge of both **3**

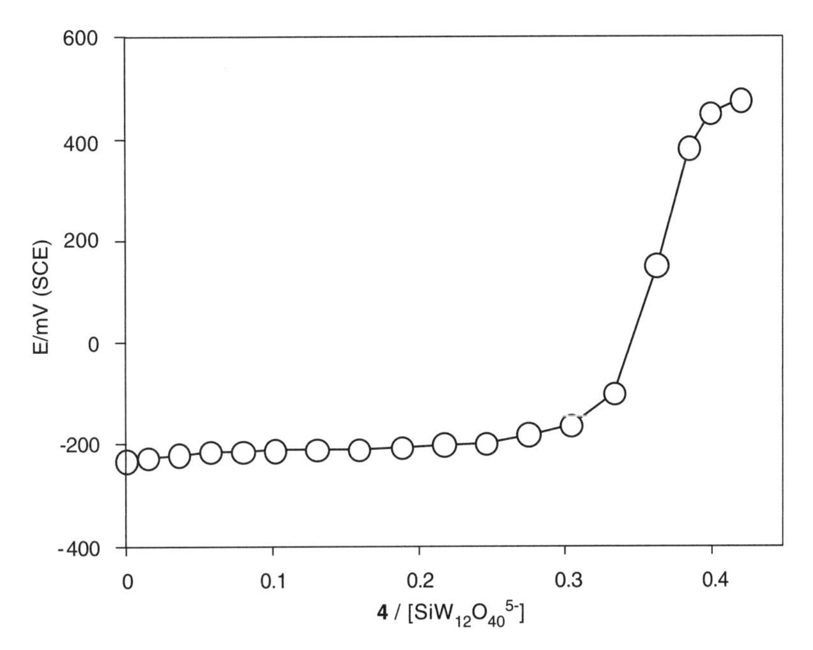

FIG. 9. The $SiW_{12}O_{40}^{5-}/SiW_{12}O_{40}^{4-}$ couple (initial solution concentration, 0.87 mM $SiW_{12}O_{40}^{5-}$) on addition of a 1 mM stock solution of **4**. Conditions, 25 mM $Na(CH_2ClCO_2)$ buffer (pH 3.0), Ar atmosphere, 25 °C.

and **4** relative to an authentic Au(III) complex, Au(acetate)$_3$, also implicate the presence of Au(III) in the LTMO complexes (*44*). The further comparison of the spectra of **3** and **4** to those of Au(I) compounds shows that they are quite different and the rising part of the edges of **3** and **4** is much closer in energy to Au(acetate)$_3$ than to any of the Au(I) complexes. Fourth, the ultra-low temperature (17 K) optical spectrum of **4** in water–glycerol glasses is quite different than the optical spectrum of the d^6 Pd(IV)-oxo complex and similar to that predicted for a d^8 electronic configuration in a six-coordinate C_{2v} structure like that in **3** and **4** (*44*). The two electronic absorption bands at 21,440 cm^{-1} (466 nm, band 1) and 25,310 cm^{-1} (372 nm, band 2) are thus tentatively assigned as $xy \rightarrow z^2$ ($^1A_1 \rightarrow {}^3A_2$) for band 1 and $xy \rightarrow z^2$ ($^1A_1 \rightarrow {}^1A_2$) for band 2. Importantly, the experimentally determined $^1A_2 \rightarrow {}^3A_2$ splitting of $\sim$4000 cm^{-1} can be compared with an $\sim$4000 cm^{-1} $^1E \rightarrow {}^3E$ splitting in the ligand field spectrum of appropriate d^2 analogues such as the $[MoOL_4Cl]^{1+}$ compounds (*50*). This provides further support for our assignment of the lowest energy ligand field transitions in **4** as arising from exchange-split triplet and singlet $xy \rightarrow z^2$ excitations. Fifth, all the NMR experiments (chemical shifts and peak width in ^{31}P and ^{17}O NMR) as well as magnetic studies show that the Au-oxo complexes **3** and **4** are diamagnetic. These collective five lines of experimental evidence are consistent with d^8 Au(III) centers in these LTMO complexes.

Computational studies conducted on $O{=}Au(OH_2)(OH)_4^{n-}$, $O{=}Au(W_2O_9H_4)_2^{n-}$, and $O{=}Au(OH_2)(W_2O_9H_4)_2^{n-}$, where $n{=}1$ and 3 (with Au(V) and Au(III) centers, respectively) to model the ligand environments of Au in **3** and **4** are inconclusive because the ligand environment of the Au centers is unstable indicating that these simple models are not adequately close to the actual polytungstate ligands in **3** and **4** to be defensible (*44*). We also conducted (CASSCF and MRSD-CI) computational investigations of AuO, AuO^+, AuO^{2+}, AuO^{3+}, and of $(NC)AuO^{q+}$ ($q = 0$, 1 and 2) (*44*). The electronic structures and oxo dissociation energies from these calculations, like the experimental data, indicate that the Au-oxo units in **3** and **4** have multiple bond character similar to $(AuC)^+$ and such species calculated by the Pyykkö research group (*51,52*).

Density Functional Theory, DFT (B3LYP), CASSCF (Complete Active-State Self-Consistent Field) and MRSD-CI (Multi-Reference Single-Double Correlation Interaction) calculations on the diatomic units AuO, AuO^+, AuO^{2+} and AuO^{3+} clearly show that stability of Au–O bond reduces in this order. This trend is consistent with the molecular orbital diagram of AuO molecule presented in Fig. 10.

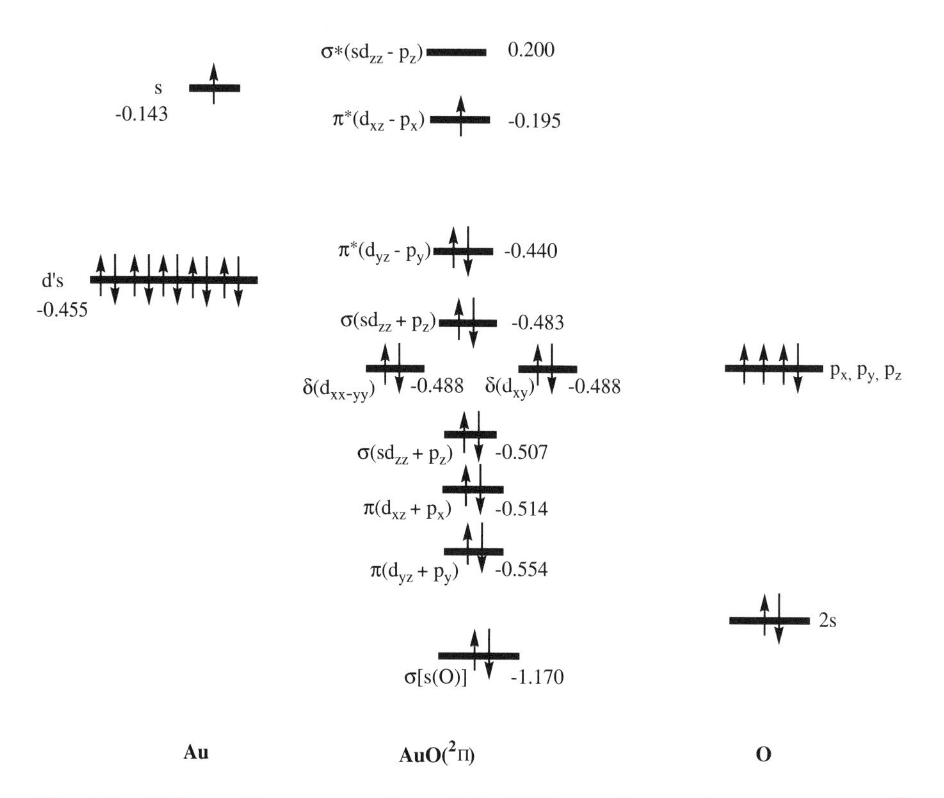

FIG. 10. Schematic presentation of the MO-correlation diagram for the $^2\Pi$ electronic state of the AuO molecule (*44*).

As seen from this figure, the ground electronic $^2\Pi$ state of AuO with Au(II) has the following electronic configuration: $[\pi(d_{yz}+p_y)]^2$ $[\pi(d_{xz}+p_x)]^2[\sigma(sd_{zz})]^2[\delta(d_{xy})]^2[\delta(d_{xx-yy})]^2[\sigma(sd_{zz}+p_z)]^2[\pi^*(d_{yz}-p_y)]^2[\pi^*$ $(d_{xz}-p_x)]^1$ $[\sigma^*(sd_{zz}-p_z)]^0$. The MRSD-CI calculated Au–O bond distance is 1.930 Å. This is in a good agreement with the experimental value of 1.912Å and also in agreement with previous computational studies of AuO. The Au–O bond dissociation energy is 31.7 kcal/mol calculated at the MRSD-CI level. This is also consistent with the previous computational investigations.

The energy for removal of an electron from the $\sigma(sd_{zz}+p_z)$ orbital of AuO is −0.483 a.u. This is principally the valence s-orbital of the Au atom. The resulting cation, $Au^{III}O^+$ has a +1 charge located at the Au center. The computational data indicate that $Au^{III}O^+$ has a $^3\Pi$ ground electronic state with the following electronic configuration: $[\sigma(sd_{zz}+p_z)]^2[\pi(d_{yz}+p_y)]^2[\pi(d_{xz}+p_x)]^2[\delta(d_{xy})]^2[\delta(d_{xx-yy})]^2[\pi^*(d_{yz}-p_y)]^2[\sigma(sd_{zz}+p_z)]^1[\pi^*(d_{xz}-p_x)]^1[\sigma^*(sd_{zz}-p_z)]^0$ (*44*). The $^3\Sigma$ state with the $[\sigma(sd_{zz}+p_z)]^2[\pi^*(d_{yz}-p_y)]^1[\pi^*(d_{xz}-p_x)]^1[\sigma^*(sd_{zz}-p_z)]^0$ orbital configuration has a very similar energy. The calculations demonstrate that AuO^+, where Au is in Au(III) oxidation state, exists and is stable relative to the Au^++O dissociation limit. The increase in oxidation state of Au (+2 to +3) slightly reduces the stability of the AuO moiety. The Au–O bond distance in $Au^{III}O^+$, 1.981 Å, is slightly greater than that for AuO. In addition, the Au^+–O bond energy, 27.8 kcal/mol, is ~4 kcal/mol smaller than in $Au^{II}O$. It should be noted that the lowest singlet $^1\Sigma$ state of $Au^{III}O^+$ has multi-determinant character. The $[\pi^*(d_{yz}-p_y)]^2[\pi^*(d_{xz}-p_x)]^0$ and $[\pi^*(d_{yz}-p_y)]^0[\pi^*(d_{xz}-p_x)]^2$ configurations (44% weight each) are the largest contributors to the total wavefunction.

Ionization of $Au^{III}O^+$ generates AuO^{2+}, which is stable on the potential energy surface in its ground electronic state ($^4\Sigma$) with the $[\sigma(sd_{zz}+p_z)]^2[\pi(d_{yz}+p_y)]^2[\pi(d_{xz}+p_x)]^2[\delta(d_{xy})]^2[\delta(d_{xx-yy})]^2[\sigma^*(sd_{zz}-p_z)]^1$ $[\pi^*(d_{yz}-p_y)]^1[\pi^*(d_{xz}-p_x)]^1$ electron configuration (*44*). The calculated Au–O bond distance in AuO^{2+} is 2.10 Å. Dissociation of AuO^{2+} into Au^{2+}+O and into Au^++O^+ is endothermic by 58.3 kcal/mol, and exothermic by 51.7 kcal/mol, respectively. These electronic, geometric and energetic data indicate that $Au^+(O^+)$ is the lowest energy form of AuO^{2+}. Further, all the doubly occupied σ-, π- and δ-MOs are d atomic orbitals of Au^+ and all three singly occupied orbitals of the $^4\Sigma$ ground state of AuO^{2+} are p atomic orbitals of O^+. There is only one true Au–O bonding orbital: $\sigma(sd_{zz}+p_z)$. The aforementioned conclusions about the electronic structure of AuO^{2+} are in accord with the calculated (and experimental) ionization potentials of Au^+ and O, which are 19.7 (20.5) and 13.2 (13.614) eV, respectively.

The minima on the potential energy surface of the reaction Au^{2+}+O^+ corresponds to the $^3\Sigma$ state of AuO^{3+}. This species has the electron configuration $[\pi(d_{yz}+p_y)]^2[\pi(d_{xz}+p_x)]^2[\delta(d_{xy})]^2[\delta(d_{xx-yy})]^2$ $[\sigma(sd_{zz}+p_z)]^2[\pi^*(d_{yz}-p_y)]^1[\pi^*(d_{xz}-p_x)]^1[\sigma^*(sd_{zz}-p_z)]^0$. The minimum corresponding to AuO^{3+} on the potential energy surface is very shallow despite the finding that the Au–O bond distance is quite short (1.90 Å). AuO^{3+} dissociates to Au^{2+}+O^{+} with the release of a very substantial energy (264.4 kcal/mol). This minimum is 150.4 kcal/mol more stable than the Au^{3+}+O dissociation limit. Again, these findings are in accord with the calculated ionization potentials of Au^{2+} and O: 32.9 eV (758.67 kcal/mol) and 13.2 eV (the experimental value is 13.614 eV) (304.39 kcal/mol), respectively.

The Au(IV) and Au(V) units, AuO^{2+} and AuO^{3+}, are highly unstable with respect to loss of O^{+} (*44*). Nonetheless, these species are stabilized under particular limiting conditions. First, ligation of Au by negatively charged σ-donating ligands will reduce the Coulomb repulsion between the positively charged Au and O centers. Second, ligation of Au by strong π-electron withdrawing ligands will delocalize the π^* electrons in AuO units. Third, (O^{+})...X (non-covalent) interactions involving AuO units embedded in the surfaces can render O^{+} dissociation less favorable. These hypotheses are supported by calculations on the complexes $(NC)AuO^{q+}$ ($q = 0$, 1 and 2). (NC)AuO, formally containing Au(III) is stable with a 1.956 Å Au–O bond distance and a triplet ground state. Thus ligation of CN^{-} to AuO^{+} reduces the Au–O bond distance and increases the Au–O bond energy from 27.8 kcal/mol in AuO^{+} to 31.0 kcal/mol in (NC)AuO.

In addition, the complexes $(NC)Au^{IV}O^{+}$ and $(NC)Au^{V}O^{2+}$ are more stable with respect to both O-atom and O-cation dissociation limits compared to the corresponding non-ligated AuO^{2+} and AuO^{3+} units. As per historical experimental and computational studies on many such species, the presence of σ-electron donating and π-electron withdrawing ligands generally stabilizes the high oxidation states of Au in AuO fragments.

The structures (geometrical and electronic) and physicochemical properties of the LTMO complexes are only now adequately established to facilitate interpretable reactivity studies. Parallel investigations are currently being conducted in both aqueous media using alkali metal salts of the complexes and in organic media (e.g., acetonitrile) using tetralkylammonium salts of the complexes. Thus far oxo transfer from the Pd-oxo and Au-oxo complexes to multiple organic substrates has been documented. This work will be reported subsequently.

III. Open-Framework Materials that Catalyze Aerobic Oxidations

Materials with controlled pore sizes and functionality, particularly in three dimensions would have many uses (*53–59*). Numerous totally inorganic microporous and mesoporous materials have been subject of thousands of papers, and applications of the former (e.g., zeolites) have a sizable impact on the global economy at present (myriad uses from production of gasoline to a host of chemicals) (*60–66*). However, the use of open-framework coordination polymers and organic–inorganic composite porous materials including metal-organic frameworks (MOFs) for applications in catalysis has been problematical for a few reasons. While diversity, synthetic tunability and high porosity of these materials are highly attractive, their lack of adequate stability under most conditions relevant to industrial catalysis is a fairly pervasive concern. Secondary concerns for the use of MOFs in catalysis include the typical lack of free coordination positions on d-electron metal active sites in these structures and adequate mass transport for catalysis conducted under low and ambient temperature conditions, the conditions that are likely to minimize the oxidative instability concerns intrinsic to organic linkers. A key feature underlying the proven utility of zeolites and other microporous totally inorganic solids in most industrial catalysis and processing applications is their ability to withstand oxidative degradation at higher temperatures (typically 200–400 °C) while exposed to O_2/air. Hydrolytic stability is very important in some industrial catalysis and processing applications but generally less so than oxidative stability. Concerns about the oxidative instability of organic ligands, molecular connectors and organic structures in general have been a significant focus of homogeneous catalytic oxidation investigators for three decades. Heteropoly acid and other POM derivatives were demonstrated to offer the active site tunability (control of geometric and electronic structure) advantages of homogeneous oxidation catalysts with the stability advantages of zeolites, metal oxides and other totally inorganic and thus oxidatively resistant species. POMs substituted with particular d-electron metal centers are oxidatively and hydrolytically stable analogues of highly selective homogeneous and biological catalysts including hemes (and more generally metalloporphyrins) (*6,67–79*).

The attractive (*80*) features of MOFs and similar materials noted above for catalytic applications have led to a few reports of catalysis by these systems (*81–89*), but to date the great majority of MOF applications have addressed selective sorption and separation of gases (*54–57,59,80,90–94*). Most of the MOF catalytic applications have involved hydrolytic processes and several have involved enantioselective processes. Prior to our work, there were only two or three reports of selective oxidation processes catalyzed by MOFs. Nguyen and Hupp reported an MOF with chiral covalently incorporated (salen)Mn units that catalyzes asymmetric epoxidation by iodosylarenes (*95*), and in a very recent study, Corma and co-workers reported aerobic alcohol oxidation, but no mechanistic studies or discussion was provided (*89*).

It is a specific and practically oriented challenge to materials and other chemists to design and realize microporous or mesoporous structures that take up (sequester) and then catalyze aerobic (air-based) oxidation of pore-trapped target reactants under ambient conditions. Major uses of materials that trap and decontaminate using

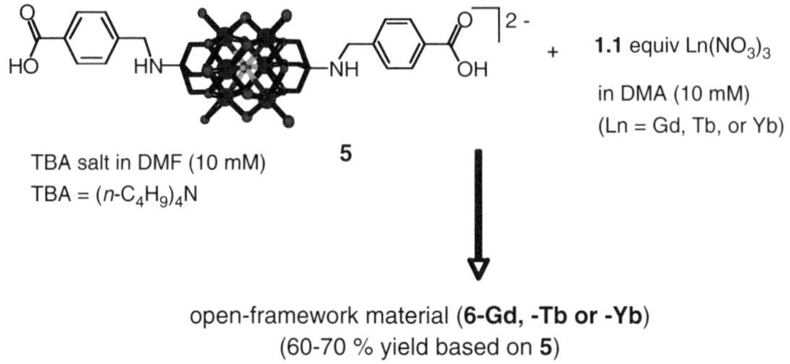

HO HN NH OH 2 -

\+ **1.1** equiv $Ln(NO_3)_3$
in DMA (10 mM)
(Ln = Gd, Tb, or Yb)

TBA salt in DMF (10 mM)
TBA = $(n\text{-}C_4H_9)_4N$

5

open-framework material (**6-Gd, -Tb or -Yb**)
(60-70 % yield based on **5**)

FIG. 11. Synthesis of an open-framework material, **6** (structure of **6-Gd** given in Figs. 12 and 13). Condensation of a bis(triester)hexavanadate unit with the pendant triester groups terminating in carboxylic acids, **5**, with lanthanide ions form the open-framework catalytic materials.

the ambient environment would include purification of air in civilian, military and care/hospital environments and decontamination of toxic agents. We report here initial studies of structures that bind and catalyze the oxidative removal of deleterious and/or odorous compounds using only the ambient environment (air at atmospheric pressure and room temperature).

Our approach is outlined in Fig. 11. It is based on linking catalytically active POM-based structural units with lanthanide ions into porous MOF-like structures (*96*). First, the POM-centered linking unit with two juxtaposed organic groups (**5** in Fig. 11) is prepared. The organic groups in **5** are attached to the POM moiety (formally $[V_6O_{19}]^{8-}$) by triester groups that are quite robust (thermally stable to 100 °C and hydrolytically stable even in water) (*96–98*). We demonstrated that the bis(triester)V_6 unit, **5**, alone catalyzes O_2-based oxidations of some organic molecules under mild conditions and in the absence of a reducing agent. These are generically attractive features in catalytic oxidation and green chemistry. We reasoned that catalytic activity exhibited by **5** in solution should be translatable to such materials, including possible microporous materials, containing **5**. The second and final step entails reaction of **5** with a lanthanide ion at high dilution. This does indeed form crystalline open-framework materials. Separate reactions with Gd(III), Tb(III) or Yb(III) all successfully led to these materials (**6-Gd**, **6-Tb** and **6-Yb**, respectively) in high yields (*96*).

Figs. 12 and 13 show the crystal structure of **6-Gd**. Four of the linear POM-catalyst dicarboxylic acid units, **5**, are linked by di-lanthanide paddle wheel junctions (Fig. 12) into the open-framework material (*96*). Fig. 13 shows the large channels in **6-Gd**. These are filled with dimethylformamide (DMF) molecules that are hard to remove (the boiling point of DMF at 1.0 atmosphere = 151 °C). Thus while the solvent-accessible internal volume of **6-Gd** is 50.5% of the crystal

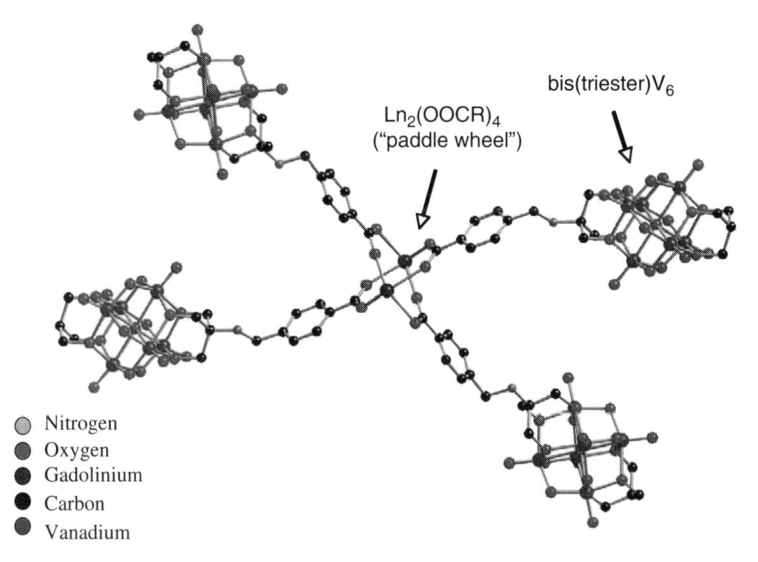

FIG. 12. A portion of the X-ray crystal structure of **6-Gd** showing one $Gd_2(OOCR)_4$ paddle wheel junction connecting four triesterified V_6 polyanions. (Only one of the triesterbenzoate moieties of the two bound to each polyanion is shown.)

volume based on PLATON calculations, standard surface area measurements indicate very low porosity. The surface area of **6-Tb** from BET adsorption isotherms is as follows: N_2 at $78\,K = 2\,m^2\,g^{-1}$; CO_2 at $273\,K = 30\,m^2\,g^{-1}$ (*96*). The fact the measurable surface area is highly dependent on temperature indicates that gas sorption is an activated process, i.e., more chemisorption than physical adsorption, and this doubtless arises because there is marked activation energy for displacing the pore-occupying DMF molecules.

Treatment of the open-framework materials at high vacuum and 70 °C results in collapse of the pores that are readily visible in the X-ray crystal structure (Fig. 13) and a nonporous structure (surface area $< 1\,m^2\,g^{-1}$) (*96*).

The open-framework materials catalyze peroxide-based oxidations such as sulfoxidation, but significantly more noteworthy, **6-Tb** catalyzes O_2-based oxidations and does so at ambient or nearly ambient temperature (*96*). A case in point is the oxidation of thiols, a common odorant in human environments and a mildly toxic class of compounds, to disulfides that have almost no odor and are less toxic. The stoichiometry for oxidation of a representative substrate, *n*-propane thiol, was established to be that in Eq. (1). After 30 days at 45 °C, **6-Tb** produces *ca.* 19 turnovers of disulfide product based on the bis(trisester)V_6 groups in the open-framework material, *using only ambient air as the oxidant.*

$$2\,PrSH + 0.5\,O_2 \rightarrow PrSSPr + H_2O \qquad (1)$$

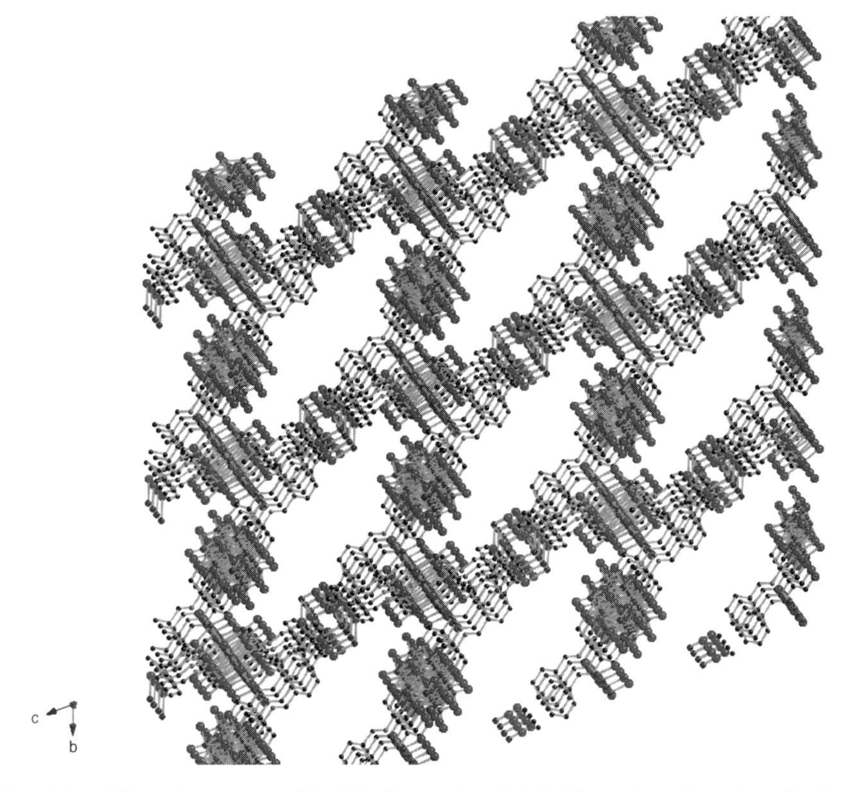

FIG. 13. The structure of **6-Gd** along the (*100*) direction showing the large channels (cross-section = *ca*. 28 × 20 Å) pores. (H atoms and DMF molecules omitted for clarity.)

No efforts have been made yet to optimize catalysis of Eq. (1) or any other attractive aerobic oxidation process catalyzed by these materials.

Although it is too early in the design and study of open-framework POM-based materials for structure–activity relationships, stability, mechanisms and other intellectually and/or developmentally important issues to be adequately understood, it is evident that such materials could be of significant interest in the context of air purification and detoxification processes. The concept of generating a family of tunable robust catalytic connectors exemplified by **5** and then assembling complex functional materials using these connectors, including microporous or nanoporous aerobic oxidation catalysts, is a general one. A key current challenge is to design open-framework POM-based materials that are sufficiently robust to permit removal of high-boiling pore-blocking solvent molecules that may be required in the synthesis of such materials in the future. This alone could well lead to catalysts with activities for aerobic reactions that are at least an order of magnitude higher than those exhibited by **6-Tb** and **6-Yb** where the pores are only slightly emptied of the high-boiling solvent molecules during catalytic turnover.

A related research effort was also reported at this Erlangen, Germany, conference involving the successful preparation of extraordinarily reactive catalysts for aerobic oxidations of several kinds, including the oxidation of aldehydes to acids and sulfides to sulfoxides. These catalysts are significantly more reactive than others reported recently for such processes (*99–103*). Aldehydes, including acetaldehyde and formaldehyde, are some of the most toxic compounds found in indoor air generally, and the chemical warfare agent, mustard, is a sulfide. These catalysts are generated when particular Cu- (or Cu- and Fe-) substituted polytungstates are combined with NO_x ligands in appropriate nonpolar media. While it is likely that the catalysts contain terminal NO_x ligands on the Cu and/or Fe center(s), even this inference is uncertain because all work thus far to determine the structures of these complex POM derivatives, including marked efforts to garner X-ray crystal structures or insightful NMR spectra, have been unsuccessful. Several X-ray crystal structures to date have all been too disordered (a common difficulty encountered with substituted α-Keggin POM derivatives) to distinguish the terminal ligands or other potential key features, while significant paramagnetism has rendered all NMR spectra comparably uninformative. Nonetheless, the activities of these latter Cu/FePOM-NO_x-based catalysts certainly call for additional careful work to elucidate the structural, physicochemical and reactivity properties of these compounds. Appropriate experiments are underway and will be reported subsequently.

Acknowledgments

Craig L. Hill thanks the U.S. Army Research Office (W911NF-05-1-0200) for funding the research on catalytic network materials, and the U.S. National Science Foundation (CHE-0553581) and U.S. Department of Energy (DE-FG02-03-ER15461 and DE-FG02-07ER15906) for funding the research on LTMO complexes and O_2 activation. Martin L. Kirk acknowledges the National Institutes of Health (GM-057378) for funding work related to metal-oxo catalysis. SSRL operations are funded by the U.S. DOE-BES and the SSRL SMB program by the NIH NCRR BTP and DOE BER.

References

1. Pope, M. T.; Müller, A. *Angew. Chem. Int. Ed.* **1991**, *30*, 34–48.
2. Pope, M. T.; Müller, A. (Eds.) *"Polyoxometalate Chemistry from Topology via Self-Assembly to Applications"*; Kluwer Academic Publishers: Dordrecht, **2001**.
3. Borrás-Almenar, J. J.; Coronado, E.; Müller, A.; Pope, M. T. *"Polyoxometalate Molecular Science"*; vol. 98; Kluwer Academic Publishers: Dordrecht, **2003**.

4. Hill, C. L. (Ed.) "Special Thematic Issue on Polyoxometalates"; vol. 98, No. 1; **1998**.
5. Pope, M. T. In: "*Comprehensive Coordination Chemistry II: From Biology to Nanotechnology*"; vol. 4; Ed. Wedd, A.G.; Elsevier: Oxford, UK, **2004**, pp. 635–678.
6. Hill, C. L. In: "*Comprehensive Coordination Chemistry-II: From Biology to Nanotechnology*"; Vol. 4; Ed. Wedd, A.G.; Elsevier: Oxford, UK, **2004**, pp. 679–759.
7. Nugent, W. A.; Mayer, J. M. "*Metal-Ligand Multiple Bonds*"; Wiley: New York, **1998**.
8. Holm, R. H. *Chem. Rev.* **1987**, *87*, 1401–1449.
9. Holm, R. H.; Donahue, J. P. *Polyhedron* **1993**, *12*, 571–589.
10. Parkin, G. In: "*Prog. Inorg. Chem*"; Vol. 47; Ed. Karlin, K.D.; Wiley: New York, **1998**, pp. 1–165.
11. Waterman, R.; Hillhouse, G. L. *J. Am. Chem. Soc.* **2003**, *125*, 13350–13351.
12. Kogut, E.; Wiencko, H. L.; Zhang, L.; Cordeau, D. E.; Warren, T. H. *J. Am. Chem. Soc.* **2005**, *127*, 11248–11249.
13. Ison, E. A.; Cessarich, J. E.; Travia, N. E.; Fanwick, P. E.; Abu-Omar, M. M. *J. Am. Chem. Soc.* **2007**, *129*, 1167–1178.
14. Cowley, R. E.; Bontchev, R. P.; Sorrell, J.; Sarracino, O.; Feng, Y.; Wang, H.; Smith, J. M. *J. Am. Chem. Soc.* **2007**, *129*, 2424–2425.
15. MacBeth, C. E.; Golombek, A. P., Jr.; Young, V; Yang, C.; Kuczera, K.; Hendrich, M. P.; Borovik, A. S. *Science* **2000**, *289*, 938–941.
16. Rohde, J.-U.; In, J.-H.; Lim, M. H.; Brennessel, W. W.; Bukowski, M. R.; Stubna, A.; Münck, E.; Nam, W.; Que, L., Jr. *Science* **2003**, *299*, 1037–1039.
17. Green, M. T.; Dawson, J. H.; Gray, H. B. *Science* **2004**, *304*, 1653–1656.
18. Bukowski, M. R.; Koehntop, K. D.; Stubna, A.; Bominaar, E. L.; Halfen, J. A.; Münck, E.; Nam, W.; Que, L., Jr. *Science* **2005**, *310*, 1000–1002.
19. Ostovic, D.; Bruice, T. C. *Acc. Chem. Res.* **1992**, *25*, 314–320.
20. Groves, J. T.; Lee, J.; Marla, S. S. *J. Am. Chem. Soc.* **1997**, *119*, 6269–6273.
21. Dubé, C. E.; Wright, D. W.; Pal, S.; Bonitatebus, P. J.; Armstrong, W. H. *J. Am. Chem. Soc.* **1998**, *120*, 3704–3716.
22. Shirin, Z.; Hammes, B. S.; Victor, G.; Young, J.; Borovik, A. S. *J. Am. Chem. Soc.* **2000**, *122*, 1836–1837.
23. Vrettos, J. S.; Brudvig, G. W. *Phil. Trans. R. Soc. Lond. B* **2002**, *357*, 1395–1405.
24. Visser, S. P. d.; Kumar, D.; Neumann, R.; Shaik, S. *Angew. Chem. Int. Ed.* **2004**, *43*, 5661–5665.
25. Khenkin, A. M.; Kumar, D.; Shaik, S.; Neumann, R. *J. Am. Chem. Soc.* **2006**, *128*, 15451–15460.
26. Hay-Motherwell, R. S.; Wilkinson, G.; Hussain-Bates, B.; Hursthouse, M. B. *Polyhedron* **1993**, *12*, 2009–2012.
27. Jacobi, B. G.; Laitar, D. S.; Pu, L.; Wargocki, M. F.; DiPasquale, A. G.; Fortner, K. C.; Schuck, S. M.; Brown, S. N. *Inorg. Chem.* **2002**, *41*, 4815–4823.
28. Anderson, T. M.; Neiwert, W. A.; Kirk, M. L.; Piccoli, P. M. B.; Schultz, A. J.; Koetzle, T. F.; Musaev, D. G.; Morokuma, K.; Cao, R.; Hill, C. L. *Science* **2004**, *306*, 2074–2077.
29. Anderson, T. M.; Cao, R.; Slonkina, E.; Hedman, B.; Hodgson, K. O.; Hardcastle, K. I.; Neiwert, W. A.; Wu, S.; Kirk, M. L.; Knottenbelt, S.; Depperman, E. C.; Keita, B.; Nadjo, L.; Musaev, D. G.; Morokuma, K.; Hill, C. L. *J. Am. Chem. Soc.* **2005**, *127*, 11948–11949. Addition/Correction: *J. Am. Chem. Soc.* **2008**, *130*, 2877–2877.
30. Filowitz, M.; Ho, R. K. C.; Klemperer, W. G.; Shum, W. *Inorg. Chem.* **1979**, *18*, 93–103.
31. Filowitz, M.; Klemperer, W. G.; Messerle, L.; Shum, W. *J. Am. Chem. Soc.* **1976**, *98*, 2345–2346.
32. Duncan, D. C.; Hill, C. L. *Inorg. Chem.* **1996**, *35*, 5828–5835.

33. Duncan, D. C.; Hill, C. L. *J. Am. Chem. Soc.* **1997**, *119*, 243–244.
34. Boykin, D. W. (Ed.) "*^{17}O NMR Spectroscopy in Organic Chemistry*"; CRC Press: Boca Raton, **1991**.
35. Bi, L.-H.; Reicke, M.; Kortz, U.; Keita, B.; Nadjo, L.; Clark, R. J. *Inorg. Chem.* **2004**, *43*, 3915–3920.
36. Bi, L.-H.; Kortz, U.; Keita, B.; Nadjo, L.; Borrmann, H. *Inorg. Chem.* **2004**, *43*, 8367–8372.
37. Bi, L.-H.; Kortz, U.; Keita, B.; Nadjo, L.; Daniels, L. *Eur. J. Inorg. Chem.* **2005**, 3034–3041.
38. Valden, M.; Lai, X.; Goodman, D. W. *Science* **1998**, *281*, 1647–1650.
39. Bond, G. C.; Thompson, D. T. *Gold Bulletin* **2000**, *33*, 41–51.
40. Chen, M. S.; Goodman, D. W. *Science* **2004**, *306*, 252–255.
41. Kim, W. B.; Voitl, T.; Rodriguez-Rivera, G. J.; Dumesic, J. A. *Science* **2004**, *305*, 1280–1283.
42. Guzman, J.; Carrettin, S.; Fierro-Gonzalez, J. C.; Hao, Y.; Gates, B. C.; Corma, A. *Angew. Chem. Int. Ed.* **2005**, *44*, 4778–4781.
43. Hutchings, G. J.; Haruta, M. (Eds.) "*Catalysis by Gold*"; vol. 291; Elsevier: New York, **2005**.
44. Cao, R.; Anderson, T. M.; Piccoli, P. M. B.; Schultz, A. J.; Koetzle, T. F.; Geletii, Y. V.; Slonkina, E.; Hedman, B.; Hodgson, K. O.; Hardcastle, K. I.; Fang, X.; Kirk, M. L.; Knottenbelt, S.; Kögerler, P.; Musaev, D. G.; Morokuma, K.; Takahashi, M.; Hill, C. L. *J. Am. Chem. Soc.* **2007**, *129*, 11118–11133.
45. Gouzerh, P.; Proust, A. *Chem. Rev.* **1998**, *98*, 77–112.
46. Jones, P. G.; Rumpel, H.; Schwarzmann, E.; Sheldrick, G. M.; Paulus, H. *Acta Cryst.* **1979**, *B35*, 1435–1437.
47. Cinellu, M. A.; Minghetti, G.; Pinna, M. V.; Stoccoro, S.; Zucca, A.; Manassero, M. *Chem. Comm.* **1998**, 2397–2398.
48. Cinellu, M. A.; Minghetti, G.; Pinna, M. V.; Stoccoro, S.; Zucca, A.; Manassero, M. *J. Chem. Soc., Dalton Trans.* **2000**, 1261–1265.
49. Cinellu, M. A.; Minghetti, G.; Cocco, F.; Stoccoro, S.; Zucca, A.; Manassero, M. *Angew. Chem. Int. Ed.* **2005**, *44*, 6892–6895.
50. Re, R. E. D.; Hopkins, M. D. *Inorg. Chem.* **2002**, *41*, 6973–6985.
51. Barysz, M.; Pyykkö, P. *Chem. Phys. Lett.* **1998**, *285*, 398–403.
52. Pyykkö, P.; Riedel, S.; Patzschke, M. *Chem. Eur. J.* **2005**, *11*, 3511–3520.
53. Barton, T. J.; Bull, L. M.; Klemperer, W. G.; Loy, D. A.; McEnaney, B.; Misono, M.; Monson, P. A.; Pez, G.; Scherer, G. W.; Vartuli, J. C.; Yaghi, O. M. *Chem. Mater.* **1999**, *11*, 2633–2656.
54. Cussen, E. J.; Claridge, J. B.; Rosseinsky, M. J.; Kepert, C. J. *J. Am. Chem. Soc.* **2002**, *124*, 9574–9581.
55. Eddaoudi, M.; Kim, J.; Rosi, N.; Vodak, D.; Wachter, J.; O'Keeffe, M.; Yaghi, O. M. *Science* **2002**, *295*, 469–472.
56. Kitagawa, S.; Kitaura, R.; Noro, S.-i. *Angew. Chem. Int. Ed.* **2004**, *43*, 2334–2375.
57. Bradshaw, D.; Claridge, J. B.; Cussen, E. J.; Prior, T. J.; Rosseinsky, M. J. *Acc. Chem. Res.* **2005**, *38*, 273–282.
58. Ockwig, N. W.; Delgado-Friedrichs, O.; O'Keeffe, M.; Yaghi, O. M. *Acc. Chem. Res.* **2005**, *38*, 176–182.
59. Matsuda, R.; Kitaura, R.; Kitagawa, S.; Kubota, Y.; Belosludov, R. V.; Kobayashi, T. C.; Sakamoto, H.; Chiba, T.; Takata, M.; Kawazoe, Y.; Mita, Y. *Nature* **2005**, *436*, 238–241.
60. Csicsery, S. M. *Pure Appl. Chem.* **1986**, *58*, 841–856.
61. Sheldon, R. A. *Chemtech* **1991**, *21*, 566–576.
62. Davis, M. E. *Acc. Chem. Res.* **1993**, *26*, 111–115.
63. Hattori, H. *Chem. Rev.* **1995**, *95*, 537–558.
64. Montero, M. L.; Voigt, A.; Teichert, M.; Uson, I.; Roesky, H. W. *Angew. Chem. Int. Ed.* **1995**, *34*, 2504–2506.

65. Neumann, R.; Levin-Elad, M. *J. Catal.* **1997**, *166*, 206–217.
66. Davis, M. E. *Nature* **2002**, *417*, 813–821.
67. Hill, C. L.; Brown, R. B., Jr. *J. Am. Chem. Soc.* **1986**, *108*, 536–538.
68. Neumann, R.; Abu-Gnim, C. *J. Chem. Soc. Chem. Comm.* **1989**, 1324–1325.
69. Hill, C. L. In: *"Activation and Functionalization of Alkanes"*; Ed. Hill, C.L.; Wiley: New York, **1989**, pp. 243–279.
70. Neumann, R.; Abu-Gnim, C. J. *J. Am. Chem. Soc.* **1990**, *112*, 6025–6031.
71. Lyon, D. K.; Miller, W. K.; Novet, T.; Domaille, P. J.; Evitt, E.; Johnson, D. C.; Finke, R. G. *J. Am. Chem. Soc.* **1991**, *113*, 7209–7221.
72. Mansuy, D.; Bartoli, J.-F.; Battioni, P.; Lyon, D. K.; Finke, R. G. *J. Am. Chem. Soc.* **1991**, *113*, 7222–7226.
73. Hill, C. L.; Prosser-McCartha, C. M. *Coord. Chem. Rev.* **1995**, *143*, 407–455.
74. Okuhara, T.; Mizuno, N.; Misono, M. *Advan. Catal.* **1996**, *41*, 113–252.
75. Neumann, R. *Prog. Inorg. Chem.* **1998**, *47*, 317–370.
76. Okuhara, T.; Mizuno, N.; Misono, M. *Appl. Catal. A* **2001**, *222*, 63–77.
77. Moffat, J. B. *"Metal-Oxygen Clusters: The Surface and Catalytic Properties of Heteropoly Oxometalates"*; vol. 9; Kluwer Academic/Plenum Publishers: New York, **2001**.
78. Kozhevnikov, I. V. *"Catalysis by Polyoxometalates"*; vol. 2; Wiley: Chichester, England, **2002**.
79. Alsters, P. L.; Witte, P. T.; Neumann, R.; Rozner, D. S.; Adam, W.; Zhang, R.; Reedijk, J.; Gamez, P.; Elshof, J. E. t.; Chowdhury, S. R. *Preprints – Amer. Chem. Soc., Division of Petroleum Chemistry* **2007**, *52*, 208–212.
80. Li, H.; Eddaoudi, M.; O'Keeffe, M.; Yaghi, O. M. *Nature* **1999**, *402*, 276–279.
81. Allen, A., Jr.; Manke, D. R.; Lin, W. *Tetrahedron Lett.* **2004**, *41*.
82. Janiak, C. *Dalton Trans.* **2003**, 2781–2804.
83. Ngo, H. L.; Lin, W. *Top. in Catal.* **2005**, *34*, 85–92.
84. Wu, C.-D.; Hu, A.; Zhang, L.; Lin, W. *J. Am. Chem. Soc.* **2005**, *127*, 8940–8941.
85. Alaerts, L.; Séguin, E.; Poelman, H.; Thibault-Starzyk, F.; Jacobs, P. A.; Vos, D. E. D. *Chem. Eur. J.* **2006**, *12*, 7353–7363.
86. Yamada, Y. M. A.; Maeda, Y.; Uozumi, Y. *Org. Lett.* **2006**, *8*, 4259–4262.
87. Wu, C.-D.; Lin, W. *Angew. Chem. Int. Ed.* **2007**, *46*, 1075–1078.
88. Xamena, F. X. L. i.; Corma, A.; Garcia, H. *J. Phys. Chem.* **2007**, *111*, 80–85.
89. Xamena, F. X. L. i.; Abad, A.; Corma, A.; Garcia, H. *J. Catal.* **2007**, *250*, 294–298.
90. Bradshaw, D.; Prior, T. J.; Cussen, E. J.; Claridge, J. B.; Rosseinsky, M. J. *J. Am. Chem. Soc.* **2004**, *126*, 6106–6114.
91. Matsuda, R.; Kitaura, R.; Kitagawa, S.; Kubota, Y.; Kobayashi, T. C.; Horike, S.; Takata, M. *J. Am. Chem. Soc.* **2004**, *126*, 14063–14070.
92. Wu, C.-D.; Lin, W. *Angew. Chem. Int. Ed.* **2005**, *44*, 1958–1961.
93. Hermes, S.; Schröter, M.-K.; Schmid, R.; Khodeir, L.; Muhler, M.; Tissler, A.; Fischer, R. W.; Fischer, R. A. *Angew. Chem. Int. Ed.* **2005**, *44*, 6237–6241.
94. Kitagawa, S.; Noro, S.-i.; Nakamura, T. *Chem. Commun.* **2006**, 107–707.
95. Cho, S.-H.; Ma, B.; Nguyen, S. T.; Hupp, J. T.; Albrecht-Schmitt, T. E. *Chem. Comm.* **2006**, 2563–2565.
96. Han, J. W.; Hill, C. L. *J. Am. Chem. Soc.* **2007**, *129*, 15094–15095.
97. Hou, Y.; Hill, C. L. *J. Am. Chem. Soc.* **1993**, *115*, 11823–11830.
98. Zeng, H.; Newkome, G. R.; Hill, C. L. *Angew. Chem. Int. Ed.* **2000**, *39*, 1771–1774.
99. Rhule, J. T.; Neiwert, W. A.; Hardcastle, K. I.; Do, B. T.; Hill, C. L. *J. Am. Chem. Soc.* **2001**, *123*, 12101–12102.
100. Boring, E.; Geletii, Y. V.; Hill, C. L. *J. Am. Chem. Soc.* **2001**, *123*, 1625–1635.
101. Okun, N. M.; Anderson, T. M.; Hardcastle, K. I.; Hill, C. L. *Inorg. Chem.* **2003**, *42*, 6610–6612.
102. Okun, N. M.; Anderson, T. M.; Hill, C. L. *J. Am. Chem. Soc.* **2003**, *125*, 3194–3195.
103. Okun, N. M.; Tarr, J. C.; Hillesheim, D. A.; Zhang, L.; Hardcastle, K. I.; Hill, C. L. *J. Mol. Catal. A. Chem.* **2006**, *246*, 11–17.

INDEX

C

O

P

R

S

T

U

CONTENTS OF PREVIOUS VOLUMES

VOLUME 49

VOLUME 50

VOLUME 51

VOLUME 52

VOLUME 53

VOLUME 54

VOLUME 55

VOLUME 56

VOLUME 57

VOLUME 58

VOLUME 59